Lecture Notes on Coastal and Estuarine Studies

Managing Editors:
Malcolm J. Bowman
Richard T. Barber
Christopher N.K. Mooers

16

Physics of Shallow Estuaries and Bays

Edited by J. van de Kreeke

Springer-Verlag
Berlin Heidelberg New York Tokyo

Managing Editors
Malcolm J. Bowman
Marine Sciences Research Center, State University of New York
Stony Brook, N.Y. 11794, USA

Richard T. Barber
Coastal Upwelling Ecosystems Analysis
Duke University, Marine Laboratory
Beaufort, N.C. 28516, USA

Christopher N. Mooers
Dept. of Oceanography, Naval Postgraduate School
Monterey, CA 93940, USA

Contributing Editors
Ain Aitsam (Tallinn, USSR) · Larry Atkinson (Savannah, USA)
Robert C. Beardsley (Woods Hole, USA) · Tseng Cheng-Ken (Qingdao, PRC)
Keith R. Dyer (Merseyside, UK) · Jon B. Hinwood (Melbourne, AUS)
Jorg Imberger (Western Australia, AUS) · Hideo Kawai (Kyoto, Japan)
Paul H. Le Blond (Vancouver, Canada) · Akira Okuboi (Stony Brook, USA)
William S. Reebourgh (Fairbanks, USA) · David A. Ross (Woods Hole, USA)
S.- Sethuraman (Raleigh, USA) · John H. Simpson (Gwynedd, UK)
Absornsuda Siripong (Bangkok, Thailand) · Robert L. Smith (Covallis, USA)
Mathis Tomaczak (Sydney, AUS) · Paul Tyler (Swansea, UK)

Editor

J. van de Kreeke
University of Miami
Rosenstiel School of Marine and Atmospheric Science
Division of Applied Marine Physics
4600 Rickenbacker Causeway
Miami, Florida 33149-1098, USA

ISBN 3-540-96328-6 Springer-Verlag Berlin Heidelberg New York Tokyo
ISBN 0-387-96328-6 Springer-Verlag New York Heidelberg Berlin Tokyo

This work is subject to copyright. All rights are reserved, whether the whole or part of the material is concerned, specifically those of translation, reprinting, re-use of illustrations, broadcasting, reproduction by photocopying machine or similar means, and storage in data banks. Under § 54 of the German Copyright Law where copies are made for other than private use, a fee is payable to "Verwertungsgesellschaft Wort", Munich.

© by 1986 Springer-Verlag New York, Inc.
Printed in Germany

Printing and binding: Beltz Offsetdruck, Hemsbach/Bergstr.
2131/3140-543210

CONTENTS

Contributors		V
Acknowledgement		IX
I	**Introduction**	1
II	**Large Scale Mixing**	5
	Large-Scale Mixing Processes in a Partly Mixed Estuary *G. Abraham, P. de Jong & F.E. van Kruiningen*	6
	Circulation and Salt Flux in a Well Mixed Estuary *B. Kjerfve*	22
	Currents and Salinity Transport in the Lower Elbe Estuary: Some Experiences from Observations and Numerical Solutions *K.C. Duwe & J. Sündermann*	30
III	**Residual Currents**	41
	Generalised Theory of Estuarine Dynamics *D. Prandle*	42
	Synoptic Observations of Salinity, Suspended Sediment and Vertical Current Structure in a Partly Mixed Estuary *R.J. Uncles, R.C.A. Elliott, S.A. Weston, D.A. Pilgrim, D.R. Ackroyd, D.J. McMillan & N.M. Lynn*	58
	Analysis of Residual Currents Using a Two-Dimensional Model *Po-Shu Huang, Dong-Ping Wang & T.O. Najarian*	71
	Residual Currents, A Comparison of Two Modelling Approaches *H. Gerritsen*	81
	On Lagrangian Residual Ellipse *R.T. Cheng, Shizuo Feng & Pangen Xi*	102
	A Simulation of the Residual Flow in the Bohai Sea *Zhang Shuzhen, Wang Huatong, Feng Shizuo, Xi Pangen*	114

	Principal Differences Between 2D- and Vertically Averaged 3D-Models of Topographic Tidal Rectification *J.T.F. Zimmerman*	120
IV	**Low-Frequency Motions**	131
	Subtidal Current Variability in the Lower Hudson Estuary *R.E. Wilson & R.J. Filadelfo*	132
	Subtidal Exchanges Between Corpus Christi Bay and Texas Inner Shelf Waters *N.P. Smith*	143
V	**Coastal Circulation**	153
	Water Circulation in a Topographically Complex Environment *E. Wolanski*	154
	Roles of Large Scale Eddies in Mass Exchange Between Coastal and Oceanic Zones *Sotoaki Onishi*	168
	Coastal Circulation in the Key Largo Coral Reef Marine Sanctuary *T.N. Lee*	178
	Effects of Coastal Boundary Layers on the Wind-Driven Circulation in Shallow Sea *Su Jilan*	199
VI	**Sediment Transport**	209
	Suspended Sediment Transport in Rivers and Estuaries *M. Markofsky, G. Lang & R. Schubert*	210
	Tide-Induced Residual Transport of Fine Sediment *J. Dronkers*	228
	Hydrodynamic Controls on Sediment Transport in Well-Mixed Bays and Estuaries *D.G. Aubrey*	245
	Sediment-Driven Density Fronts in Closed End Canals *Chung-Po Lin & A.J. Mehta*	259
	Index	277

CONTRIBUTORS

G. ABRAHAM
Delft Hydraulics Laboratory
P.O. Box 177
2600 MH Delft
The Netherlands

D.R. ACKROYD
Department of Marine Science
Plymouth Polytechnic
Drake Circus
Plymouth, PL4 8AA
U.K.

D.G. AUBREY
Woods Hole
 Oceanographic Institution
Woods Hole, MA 02543
U.S.A.

R.T. CHENG
U.S. Geological Survey WRD
345 Middlefield Road MS/496
Menlo Park, CA 94025
U.S.A.

Lin CHUNG-PO
University of Florida
Department of Coastal
 & Oceanographic Engineering
336 Weil Hall
Gainesville, Fl 32611
U.S.A.

J. DRONKERS
Rijkswaterstaat
van Alkemadelaan 400
The Hague
The Netherlands

K.C. DUWE
Institut für Meereskunde
University of Hamburg
Hasencleverstr. 27E, D-2000
Hamburg 74
Federal Republic of Germany

R.C.A. ELLIOT
Natural Environment Research Council
Institute for Marine Environmental
 Research
Prospect Place, The Hoe
Plymouth, PL1 3DH
U.K.

Shizuo FENG
Shandong College of Oceanography
P.O. Box 90,
Qingdao, Shandong
Peoples Republic of China

R.J. FILADELFO
Marine Sciences Research Center
State University of New York
Stony Brook, NY 11794
U.S.A.

H. GERRITSEN
Delft Hydraulics Laboratory
P.O. Box 177
2600 MH Delft
The Netherlands

Po-Shu HUANG
Najarian & Associates Inc.
1 Industrial Way West
Eatontown, New Jersey 07724
U.S.A.

P. DE JONG
Delft Hydraulics Laboratory
P.O. Box 177
2600 MH Delft
The Netherlands

B. KJERFVE
Belle W. Baruch Institute
 for Marine Biology
 and Coastal Research
University of South Carolina
Columbia, SC 29208
U.S.A.

F.E. van Kruiningen
Delft Hydraulics Laboratory
P.O. Box 177
2600 MH Delft
The Netherlands

G. Lang
Institut für Strömungsmechanick
Hannover University
Callinstrasse 32
3000 Hannover 1
Federal Republic of Germany

T.N. Lee
Div. of Meteorology and
  Physical Oceanography
RSMAS, University of Miami
4600 Rickenbacker Causeway
Miami, Florida 33149
U.S.A.

N.M. Lynn
Department of Nuclear Science
  and Technology
Royal Naval College
Greenwich, London SE10 9NN
U.K.

M. Markofsky
Institut für Strömungsmechanick
Hannover University
Callinstrasse 32
3000 Hannover 1
Federal Republic of Germany

D.J. McMillan
Department of Nuclear Science
  and Technology
Royal Naval College
Greenwich, London, SE10 9NN
U.K.

A.J. Mehta
University of Florida
Department of Coastal
  & Oceanographic Engineering
336 Weil Hall
Gainesville, Fl 32611
U.S.A.

T.O. Najarian
Najarian & Associates Inc.
1 Industrial Way West
Eatontown, New Jersey 07724
U.S.A.

Sotoaki Onishi
Civil Engineering Department
Science University of Tokyo
Noda City, Chiba Pref.
Japan 278

D.A. Pilgrim
Department of Marine Science
Plymouth Polytechnic
Drake Circus
Plymouth, PL4 8AA
U.K.

D. Prandle
Institute of Oceanographic
  Sciences
Bidston Observatory,
Birkenhead, Merseyside, L43 7RA
U.K.

R. Schubert
Institut für Strömungsmechanick
Hannover University
Callinstrasse 32
3000 Hannover 1
Federal Republic of Germany

Ned. P. Smith
Harbor Branch Foundation Inc.
RR#1, Box 196
Ft. Pierce, Fl 33450
U.S.A.

Jilan Su
Second Inst. of Oceanography
Nat. Bur. of Oceanography
P.O. Box 75
Hangzhou, Zhejiang 310005
People's Republic of China

J. SÜNDERMANN
Institut für Meereskunde
University of Hamburg
Hasencleverstr. 27E, D-2000
Hamburg 74
Federal Republic of Germany

R.J. UNCLES
Institute for Marine
  Environmental Research
Prospect Place, The Hoe
Plymouth PL1 3DH
U.K.

Huatong WANG
Shandong College of Oceanography
P.O. Box 90
Qingdao, Shandong
People's Republic of China

Dong-Ping WANG
Najarian & Associates Inc.
1 Industrial Way West
Eatontown, New Jersey 07724
U.S.A.

S.A. WESTON
Natural Environment Research Council
Institute for Marine
  Environmental Research
Prospect Place, The Hoe
Plymouth PL1 3DH
U.K.

R.E. WILSON
Marine Sciences Research Center
State University of New York
Stony Brook, NY 11794
U.S.A.

E. WOLANSKI
Australian Institute of
  Marine Science
PMB No.3, Townsville
M.S.O., Queensland, 4810
Australia

Pangen XI
Shandong College of
  Oceanography
P.O. Box 90
Qingdao, Shandong
People's Republic of China

Shuzhen ZHANG
Shandong College of Oceanography
P.O. Box 90
Qingdao, Shandong
People's Republic of China

J.T.F. ZIMMERMAN
Netherlands Institute
  of Sea Research
1 Industrial Way
P.O. Box 59
1790 AB Texel
The Netherlands

ACKNOWLEDGEMENT

These Lecture Notes are composed of papers presented at the Symposium Physics of Shallow Estuaries and Bays, held in Miami, Fla, 29-31 August 1984. The symposium was sponsored by

the Rosenstiel School of Marine and
Atmospheric Science,
University of Miami
and
the Coastal Engineering Research Council
of the American Society of Civil Engineers

Financial support was received from

the National
Oceanic and Atmospheric Administration,
Office of Sea Grant, U.S. Department of Commerce,

and

the National Science Foundation under Grant no CEE-8403878

I am grateful to all those who assisted in the organization of the symposium. In particular I would like to thank Ms. Mildred Corderelle who took care of the voluminous correspondence associated with such an effort. The papers of the symposium were selected with the help of Mr. J.F. Festa, Dr. G. Han, Dr. D.V. Hansen. Dr. T.N. Lee and Dr. J.D. Wang, who also assisted in editorial duties.

The appearance of this book is primarily due to facilities made available by the Center for Mathematics and Computer Science, Amsterdam. I am particularly indebted to Mrs. Linda M. Brown and Mrs. Nada Mitrovic of CMCS who were responsible for the layout of these Lecture Notes and prepared the manuscripts in type-set form. I express my thanks to Mrs. Emmy L. Goldbach who carefully checked all corrections and took care of all those loose ends accompaying the preparation of this volume.

I

INTRODUCTION

Introduction

J. van de Kreeke

Rosenstiel School of Marine
and Atmospheric Science
University of Miami, USA

Easy access to the sea and safe anchorage have made many estuaries the location of early settlements. Initially, land for further development was abundant, depths were large enough to accommodate the small vessels, and except for some local degradation water quality was little affected. However, as the population increased and industries were added, estuary basins were modified to better suit the demands

- land was reclaimed to allow industry to expand and to create waterfront residential areas
- navigation channels and harbors were dredged to accommodate larger vessels
- causeways and bridges were constructed to ease traffic
- marinas were built to harbor recreational vessels.

The changes in the estuary basins together with modifications of the run-off associated with upland development have the potential of unfavorably affecting the circulation. This together with increased pollutant input could result in a general decline in water quality. For example in South Florida, the construction of irrigation and water control projects has drastically changed the freshwater inflow in many bays and lagoons; the gradual inflow of the past has been replaced by the release of large quantities of freshwater in a short period of time. This led to a change in the circulation and to large variations in salinity which can only be tolerated by a limited number of species. In addition, the vertical stratification during periods of higher freshwater inflow causes a reduced vertical exchange and therefore a reduced oxygen concentration in the bottom layers. The construction of navigation channels, harbors, causeways and bridges, in the lagoons separating the mainland and the barrier islands limits the horizontal exchange and encourages the formation of pockets of stagnant water.

Properly incorporating environmental aspects in engineering modifications of the estuary, requires a thorough knowledge of, among other things, its physical oceanography. Knowledge of tides, currents and circulation is not only needed to correctly dimension the various structures but also, together with knowledge of marine biological and chemical features serves to evaluate the environmental consequences. It is only then that we can weigh various conflicting requirements and prevent the destruction of those assets for which we selected to live near the estuary in the first place.

To evaluate the state of the art in physical oceanography of estuaries and to suggest research priorities, a symposium was organized at the Rosenstiel School of Marine and Atmospheric Science, Miami, on August 29-31, 1984. Emphasis was on shallow estuaries, bays and lagoons. Special attention was given to topics of direct interest to the coastal (estuarine) engineering community. Papers presented at the symposium are presented in these Lecture Notes. The Lecture Notes are divided in five sections each containing a somewhat coherent set of papers.

The section *Large Scale Mixing* deals with mixing phenomena having a time scale on the order of the tidal period. Examples are mixing due to phase differences at channel junctions, the presence of harbour basins along the estuary, the exchange at the mouth and the emptying and filling of flats. In *Residual Currents* the 2D-vertical structure of wind, tide- and density induced residual currents are analysed. Pros and cons are described of calculating the vertically integrated residual currents by time averaging the solution of the tidal equations and by directly solving the equations of the residual current. Differences in the depth averaged residual currents obtained from a 2D-horizontal model and the depth averaged solution of a 3D-model are discussed. A second order addition to the costumary expression relating the Lagrangian residual current to the Eulerian velocity field is presented. *Low Frequency Motions* are fluctuations in waterlevel and current having periods in excess of one day. Using waterlevel and current observations, low frequency motions in the Lower Hudson Estuary. N.Y and Corpus Christi Bay, Texas are discussed. As demonstrated for Corpus Christi Bay, low frequency motions do play a significant though supplementary (to the tide) role in the estuary-shelf exchange. The renewal rate of estuarine waters is to a large degree determined by the *Coastal Circulation* off the mouth of the estuary. Because of the shallow and often complicated topography, the proximity of strong oceanic currents (Gulf stream, Kuroshio and Oyashio Current) and density gradients, circulation in this region in general is complicated and highly variable. In this section examples of the formation of tidal jets and eddiers associated with islands and interacting tidal- and oceanic currents are presented. The complicated velocity and temperature field in a coral reef area is described. Coastal circulation on the one hand constitutes the boundary condition for the estuary and on the other hand that for the shelf circulation. The formulation of the boundary condition when calculating wind-driven currents on the shelf is discussed. The section *Sediment Transport* focusses on suspended sediment transport. A review of analytical solutions for the vertical distribution of suspended sediment in various flow fields and for different bottom boundary conditions is presented. The tidal distortion and especially the asymmetry in the tidal curves about High- and Low Water Slack as possible mechanisms for suspended sediment transport are discussed. Motions of density fronts associated with suspended sediment (turbidity currents) are analysed.

From the papers in these Lecture Notes and discussions at the symposium a number of research priorities emerged.

- Advances in our understanding of large scale mixing processes, residual currents and suspended sediment transport are hampered by a limited knowledge of the small scale turbulent exchange processes. In particular research should focus on the reduction and the ebb-flood asymmetry of the exchange in the presence of vertical stratification.
- In view of potential pollution problems focus of residual currents is on transport- and Lagrangian velocities. It might be worthwhile to investigate these current fields in terms of the Lagrangian- rather than the Eulerian form of the pertinent equations.
- Although it is generally accepted that residual flow plays an important role in the renewal of waters of shallow bays so far no device exists to accurately measure these flows. Therefore model calculation cannot be verified. The development of a "residual flow meter" seems a worthwhile project.
- The coastal boundary layer plays an important role in the renewal of estuarine waters. Because of its complexity the hydrodynamics of this region has been somewhat neglected. More fundamental research is needed.

- Very little is known about the many processes that determine suspended sediment transport. In addition to the earlier mentioned vertical turbulent exchange research should focus on the physical and chemical properties of the sediment, the effect of the sediment concentration on setting velocities and erosion - deposition mechanisms in the bottom boundary layer..

Finally the reader will probably find these Lecture Notes less coherent than hoped for; unfortunately authors tend to expand prescribed limits. On the other hand this might lead to pleasant surprises as papers do contain valuable information on topics other than those suggested by the section headings.

II

LARGE SCALE MIXING

Large-scale Mixing Processes in a Partly Mixed Estuary

G. Abraham, P. de Jong & F.E. van Kruiningen

Delft Hydraulics Laboratory,
The Netherlands

ABSTRACT

This paper presents experimental evidence for the considerable effect of the tidal phase difference between two estuary channels on the longitudinal density distribution in the Rotterdam Waterway, a partly mixed estuary. It further shows the effect of harbour basins along the estuary on the salt intrusion to be caused by two mechanisms which in the Rotterdam Waterway are counteracting: the effect of the basins on tidal flow and the temporary storage of salt water in the basins during part of the tidal cycle.

1. Introduction

This paper deals with the salt intrusion in the Rotterdam Waterway, a partly mixed estuary. It describes large-scale advective mixing processes induced by the large-scale geometry of the estuary.

One large-scale mixing process considered is the tidal phase difference between two estuary channels. The tidal phase difference induces advective processes which have a considerable effect on the longitudinal salt distribution. Because of the tidal phase difference and because of a different channel geometry, both estuary channels react in a different manner to variations of the tidal amplitude. In the one estuary channel the salt intrusion at high water slack decreases with increasing tidal amplitude, while in the other channel the smallest intrusion at high water slack occurs with normal tide.

Another large-scale mixing process considered is the effect of harbour basins along the estuary. The basins lead to two mechanisms which in the Rotterdam Waterway are counteracting. The tide and density induced exchange flows cause salt water to be temporarily stored in the basins during part of the tidal cycle. This temporary storage tends to increase salt intrusion. With the harbour basins tidal action is stronger than it would be without. Consequently with the harbour basins turbulent mixing is stronger in the main estuary channels than it would be without. This effect of the basins tends to reduce salt intrusion.

The combined effect of the considered large scale processes depends on the integrated effect of small scale turbulent mixing. The implications of this finding for Rotterdam Waterway salt intrusion modelling are given.

2. General characteristics of Rotterdam waterway

The Rotterdam Waterway Estuary, which is represented in Fig. 1, is formed by the New Meuse, the Old Meuse and the New Waterway. Fresh water, which flows into the estuary through the New Meuse and the Old Meuse, issues into the North Sea through the New Waterway.

In the Rotterdam Waterway the salt intrusion is influenced by large-scale and small-scale transport processes. On a large scale these processes are primarily advective ones, and include tidal action, fresh water discharge and gravitational circulation. On a small scale these processes are primarily turbulent ones, and include the turbulent transport of momentum and mass. Both the large-scale and the small-scale processes depend on the geometry of the estuary. The combined effect of the large-scale processes depends on the integrated effect of the small-scale turbulent transports. The junction of the New Meuse and the Old Meuse is an important feature of the estuary. There is a phase difference between the tidal velocities in the New Meuse and those in the Old Meuse. This phase difference acts as a large scale mixing mechanism.

Several harbour basins are located along the estuary, and are openly connected to it; the Europoort harbour basins connect with the New Waterway at 1033 km, the Botlek harbour basins connect with the New Waterway at 1014 km, and several harbour basins connect with the New Meuse between 1001 km and 1012 km (see Figure 1 and Table 1). The basins lead to phase effects. In addition, density currents of the lock exchange flow-type (SCHIJF and SCHÖNFELD, 1953) cause relatively salt water to flow into the harbour basins on the flood tide, and out of the basins on the ebb tide. Both the phase effects and the density induced exchange flows cause large scale mixing to take place in the main estuary channels.

For conditions considered in the paper, high water slack salt intrusion reaches to about 995 km in the New Meuse as well as in the Old Meuse, in each channel with respect to its own coordinate system (see Fig. 1).

Figure 1 and Table 2 give the dimensions of the channels where salt intrusion takes place. For the New Meuse distinction has to be made between a deep part from 1013 km (the junction with the Old Meuse) to 1001 km, and a shallow part upstream from 1001 km. Table 2 lists geometric and hydrodynamic characteristics of the separate channels, making a distinction between the deep part and the shallow part of the New Meuse. For the separate channels Table 2 lists the following quantities: the cross-section below mean sea level (A), the width (B), the depth below mean sea level (h), the fresh water flow rate in the conditions of small discharges which are considered in this study (Q_f), the fresh water velocity ($u_f = Q_f / A$), the maximum flood discharge (Q_0) for respectively spring tide, normal tide and neap tide, and the corresponding maximum flood velocity ($U_0 = Q_0 / A$).

The fresh water flow rate as well as the tidal velocities in the Old Meuse are larger than they are in the New Meuse, while the cross-section of the Old Meuse is smaller than that of the New Meuse. Because of these different characteristics, the salt intrudes further into the New Meuse than it does into the Old Meuse, measured from the junction of these rivers.

The tidal flow supplies the energy which causes the mixing between fresh and salt water. The energy input by the tide per unit time and per unit area is propertional to the product of the bottom shear and the tidal velocity, i.e. to the tidal velocity cubed.

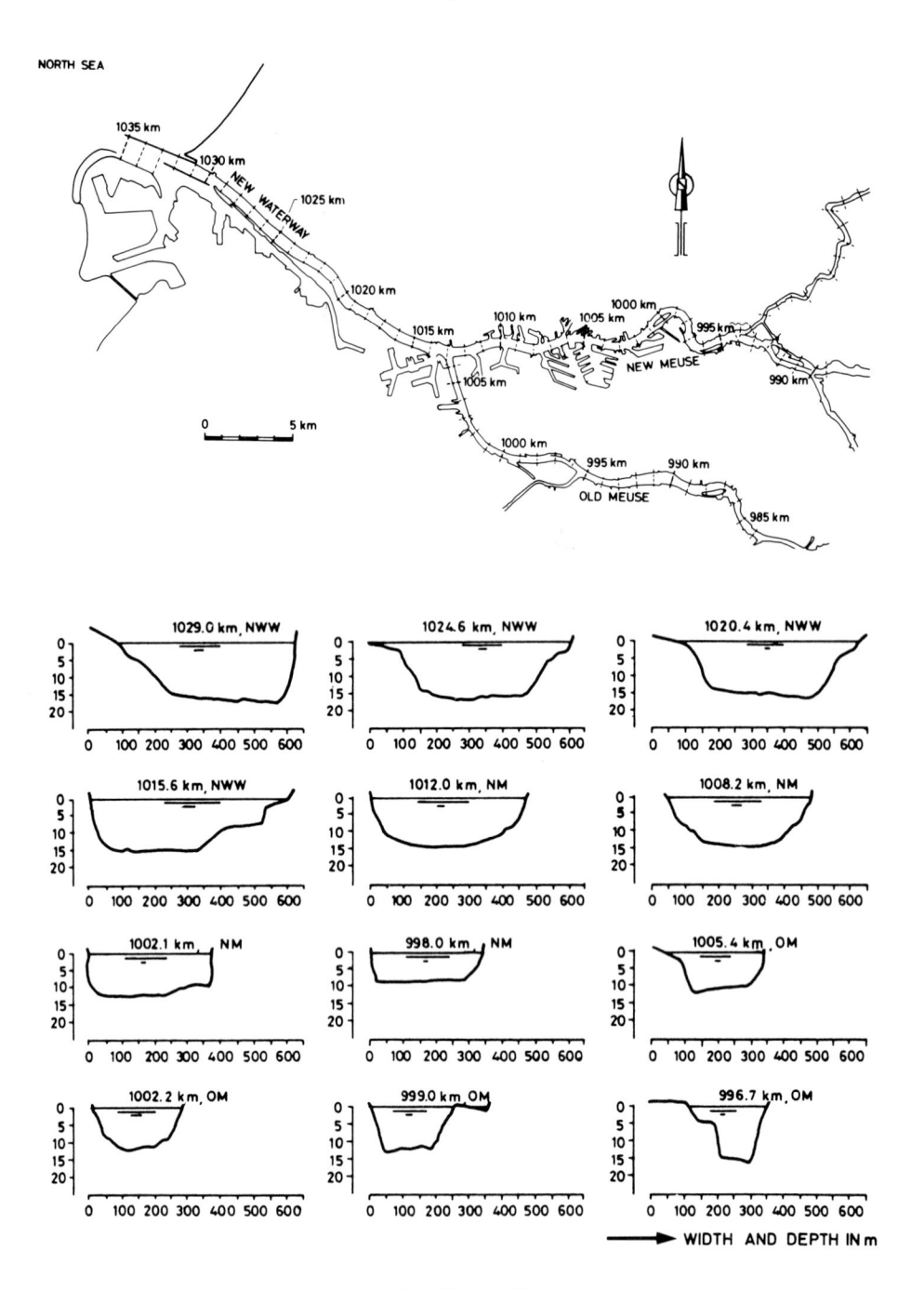

FIGURE 1. Rotterdam Waterway Estuary

Table 1 Dimensions of harbour basins

harbour basin	surface area (m^2)	location (km)
Botlek Harbour	3.6 10^6	1014
Waal Harbour	2.2 10^6	1005
Eem Harbour	1.5 10^6	1008
1st Petrol Harbour	0.5 10^6	1011
Meuse Harbour	0.6 10^6	1003
2nd Petrol Harbour	0.6 10^6	1010
Σ small harbours	3.0 10^6	1002-1012
	12.0 10^6	

Table 2 Geometric and hydrodynamic characteristics of Rotterdam Waterway (discharge of River Rhine 800 m^3/s)

existing situation								
channel	New Waterway		New Meuse		New Meuse		Old Meuse	
location (km)	1029	1015	1011	1002	999	995	1005	996
$A(m^2)$	6000	5600	5500	4200	3000	2700	2700	2300
$B(m)$	450	450	490	390	380	370	300	260
$h(m)$	13.5	12.5	11.5	10.5	8	7.5	9	9
$Q_f(m^3/s)$	640	640	210	210	210	210	430	430
$U_f(m/s)$	0.11	0.11	0.04	0.05	0.07	0.08	0.16	0.19
$Q_o(m^3/s)$ spring tide[1]	9000	8200	5600	3430	3120	2900	3000	3090
normal tide	7450	7160	4980	2830	2480	2380	2450	2560
neap tide	5200	4800	3300	2005	1870	1800	1860	1910
$U_o(m/s)$ spring tide	1.50	1.46	1.02	0.82	1.04	1.07	1.11	1.34
normal tide	1.26	1.28	0.91	0.67	0.83	0.88	0.91	1.11
neap tide	0.87	0.86	0.60	0.48	0.62	0.67	0.67	0.83
E_{local} spring tide	100%	100%	100%	100%	100%	100%	100%	100%
normal tide	59%	67%	71%	55%	51%	56%	55%	57%
neap tide	20%	20%	20%	20%	21%	25%	22%	24%
$E_{overall}$ spring tide	100%	92%	31%	16%	33%	36%	41%	71%
normal tide	59%	62%	22%	9%	17%	20%	22%	41%
neap tide	20%	19%	6%	3%	7%	9%	9%	17%

1) Δh = 2.0 m spring tide   Δh: difference between HW and LW and mouth of estuary
   = 1.6 m normal tide
   = 1.2 m neap tide

Table 2 gives the following dimensionless parameters, as a measure for the energy available to cause mixing

$$E_{local} = \frac{U_{0,local}^3}{(U_{0,local}^3)_{spring}} \tag{1}$$

$$E_{overall} = \frac{U_{0,local}^3}{(U_{0,1029}^3)_{spring}} \tag{2}$$

where the indices local, 1029 and spring refer to respectively the local value of u_0, the value at 1029 km and the value during spring tide. The parameter E_{local} is a measure for the energy available at a given location for different tidal conditions, where the local spring tide value is set at 100%. The parameter $E_{overall}$ is a measure for the energy available at different locations along the estuary, where the spring tide value at 1029 km is set at 100 %.

The values of E_{local} pertaining during a spring tide are substantially larger than the values pertaining during a neap tide. Consequently the estuary is less stratified during a spring tide than it is during a neap tide (see Fig. 3).

The estuary number, E_D, is a measure of the stratification (THATCHER and HARLEMAN, 1981). It is defined as

$$E_D = \frac{P_t}{Q_f T} \frac{U_0^2}{\frac{\Delta\rho}{\rho} gh} \sim \frac{\rho B U_0^3}{Q_f \Delta\rho g} \quad \text{(at km 1029)} \tag{3}$$

where

$\Delta\rho$: density difference between river and sea water

ρ : density of either river or sea water

P_t : volume of sea water entering the estuary on the flood tide

T : duration of tidal cycle

g : gravitational acceleration

For the conditions of small fresh water flow rates considered in this paper the estuary number ranges from 3 (for spring tide conditions) to 0.6 (for neap tide conditions). Estuary numbers of this order of magnitude refer to partly mixed conditions.

In Eq. 3 Q_f is a measure of the quantity of fresh water to be mixed in order to obtain mixed conditions, $\Delta\rho\, g$ is a measure of the force per unit volume of salt water counteracting the mixing, while $\rho B U_0^3$ is a measure of the energy supplied by the tide. As it is the latter energy which causes the mixing, E_D is a measure of the stratification.

3. Available data

Information on the salt intrusion in the Rotterdam Waterway Estuary has been derived from several series of field measurements. In addition hydraulic scale model studies and mathematical model studies have been made. A review of current Rotterdam Waterway research is given by ROELFZEMA et al (1984).

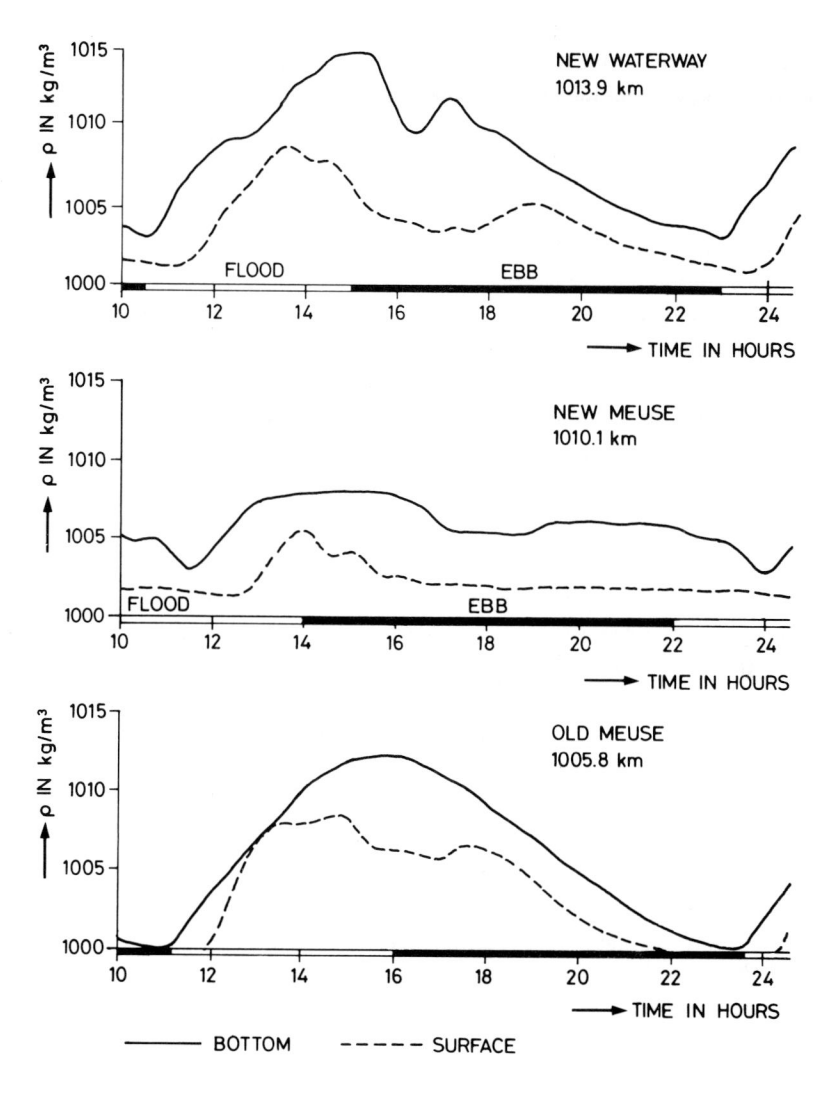

FIGURE 2   Time history of density at junction (from hydraulic scale model)

Field measurements, covering a whole tidal cycle of salt concentrations and velocities have been made on several occasions. Rather complete data sets have been obtained for different conditions with respect to the tide and the fresh water flow. The data sets show how the measured quantities vary with time, in the longitudinal direction and over the cross-section.

Analyzing the available field data by means of the decomposition method (FISCHER, 1972), it has been found that in the Rotterdam Waterway the gravitational circulation is primarily in the longitudinal direction (WINTERWERP, 1983) and that both turbulence generated at the solid boundaries and turbulence arising in the interior are to be taken into account (ABRAHAM, 1980). DRONKERS (1969) derived the effect of density gradients on the bottom shear from the field measurements.

A hydraulic scale model of the Rotterdam Waterway Estuary (vertical scale 1:64, horizontal scale 1:640, distortion 10) has been available since 1965. A survey of the type of engineering studies performed in the model is given by van REES et al (1972).

The hydraulic scale model gives a fair simulation of the physical processes which control salt intrusion, including gravitational circulation and turbulent mixing. This has been demonstrated by comparing the performance of the model with several available field data sets (BREUSERS and van Os, 1980; van der HEIJDEN et al, 1984). As this stage has not yet been reached in mathematical modelling, predictions to do with salt intrusion under new conditions are primarily derived from the hydraulic scale model.

To obtain basic knowledge on the physical processes related to salt intrusion, flume studies were made to study the intrusion in a schematized estuary of rectangular cross-section (see for example ROELFZEMA and van Os, 1978). In these studies the flume was used as a distorted hydraulic scale model of the schematized estuary.

From the flume studies, dimensionless correlations of the salt intrusion length with the determining quantities were derived (RIGTER, 1973). These correlations may not be applied to the Rotterdam Waterway as they do not incorporate the effect of the junction of the New Meuse and the Old Meuse, the different characteristics of these rivers, and the effect of the harbour basins. In addition, groins occur along part of the banks of the Rotterdam Waterway. This makes the mixing characteristics of the Rotterdam Waterway different from those of the flume, while in the flume studies salt intrusion was found to vary with the type of flow resistance (bottom roughness, side wall roughness or strips) i.e. with the mixing characteristics (ABRAHAM et al, 1975).

In one-dimensional, real time mathematical modelling of salt intrusion the effects of stratification and gravitational circulation have to be incorporated in the dispersion coefficient. The correlation between the dispersion coefficient and determining conditions is a unique characteristic of an individual estuary (ABRAHAM et al 1975, FISCHER et al, 1979, Section 7.1). For the Rotterdam Waterway no satisfactory correlation has been found as yet. Given this limitation of one-dimensional modelling, at least a two-dimensional laterally averaged ($2DV$) model is necessary to reproduce the gravitational circulation in the vertical plane through the axis of the estuary, and its effect on salt intrusion.

Laterally averaged two-dimensional modelling studies have been made by PERRELS and KARELSE (1981) to reproduce salt intrusion in the schematized estuary of rectangular cross-section. At present, their modelling technique is being applied to the Rotterdam Waterway. Tentative laterally averaged two-dimensional modelling studies of salt intrusion in the Rotterdam Waterway have been made by HAMILTON (1975) and SMITH and DYER (1979).

Given the above considerations the information on the vertical and longitudinal density distribution presented in this paper is derived from the hydraulic scale model for different tidal conditions, with the harbour basins openly connected with the Rotterdam Waterway as well as closed.

4. Effect of junction on salt intrusion

Through the New Meuse and the Old Meuse the tide penetrates into different networks of channels and wide basins. This causes phase differences between the tidal velocities in the New Meuse and in the Old Meuse. In the New Meuse the flood tide and the ebb tide begin earlier than in the New Waterway; in the New Waterway these tides begin earlier than in the Old Meuse.

The above phase differences act in various ways as large scale advective mixing mechanisms, causing contacts between water having a different salt content. The advective processes are the most pronounced when tidal excursion paths are large, and when salt intrusion is beyond the junction. The former condition is satisfied on a spring tide, the latter condition when the fresh water flow rates are small.

Figure 2 shows time histories of density at different stations in the vicinity of the junction. It illustrates some consequences of the phase differences. Late on the flood tide (from 14 hours till 15 hours) the salt content of the water, which arrives at the junction from the sea, has its peak value. At this phase of the tide, all water reaching the junction from the sea flows into the Old Meuse, not into the New Meuse, where a weak ebb flow starts to develop. This explains why in the New Meuse (1010.1 km) the density remains constant, while in the Old Meuse (1005.8 km) it continues to rise.

Early on the ebb tide (from 15 hours till 16 hours) all water, which passes the junction to flow into the New Waterway, comes from the New Meuse, where the salt content is relatively small. Later on (from 16 hours) it also comes from the Old Meuse, where the salt content is relatively large. This explains why in the New Waterway (1013.9 km) the time history of the density exhibits a peak at about 17 hours.

The above phenomena have a significant effect on the longitudinal density distribution in both the New Meuse and the Old Meuse.

5. Effect of tidal amplitude on salt intrusion

Large tidal amplitudes at the mouth of the estuary are associated with strong tidal currents, and hence with strong turbulent mixing and weak stratification. The stronger the turbulent mixing, the smaller salt intrusion tends to be. Because of this effect salt intrusion tends to become smaller with increasing tidal amplitude, in particular when the estuary is stratified.

Large tidal amplitudes at the mouth of the estuary are further associated with large tidal excursion paths. The larger the tidal excursion path, the larger salt intrusion tends to be at hight water slack. Because of this effect the salt intrusion tends to become larger with increasing tidal amplitude, in particular when the estuary is mixed.

Which of the above counteracting mechanisms prevails varies with the stratification, i.e. with the strength of the tide. For the schematized estuary this is clearly demonstrated by the fact that salt intrusion at high water slack decreases with increasing tidal amplitude when the latter is relatively small, while salt intrusion at high water slack increases with increasing tidal amplitude when the latter is relatively large. For intermediate tidal amplitudes salt intrusion at high water slack is found to have its smallest magnitude

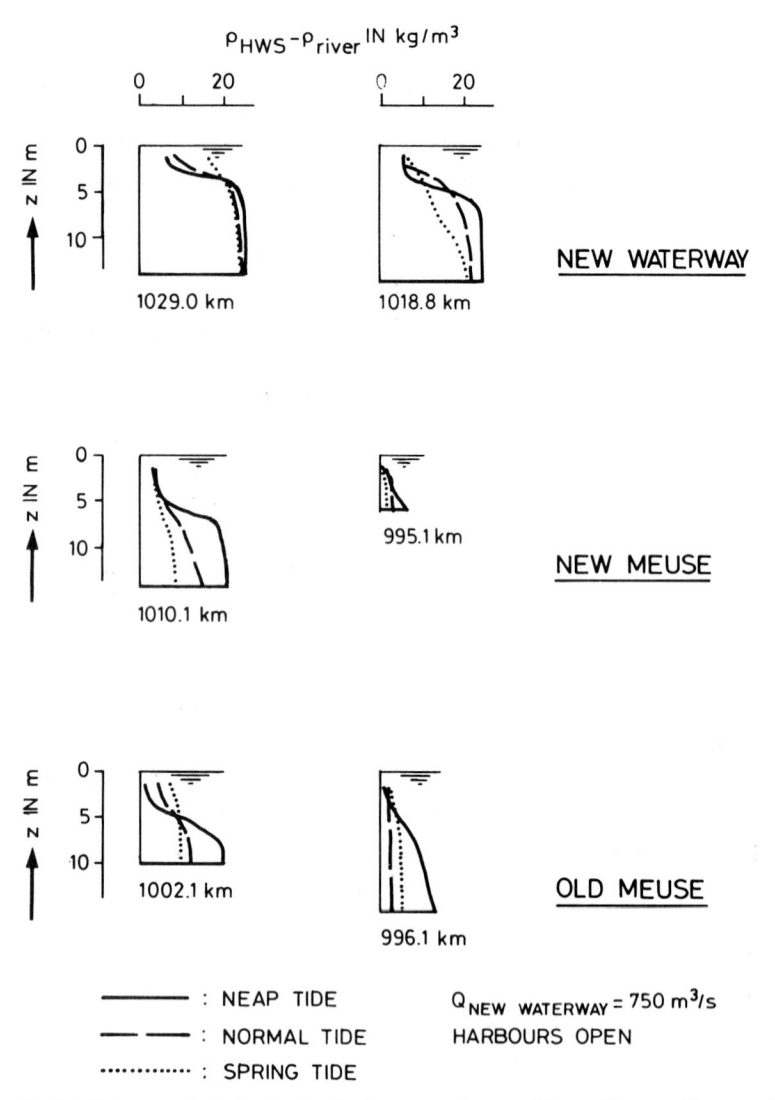

FIGURE 3. Vertical high water slack density distributions comparing conditions with neap tide, normal tide and spring tide (from hydraulic scale model)

(RIGTER 1973, Fig. 2).

The above counteracting mechanisms also play a role in the Rotterdam Waterway, though modified by the phase effects at the junction. This has been found in the hydraulic scale model, comparing vertical distributions of density measured at high water slack at different locations along the estuary. Fig. 3 shows vertical distributions of high water slack density for a spring tide, normal tide and neap tide with Δh respectively 2.0, 1.6 and 1.2 m, where Δh represents the difference between high water and low water at the mouth of the estuary. The corresponding tidal characteristics are given in Table 2.

For the New Meuse, Fig. 3 shows that salt intrusion at high water slack increases with decreasing tidal amplitude. For a normal tide it is larger than it is for a spring tide, while for a neap tide it is larger than it is for a normal tide. For the Old Meuse, Fig. 3 shows the smallest intrusion at high water slack occurring with a normal tide. This difference between the New Meuse and the Old Meuse is due to the fact that late on the flood tide there is only inflow of water from the New Waterway into the Old Meuse (see Fig. 2).

6. Effect of harbour basins on salt intrusion

6.1 Exchange of water between estuary channel and harbour basin

In a harbour basin, which is located along and in open connection with an estuary, water is stored outside the estuary channel over part of the tidal cycle. The mechanisms behind this temporary storage are tidal filling and emptying, density induced exchange flows of the lock exchange type, and to some extent eddies with a vertical axis induced by the river flow in the entrance of the basin.

Tidal filling and emptying are associated with the variation of depth within a tidal cycle. Assuming that the water surface remains level and neglecting variations of velocity with depth, the velocity through the entrance of the basin, induced by the tidal filling, v, satisfies

$$v = \frac{A \frac{dh}{dt}}{Bh} \tag{5}$$

where A is the horizontal area of the harbour basin, and where B and h are the width and depth of the entrance of the basin. The density of the water in the estuary channel varies with time. When increasing it is larger than the density of the water in the harbour basin, when decreasing it is smaller than the density of the water in the basin. This is because the mean density of the water in the basin can only increase if water from the basin is replaced by water with a comparatively large or small density from the channel by density induced exchange flows of the lock exchange type.

When $v = 0$, the density induced velocity of inflow and/or outlfow, c, is in first approximation given by (SCHIJF and SCHÖNFELD, 1953)

$$c = \frac{1}{2}(\frac{\Delta \rho}{\rho}gh)^{\frac{1}{2}} \tag{6}$$

where $\Delta \rho$ is the driving difference in density between the harbour basin and the estuary channel.

Tidal filling and emptying causes the density induced exchange currents to be reduced to, in first approximation,

$$c_n = c - |v| \tag{7}$$

where c_n is the magnitude of the net density induced exchange velocity. Fig. 4 shows the rationale behind Eq. 7.

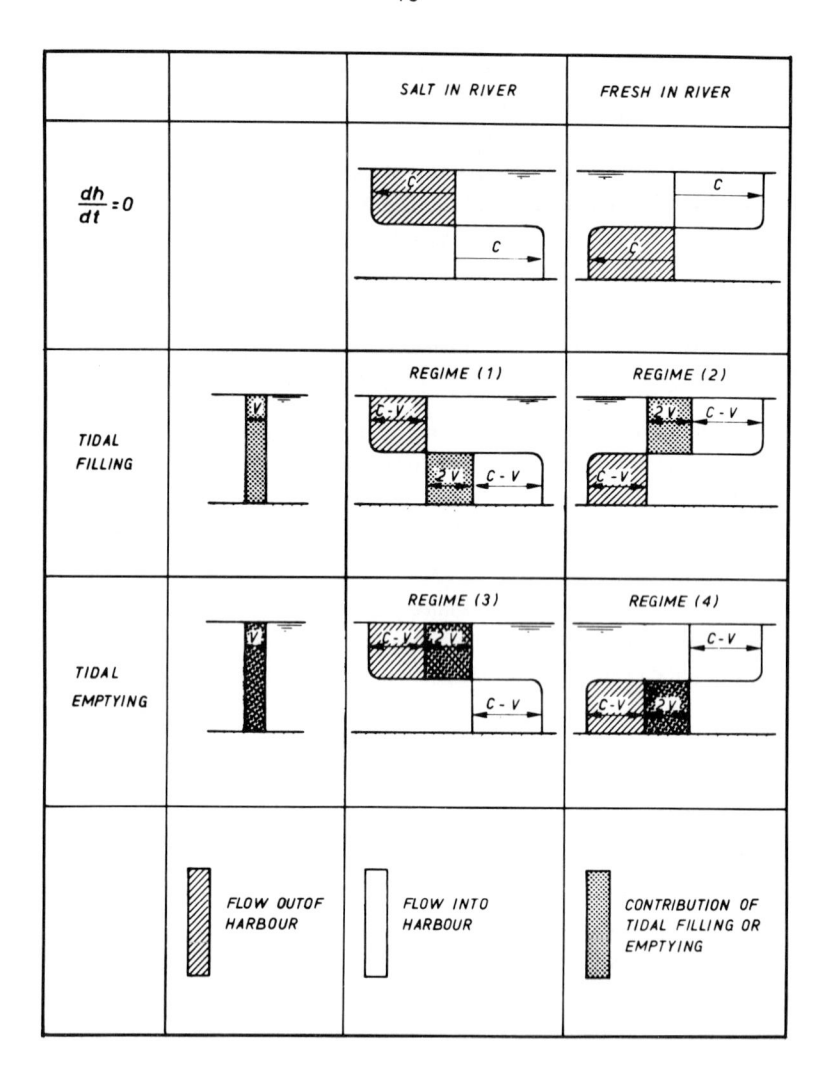

FIGURE 4. Combined effect of tidal filling and density induced exchange flows

For the Botlek Harbour, Fig. 5 shows the relative significance of tidal filling and emptying (in the figure referred to as $\partial h / \partial t$), density induced exchange flows (in the figure referred to as $\Delta \rho$) and the exchange induced by eddies in the entrance of this harbour basin (in the figure referred to as eddy). In Fig. 5 the four regimes, which are distinguished in Fig. 4, can be recognized. Fig. 5 has been obtained from field measurements.

Further details on the exchange between harbour basins and estuary channels are given by ALLEN and PRICE (1959), VOLLMERS (1976) and ROELFZEMA and OS (1978).

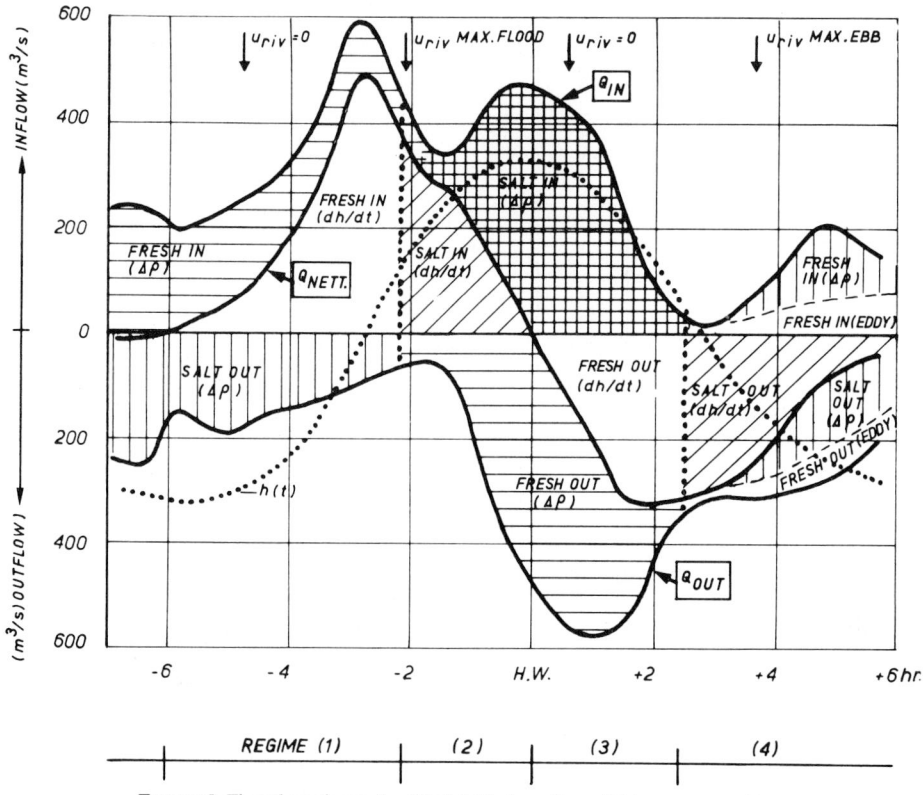

FIGURE 5. Flow through mouth of Botlek Harbour (from field measurements)

6.2. Effect of harbour basins on salt intrusion mechanisms
The harbour basins influence the salt intrusion in two ways:

Firstly, because of the above exchange mechanisms there is a temporary storage of relatively salt water in the basins on the flood tide. Therefore, they act as a source of relatively salt water at the end of the ebb tide (see Fig. 5). Because of this effect closing the basins - which is feasible in the hydraulic scale model only - would tend to reduce the salt intrusion.

Secondly, the total area of the basins is so large that tidal action is much stronger with the basins than it would be without. This applies from the area where the basins are located towards the sea.

Tables 2 and 3 quantitatively illustrate the above observation. These tables give the tidal characteristics of the separate channels, respectively when the harbour basins listed in Table 1 are in open connection with the estuary and supposed to be closed. In both tables the parameter $E_{overall}$ is related to its spring tide value at 1029 km, the harbour basins being in open connection with the estuary. This value is set at 100%. From the zone occupied by these harbours towards the sea, spring tide conditions without the basins, which are listed in Table 1 (i.e. when these basins were closed) are about the same as normal tide conditions with these basins. This can be seen from the $E_{overall}$ values listed in Table 2 and 3 for 1029 km, 1015 km and 1011 km. The above reduction when the harbour basins listed in Table 1 would not exist

channel location (km)	New Waterway		New Meuse		New Meuse		Old Meuse	
	1029	1015	1011	1002	999	995	1005	996
$A(m^2)$	6000	5000	5500	4200	3000	2700	2700	2300
$B(m)$	450	450	490	390	380	370	300	260
$h(m)$	13.5	12.5	11.5	10.5	8	7.5	9	9
$Q_f(m^3/s)$	640	640	210	210	210	210	430	430
$U_f(m/s)$	0.11	0.11	0.04	0.05	0.07	0.08	0.16	0.19
$Q_o(m^3/s)$ spring tide[1]	7900	7100	4200	3365	3015	2900	3000	3000
normal tide	6220	5780	3725	2765	2510	2415	2440	2450
neap tide	4135	3845	2570	1940	1845	1770	1700	1940
$U_o(m/s)$ spring tide	1.32	1.27	0.76	0.80	1.01	10.7	1.11	1.30
normal tide	1.04	1.03	0.68	0.66	0.84	0.89	0.90	1.07
neap tide	0.69	0.69	0.47	0.46	0.62	0.66	0.63	0.84
E_{local} spring tide	100%	100%	100%	100%	100%	100%	100%	100%
normal tide	49%	81%	72%	56%	58%	58%	53%	56%
neap tide	14%	16%	24%	19%	23%	23%	18%	27%
$E_{overall}$ spring tide	68%	61%	13%	15%	31%	36%	41%	65%
normal tide	33%	32%	9%	9%	18%	21%	22%	36%
neap tide	10%	10%	3%	3%	7%	9%	7%	18%

1) Δh = 2.0 m spring tide Δh: difference between HW and LW and mouth of estuary
    = 1.6 m normal tide
    = 1.2 m neap tide

TABLE 3. Geometric and hydrodynamic characteristics of Rotterdam Waterway (discharge of River Rhine 800 m³/s)

of tidal action, induced by closing the harbour basins, tends to increase salt intrusion into the New Waterway and into the New Meuse, and hence to increase the salt content at the junction. This follows directly from the information given in Section 5. In the Old Meuse the tidal excursion paths do not change when the harbour basins are closed. Therefore, the increase of salt content at the junction would lead to an increase of salt intrusion into the Old Meuse. Hence, both for the New Meuse and the Old Meuse the modification of the tidal action induced by closing harbour basins tends to increase salt intrusion.

Figure 6 shows vertical distributions of high water slack density for a spring tide, both when all harbour basins are in open connection with the estuary, and when the basins listed in Table 1 are closed. Figure 6 has been obtained from the hydraulic scale model. Figure 6 shows that closing the basins would increase salt intrusion in the New Meuse and reduce salt intrusion in the Old Meuse. It appears that for the New Meuse, the effect of reducing tidal action is stronger than that of eliminating the temporary storage of salt water, in particular between 1002 km and 1012 km, where the harbours are located and the reduction of tidal action is effectuated. For the Old Meuse (996.1 km) it appears that the effect of eliminating the temporary storage of salt water in the

Botlek harbour at 1014 km is stronger than that of reducing the tidal action.

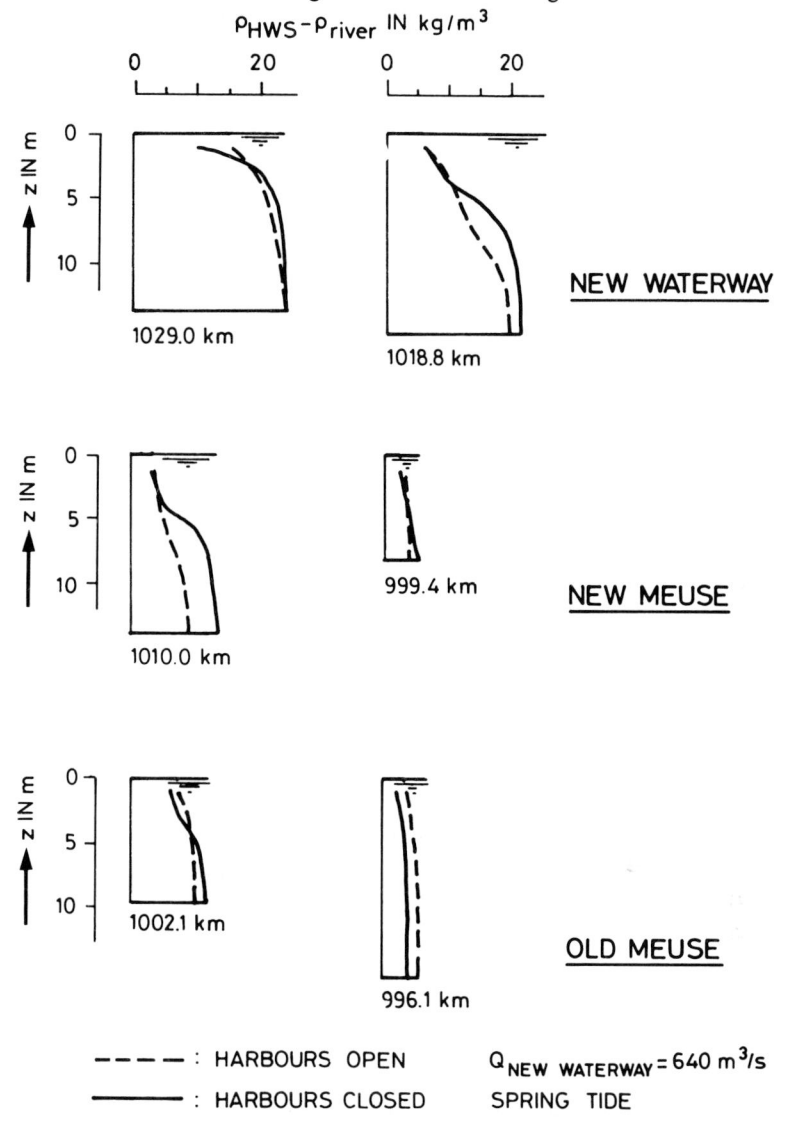

FIGURE 6. Vertical high water slack density distributions, comparing conditions with harbour basins open and closed (from hydraulic scale model)

7. Conclusions

The salt intrusion in the Rotterdam Waterway is influenced by the junction of the New Meuse and the Old Meuse and by the harbour basins. The junction and the basins represent unique large scale geometric features of the estuary.

The effect of the tide on salt intrusion is determined by two counteracting mechanisms: the effect of the tide on turbulent mixing and stratification, and the effect of tidal amplitude on the tidal excursion path. For the Rotterdam Waterway a third mechanism has to be added: the phase differences at the junction.

The effect of the harbour basins on salt intrusion depends on two mechanisms, which in the Rotterdam Waterway were found to be counteracting: the temporary storage of salt water in the basins and their effect on the tide. Through their effect on the tide the harbour basins influence turbulent mixing and the stratification of the estuary.

From a modelling perspective, the above findings imply that the combined effect of the acting large-scale processes depends on the combined effect of the small-scale turbulent transports. Therefore, the small-scale turbulent mixing and the resulting stratification have to be reproduced with sufficient accuracy in salt intrusion modelling, including the vertical dimension, whether by a hydraulic scale model or by a laterally averaged two-dimensional ($2DV$) or a three-dimensional ($3D$) mathematical model. This applies to the Rotterdam Waterway the depth of which varies little over the width and where gravitational circulation is primarily in the longitudinal direction.

Thus far, a satisfactory performance could only be demonstrated for the hydraulic scale model, comparing its performance with several available field data sets (BREUSERS and VAN OS, 1984, VAN DER HEYDEN et al, 1984). Therefore, given their stage of development, the vertical density distributions supporting this paper were derived from the hydraulic scale model rather than from a $2DV$ or $3D$ Rotterdam Waterway mathematical model.

Acknowledgement

This paper describes part of a current study (ROELFZEMA et al, 1984) commissioned by Rijkswaterstaat (Ministry of Public Works).

References

ABRAHAM, G., 1980, On internally generated estuarine turbulence, Proceedings 2nd International Symposium on Stratified Flow, Vol. 1, Trondheim, Norway, pp. 344-353.

ABRAHAM, G., KARELSE, M. & LASES, W.B.P.M., 1975, Data requirements for one-dimensional mathematical modelling of salinity intrusion in estuaries, Proceedings 16th Congress of International Association for Hydraulic Research, Vol. 3, Sao Paulo, Brasil, pp. 275-283.

ALLEN, F.H. & PRICE, W.A., 1959, Density currents and siltation in docks and tidal basins, Dock and Harbour Authority, Vol. 40, no. 465, pp. 72-76.

BREUSERS, H.N.C. & VAN OS, A.G., 1981, Physical modelling of the Rotterdamse Waterweg, Proceedings American Society of Civil Engineers, Journal of Hydraulics Division, Vol. 107, No. HY 11, pp. 1351-1370.

DRONKERS, J.J., 1969, Some practical aspects of tidal computations, Proceedings 13th Congress of International Association for Hydraulic Research, Vol. 3, Kyoto, Japan, pp. 11-20.

FISCHER, H.B., 1972, Mass transport mechanisms in partially stratified estuaries, Journal of Fluid Mechanics, Vol. 53, pp. 672-687.

FISCHER, H.B., LIST, E.J., KOH, R.C.Y., IMBERGER, J. & BROOKS, N.H., 1979, Mixing in inland and coastal waters, Academic Press, New York.

HAMILTON, P., 1975, Numerical model for the vertical circulation of tidal estuaries and its application to the Rotterdam Waterway, Geophysical Journal, Royal Astrological Society, Vol. 40, pp. 1-21.

PERRELS, P.A.J. & KARELSE, M., 1981, A two-dimensional, laterally averaged model for salt intrusion in estuaries, Chapter 13 of Transport models for inland and coastal waters, Fischer, H.B., editor, Academic Press, New York.

RIGTER, B.P., 1973, Minimum length of salt intrusion in estuaries, Proceedings American Society of Civil Engineers, Journal of Hydraulics Division, Vol. 99, No. HY 9, pp. 1475-1496.

ROELFZEMA, A. & VAN OS, A.G., 1978, Effects of harbours on salt intrusion in estuaries, 6th International Conference on Coastal Engineering, Proceedings, Vol. 3, Hamburg, part 4, Chapter 172, 2810-2826.

ROELFZEMA, A., KARELSE, M., STRUYK, A.J. & ADRIAANSE, M., 1984, Waterquantity and quality research for the Rhine Meuse Estuary, Paper presented at 19th Conference on Coastal Engineering, Houston, Texas.

SCHIJF, J.B. & SCHÖNFELD, J.C., 1953, The motion of salt and fresh water, Proceedings Minnesota International Hydraulics Convention, Joint Meeting of International Association for Hydraulic Research and Hydraulics Division of American Society of Civil Engineers, pp. 321-333.

SMITH, T.J. & DYER, K.R., 1981, Mathematical modelling of circulation and mixing in estuaries, in Mathematical modelling of turbulent diffusion in the environment, Harris, C.J., editor, Academic Press, London, pp. 301-341.

THATCHER, M.L. & HARLEMAN, D.R.F., 1981, Long-term salinity calculation in Delaware Estuary, Proceedings American Society of Civil Engineers, Journal of Environmental Engineering Division, Vol. 107, No. EE1, pp. 11-27.

VAN DER HEIJDEN, H.N.M.C., DE JONG, P., KUYPER, K. & ROELFZEMA, A., 1984, Calibration and adjustment procedures for the Rhine-Meuse estuary scale model, Paper presented at 19th Conference on Coastal Engineering, Houston, Texas.

VAN REES, A.J., VAN DER KUUR, P. & STROBAND, H.J., 1972, Experience with tidal salinity model Europoort, 13th International Conference on Coastal Engineering, Vancouver, Proceedings, Vol. III, Chapter 135, American Society of Civil Engineers, New York, (1973), pp. 2355-2378.

VOLLMERS, H.J., 1976, Harbour inlets on tidal estuaries, 15th International Conference on Coastal Engineering, Proceedings, Vol. II, Hawaii, pp. 1854-1867.

WINTERWERP, J.C., 1983, Decomposition of the mass transport in narrow estuaries, Estuarine, Coastal and Shelf Science, Vol. 16, pp. 627-638.

Circulation and Salt Flux in a Well Mixed Estuary

Björn Kjerfve

Belle W. Baruch Institute for Marine Biology
and Coastal Research, University of South
Carolina, U.S.A.

ABSTRACT

The North Inlet, South Carolina, salt marsh estuary is dominated by tidal and coastal far-field forcing. Fresh water input is usually insignificant, explaining the absence of gravitational estuarine circulation. The circulation can be simulated from an instantaneous momentum balance of local longitudinal acceleration, pressure gradient, bottom friction, and viscous drag. Systematic deviations from the simulated solution indicate that the non-linear field acceleration is also an important circulation factor. In general, net currents flow in opposite directions on different sides of the same estuarine cross-section. The net longitudinal salt transport is largely a balance between ebb-directed advective salt flux and flood-directed dispersive tidal flux due to triple-correlation between tide, velocity, and salinity. The salt flux is in general not in steady state from one tidal cycle to the next and there is a pronounced transverse circulation contribution to the mean advective salt flux.

1. Introduction

The North Sea Inlet system in South Carolina consists of 32.2 km² of *Spartina alterniflora* salt marshes, intersected by numerous, interconnecting, winding estuarine creeks (Fig. 1). It is typical of many of the estuarine systems located between Cape Hatteras and Cape Canaveral on the Atlantic coast of the southeastern United States.

Because of the role that salt marshes might play in support of high productivity in coastal waters (NIXON, 1980) the North Inlet system has since the mid 1970's been the focus of intensive field studies. Foremost of these, material fluxes were measured directly in an experimental cross-section (Fig. 1) by simultaneously making velocity measurements and drawing water samples. The main objective was to assess the extent to which organic matter and nutrients (principally various forms of carbon, nitrogen, and phosphorus) are exchanged between the coastal ocean and a salt marsh estuary, which is relatively undisturbed by man's impact. Some of the results of this so called *Outwelling Project* are now available (CHRZANOWSKI et al., 1982; WHITING et al., in Press).

The flux measurements were made by cross-correlating velocity and material concentrations, based on measurements from three stations in the experimental cross-section every meter and 1.5 lunar hours for 32 tidal cycles spread over a year. Preliminary calibration studies with as many as 11 cross-sectional stations indicated that 3 stations in a section would be an optimum choice, considering both logistics and the potential error (KJERFVE et al 1981). One conclusion was that the fluxes are considerably more sensitive to velocity variations as compared to concentration changes. In general, a good understanding of the hydrodynamic characteristics of the system is essential in minimizing the number of stations, velocity measurements, and nutrient analyses, and thus the cost of carrying out a field study.

The objective of this paper is to provide a description of the hydrodynamic characteristics of the North Inlet system, focusing on circulation and the longitudinal salt balance. Presumably, many other salt marsh estuaries with characteristics similar to North Inlet will exhibit the same dynamic behavior.

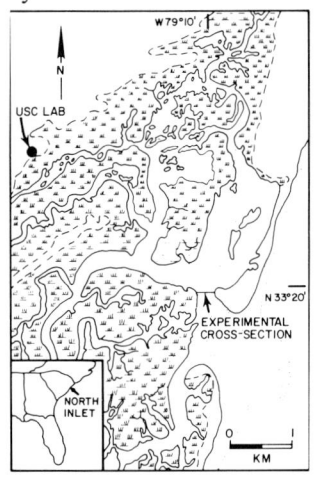

FIGURE 1. North Inlet area map showing the experimental cross-section where flux measurements were made and the location of the University of South Carolina Belle W. Baruch Coastal Field laboratory.

2. Description

The North Inlet system is tidally dominated. It is characterized by a semidiurnal tide with an average range of 1.5 m. The range varies from 0.9 m on a neap tide to 2.5 m on an extreme spring tide. The associated peak tidal currents are usually 1.4 m/s but have been measured to exceed 2.3 m/s. The tidal form number (DEFANT 1960) is defined as the ratio of the main semidiurnal ($M_2 + S_2$) to the main diurnal ($K_1 + O_1$) constituent amplitudes and measures 0.24 for North Inlet.

As the tidal creeks are very shallow, only 3 m on the average and nowhere deeper than 8 m, the tidal range constitutes a significant fraction of the total water depth. As a result, the tidal currents are relatively uniform from the surface to a distance very close to the bottom. Approximately 40% of the total water volume leaves the estuarine system on each ebbing tide. Thus, the average hydrodynamic residence time is extremely short, only 15.5 hours.

Because of limited freshwater runoff from the adjacent forest drainage basin, the salinity is usually high, in the range from 31-34 ppt. However, subsequent to periods of intense rainfall, the salinity of the inner ends of the marsh creeks may experience salinities as low as 4 ppt for up to a couple of days at the time. Estimates of the mean freshwater supply range from 1 to 5 m^3/s.

Because of shallow water depths and intense tidal mixing, the North Inlet creeks are well-mixed and almost vertically homogeneous. Based on several thousand vertical profiles of velocity and salinity collected since 1974, the North Inlet estuarine system may be classified as *la* according to HANSEN and RATTRAY (1966). This implies that gravitational circulation is absent and vertical stratification extremely weak. Dynamically, this place North Inlet among coastal lagoons.

3. Circulation

KJERFVE (1978, 1979) has described the salient characteristics of the net circulation in the North Inlet system. The main feature of the circulation is the net lateral reversal of currents in each estuarine cross-section. Net ebb flow dominates from surface to bottom in the deepest part of the cross-section. However, on the other side of the cross-section, net flood-directed currents dominated from surface to bottom. Gravitational vertical-longitudinal circulation is absent in North Inlet on account of the very limited fresh water supply and shallow water depths. PRITCHARD (1955) refers to this type of a system with a lateral-longitudinal circulation as a *type C* estuary.

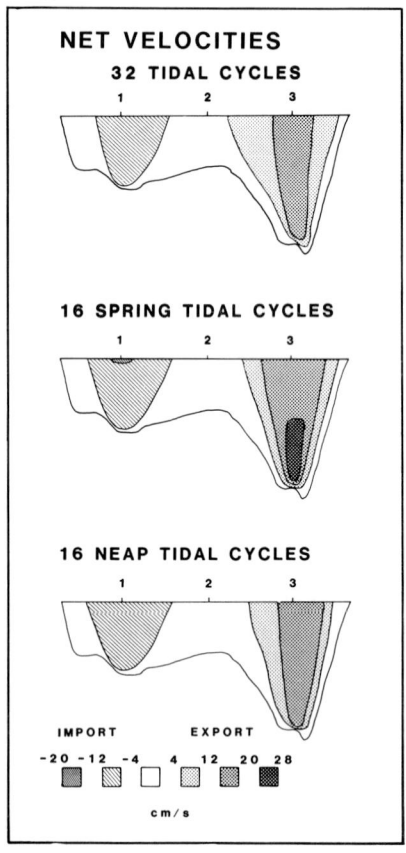

FIGURE 2. Time-averaged isotachs on velocity profiles made simultaneously at 3 stations every 1.5 hours for 32 tidal cycles in 1979. The 32 tidal cycles include 16 neap and 16 spring tides.

The cross-sectional distribution of net longitudinal currents is shown in Fig. 2. These results are based on measurements at three stations in the experimental cross-section (Fig. 1) for 32 tidal cycles in 1979, with vertical velocity profiles measured every 1.5 lunar hours from surface to bottom at each station. These data are descibed in greater detail by KJERFVE (submitted). In the net currents and the net isotach plots are calculated in the manner described by KJERFVE and SEIM (1984).

The intensity of the lateral-longitudinal circulation varies with the tidal forcing. During spring tides (average range 1.8m), the net ebbing and flooding currents are significantly stronger as compared to the net currents during neap tides (average range

1.1 m) (Fig. 2). Intuitively, the existence of this lateral-longitudinal circulation has implications with respect to exchange of sediments, nutrients, organic matter, larvae, etc., between the estuary and the coastal ocean.

A numerical model has been used to simulate the depth-averaged currents in the experimental cross-section (UNCLES and KJERFVE, in Press). The instantaneous longitudinal acceleration is balanced by tidal surface slopes, friction, and viscous drag, and the resulting velocities compare well with the observed pattern. The model is expressed

$$h \, \partial u / \partial t = -gh \, \partial \eta / \partial x - D|u|u + \partial(v_y h \partial u / \partial y) / \partial y \tag{1}$$

where h is the local water depth; u longitudinal current; g gravity; t time; x and y longitudinal and transverse coordinates, respectively; D a bottom friction factor; v_y transverse eddy viscosity; and the portion of the pressure gradient term, $g \partial \eta / \partial x$, is taken to oscillate over the tidal cycle around a mean value. Although this model seems to explain most of the circulation, small systematic deviations from the model point to the importance of the non-linear field acceleration term, $u \partial u / \partial x$. These deviations become especially pronounced during spring tides with increased current velocities and changing cross-sectional width during high tides.

4. Longitudinal salt flux

Instaneous vertical profiles of velocity and salinity for 10 stations across the North Inlet experimental cross-section (cf. Fig. 1) were used to compute instanteneous and time-averaged salt flux estimates.

Many authors have recently decomposed the longitudinal salt flux to determine the contribution by vertical, transverse, and cross-sectional component effects (HANSEN and RATTRAY, 1966; FISCHER, 1972; DYER, 1974; MURRAY and SIRIPONG, 1978). The contention has been that transverse effects contribute as much as vertical effects to the overall salt transport. RATTRAY and DWORSKI (1980), however, showed in a convincing way that this contention is usually an artifact of how the estuarine cross-section most often is partioned. Instead, they concluded that vertical effects dominate the mean advective salt flux and that the transverse circulation contribution is small because of poor correlation between transverse velocity and salinity variations. Further, the various vertical-lateral decomposition schemes give rise to numerous flux terms without a clear physical meaning, making interpretation difficult.

For these reasons, I have chosen to decompose the flux in the vertical for each station, and then compare the results. Differences between the stations can then be attributed to transverse circulation effects. If differences between stations are significant, they will be noticible in a system such as North Inlet and indicate the degree to which the lateral circulation contributes to the mean advective salt flux.

To evaluate the dispersion mechanisms, the net salt flux was decomposed in manner similar to BOWDEN (1963). His analysis had to be modified slightly to account for the effect of tidally varying water depth and is for this reason given below.

In general, the salt flux per cross-sectional width (F) can be calculated for each station in a cross-section as

$$F = \int_0^h us \, dz \tag{2}$$

where u is instantaneous velocity normal to the cross-section; s is salinity; h is station water depth; and z is the vertical coordinate. I have assumed that both u and s have been filtered to remove turbulent fluctuations shorter than a couple of minutes.

Experimental data indicate that longitudinal salt fluxes associad with such short-term turbulence fluctuations are small compared to other dispersion effects. Velocity and salinity are decomposed according to $u = \bar{u} + u'$ and $s = \bar{s} + s'$. The primed variables represent deviations from the depth-averaged variables, $\bar{u}$ and $\bar{s}$. Clearly, u' and s' are functions of both time and depth. The depth-averaged quantities are further decomposed into time-averaged and a time-varying components

$$\bar{u} = <\bar{u}> + U_T \tag{3}$$

$$\bar{s} = <\bar{s}> + S_T \tag{4}$$

where the time-averaging operation is designed by the bracketed terms and represent an average over one or more complete tidal cycles.

Substitution of u and s into Eq. (2) is followed by substitution of Eqs. (3) and (4) and then time-averaging the result to get

$$<F> = \frac{1}{nT} \int_0^{nT} h \{ <\bar{u}><\bar{s}> + <\bar{u}>S_T + <\bar{s}>U_T +$$

$$U_T S_T + \frac{1}{h} \int_0^h u's' dz \} dt \tag{5}$$

The left hand side of Eq. (5) may be expressed as

$$<F> = <h\overline{us}> \tag{6}$$

and Eqs. (5) can then be rewritten as

$$<F> = <h><\bar{u}><\bar{s}> + <hU_T S_T> + <\bar{u}><hS_T> +$$

$$<\bar{s}><hU_T> + <\overline{hu's'}> \tag{7}$$

Then net longitudinal salt transport per unit width can now be calculated for each station based on simultaneous vertical velocity and salinity profiles over at least one tidal cycle. The net longitudinal flux is a balance of an advective flux and a number of dispersive fluxes (cf. Fig. 3). The turbulence flux usually is several orders of magnitude smaller than the other fluxes and its derivation has not been included.

The large total flux estimates in each of the three regions of the cross-section point to the fact that the salt flux is seldom in steady state over a few tidal cycles. This is a manifestation of far-field coastal forcing of the North Inlet system.

The net salt balance in this well mixed estuary is largely made up of ebb-transport due to advective salt flux balanced by flood-directed transport due to *tidal sloshing*. *Tidal sloshing* is represented as the triple correlation between tidal depth changes, tidal currents, and tidal salinity changes, $<hU_T S_T>$, and is the dominant dispersive flux. It represents the net salt transport by oscillatory tidal currents over one or more tidal cycles. Two other tidal dispersive terms are of lesser importance. Contrary to estuaries with vertical stratification, BOWDEN (1963), the shear flux term is negligible in the well-mixed North Inlet system.

Region of Cross Section	Total Flux $<F>$	Advective Flux	$<hU_TS_T>$	Tidal Fluxes $<\bar{u}><hS_T>$	$<\bar{s}><hU_T>$	Shear Flux
1 (Flood Channel)	-327.7	-19.2	-315.0	+0.2	+6.4	-0.1
2 (Mid Region)	-190.7	+68.4	-234.7	+0.3	-24.3	-0.1
3 (Ebb Channel)	+136.2	+288.3	-127.2	0.0	-24.2	-0.7

TABLE 1. Summary of salt flux computations in the experimental North Inlet cross-section. Regions 1, 2, and 3 estimates are based on 4, 3, and 3 stations; respectively, with simultaneous measurements made each 0.5 hour for 2 tidal cycles in 1977. Each one of the above fluxes has been divided by the corresponding mean water depth. Thus, the units are ppt. cm/s. Positive fluxes correspond to ebb-transport (export) and negative fluxes to flood-transport (import).

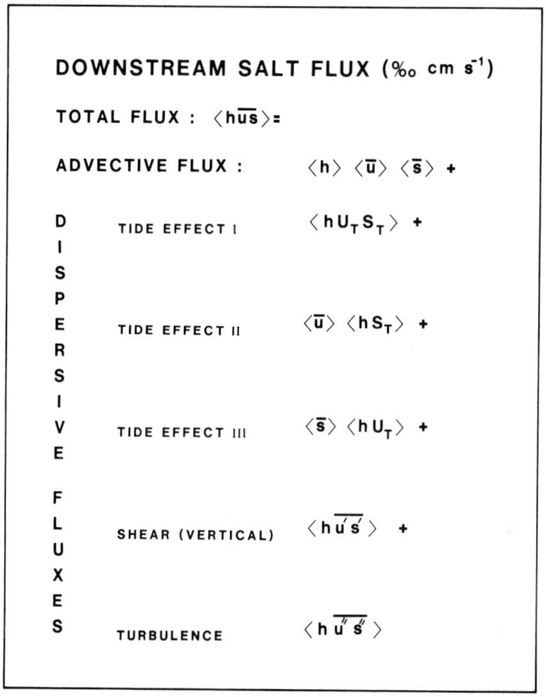

```
DOWNSTREAM SALT FLUX (‰ cm s⁻¹)

TOTAL FLUX :  ⟨hūs⟩ =

ADVECTIVE FLUX :        ⟨h⟩ ⟨ū⟩ ⟨s̄⟩ +

D
I
S       TIDE EFFECT I     ⟨hU_T S_T⟩ +
P
E
R
S       TIDE EFFECT II    ⟨ū⟩ ⟨hS_T⟩ +
I
V
E       TIDE EFFECT III   ⟨s̄⟩ ⟨hU_T⟩ +

F
L       SHEAR (VERTICAL)  ⟨hu's'⟩ +
U
X
E
S       TURBULENCE        ⟨hu″s″⟩
```

FIGURE 3. Interptation of the longitudinal salt dispersion terms.

It is interesting to note that *tidal sloshing* always acts to bring salt into the estuary, independent of the region of the cross-section. However, the advective flux varies significantly from the ebb to the flood side of the cross-section. In the ebb-channel, the advective flux is strongly positive and dominates the import of salt due to *tidal sloshing*. However, in the flood channel, the advective flux is close to zero and the total import of salt in the flood channel is largely a function of the *tidal sloshing* flux. The difference in advective fluxes from one side of the cross-section to the other points to the important role of the transverse circulation in the mean longitudinal advective salt flux.

5. Conclusions and future directions

North Inlet is a shallow well mixed estuary characterized by far-field forcing and a tidally driven net circulation. The system is in general not in steady state as evidenced by sub-tidal water level variations, net current variability, and large changes in the longitudinal salt flux from tidal cycle to tidal cycle. Classical estuarine (gravitational) circulation is absent in this type 1a estuary.

The investigation of the North Inlet estuarine system continues. The main focus is presently the extent to which adjacent coastal waters are estuarine in nature. The existence of a turbid coastal boundary layer within a couple of kilometers of the coast, separated from shelf waters by a frontal feature, and largely having water characteristics of estuarine rather than coastal characteristics, implies that studies of material transports should not necessarily be carried out at estuarine entrances. Rather, the extent to which estuaries of the Southeast export materials to the nearshore shelf waters, can probably best be assessed by estimating the fluxes across the coastal boundary layer front.

Acknowledgements

This is Contribution No. 570 from the Belle W. Baruch Institute for Marine Biology and Coastal Research. Funding for this project came from NSF grants DEB-76-83010 and DEB-81-19752.

References

BOWDEN, K.F. 1963. The mixing processes in a tidal estuary. International Journal of Air and Water Pollution 7: 343-356.

CHRZANOWSKI, T.H., STEVENSON, L.H. & SPURRIER, J.D. 1982. Transport of particulat organic carbon through the North Inlet ecosystem. Marine Ecology-Progress Series 7: 231-245.

DEFANT, A. 1960. Physical oceanography. Vol. 2. Pergamon Press. New York. 598 pp.

DYER, K.R. 1974. The salt balance in stratified estuaries. Estuarine and Coastal Marine Science 2: 273-281.

FISCHER, H.B. 1972. Mass transport mechanisms in partially stratified estuaries. Journal of Fluid Mechanics 53: 671-687.

HANSEN, D.V. 1965. Currents and mixing in the Columbia River estuary. pp. 943-955. In: Ocean Science and Ocean Engineering, Vol. 2. The Marine Technology Society, Washington, DC.

HANSEN, D.V. & RATTRAY, M., Jr. 1966. New dimensions in estuary classification. Limnology and Oceanography 11: 319-326.

KJERFVE, B. 1978. Bathymetry as an indicator of net circulation in well mixed estuaries. Limnology and Oceanography 23: 814-821.

KJERFVE, B. & SEIM, H.E. 1984. Construction of net isopleth plots in cross-sections of tidal estuaries. Journal of Marine Research 42: 503-508.

KJERFVE, B. & PROEHL, J.A. 1979. Velocity variability in a cross-section of a well-mixed estuary. Journal of Marine Research 37: 409-418

KJERFVE, B., STEVENSON, L.H., PROEHL, J.A., CHRZANOWSKI T.H. & KITCHENS, W.M. 1981. Estimation of material fluxes in an estuarine cross-section: A critical analysis of spatial measurement density and errors. Limnology and Oceanograpy 26: 325-335.

KJERFVE, B. Submitted. Water and salt fluxes in a salt marsh estuary: Hydrographic summary of Project Outwelling.

MURRAY, S.P. & SIRIPONG, A. 1978. Role of lateral gradients and longitudinal dispersion in the salt balance of a shallow well-mixed estuary. pp. 113-124. In: Estuarine Transport Processes. B. Kjerfve (ed.). University of South Carolina Press, Columbia, SC.

NIXON, S.W. 1980. Between coastal marshes and coastal waters a review of twenty years of speculation and research on the role of salt marshes in estuarine productivity and water chemistry. pp. 437-525. In: Estuarine and Wetland Processes. P. Hamilton and K.B. McDonald (eds.). Plenum Publishing Corp., New York.

PRITCHARD, D.V. 1955. Estuarine circulation patterns. Proceedings of the American Society of Civil Engineers 81: 1-11.

Rattray, M., Jr. & Dworski, J.G. 1980. Comparison of methods for analysis of the transverse and vertical circulation contributions to the longitudinal advective salt flux in estuaries. Estuarine and Coastal Marine Science 11: 515-536.

Uncles, R.J. & Kjerfve, B. In press. Transverse structure of residual flow in North Inlet, South California. Estuaries.

Whiting, G.J., McKellar, H.N., Jr., Kjerfve, B. & Spurrier, J.D. In press. Sampling and computational design of nutrient flux from a southeastern salt marsh. Estuarine, Coastal and Shelf Science.

Currents and Salinity Transport in the lower Elbe Estuary: some Experiences from Observations and Numerical Simulations

K.C. Duwe & J. Sündermann

Institut für Meereskunde, University of Hamburg
Federal Republic of Germany

ABSTRACT

Experiences from field and numerical investigations are presented on the dynamics and mass transport in a partially mixed estuary with strong tidal currents and a pronounced brackish water zone. The numerical model system used comprises a three-dimensional semi-implicit circulation model with a flux-corrected transport scheme which proved to give satisfactory results in describing circulation patterns due to the interaction of tidal currents and the horizontal density gradient caused by variable salinity. Some specific field observations are discussed and compared with numerical computations of the circulation of the Lower Elbe.

1. Introduction

The Lower Elbe estuary may be characterized as a transition zone between the limnic and marine regimes with a high tidal range and a significant freshwater discharge (see Fig. 1). Its brackish water zone varies in position and length depending on tidal amplitude, fresh water discharge and on meteorological conditions. Apart from the few situations where strong westerly winds prevail the boundary of the brackish water zone hardly extends further than 50 km inland. A seaward boundary is far more difficult to define: the relatively fresh surface water of the River Elbe is transported with the strong ebb currents into the German Bight where it mixes gradually with sea water and flows northward with the mean flow. Within this very vaguely defined area the salinity varies between 2 and 35 ppt; the purely tidally induced variation at a certain position may amount to 4 - 8 ppt. In special circumstances near front-generating zones this range may be exceeded significantly.

A tidal range of two to four meters implies strong currents (up to 2 m/s) accounting for the high transport rates through the narrow cross-section of the waterway within the short time of half a tidal period. The current system shows very complex patterns around the extensive tidal flats and in the deep tidal channels, which effect the ebb and flood currents differently. Whereas the tidal flats experience flooding only for a few hours in a tidal cycle providing the dynamics with their distinct directional characteristics, the deep river bed is important for the overall transport capacity of the River Elbe. Here large water masses are being moved back and forth influencing the distribution of transported substances and properties of the water. Strong current shear both horizontally and vertically adds to the complexity of this dynamic regime.

The discharge of the river varies normally between 400 m^3/s in September and 1200

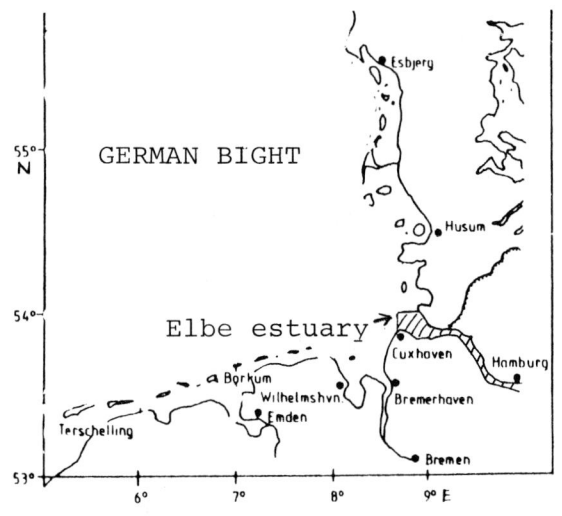

Figure 1: Location of investigated area

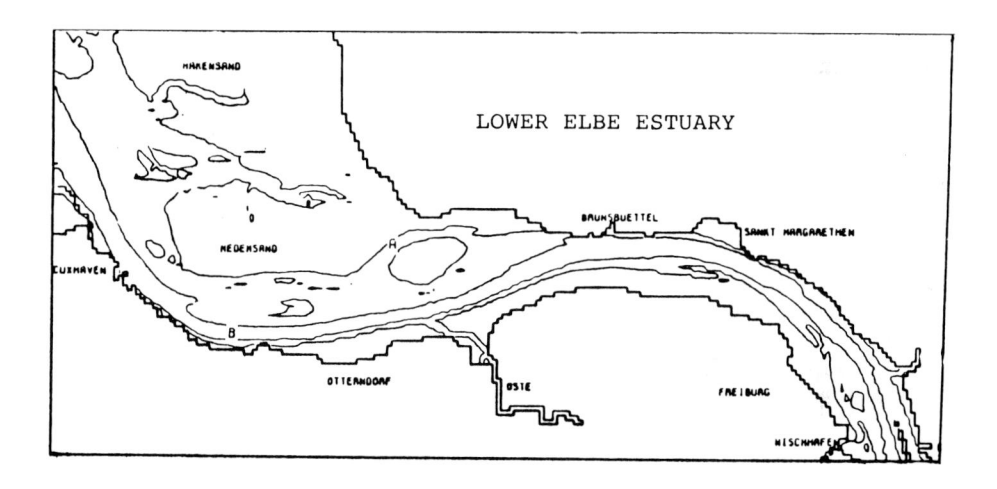

FIGURE 2: Tidal flats and waterways in the Lower Elbe estuary. Point A: Velocity and salinity measurements in Fig. 3; Point B: Velocity measurements in Fig. 5.

m^3/s in April; the additional discharges of its tributaries between Hamburg and Cuxhaven amount to 100 and 260 m^3/s respectively. River discharge not only influences the rate of salinity variation in different river sections, but also has an effect on the water level elevations. Accordingly a *mean* tide for the Elbe estuary is not easy to define especially since the water level depends also in a nonlinear way on wind direction, on the actual tidal range and the mean water level in the German Bight. Due to high current velocities further features of the Lower Elbe estuary are high rates of erosion and sedimentation as well as a very variable distribution of suspended matter. In the following experiences from field and numerical investigations are presented on the dynamics and mass transport in the Lower Elbe estuary.

2. The influence of salinity variations on the dynamics and mechanisms for the development of fronts in the Elbe

Apart from tides and wind forcing the variable mass distribution causes non-negligible currents in the transition zone between salt and fresh water. In the Elbe, currents due to varying salinity are much more important than the seasonal thermally induced circulation. The salinity variations do not only cause a modification of the pressure gradient and its corresponding accelerations, but the stratified medium also influences the character of turbulent mixing. Therefore the mass distribution has to be taken into account considering transport and mixing phenomena of passive substances, pollutants or warm water. Strong gradients of salinity have the effect of barriers for suspended matter, nutrients and even oxygen, that are most pronounced at fronts. Moreover the temporal changes of salinity have their implications for the biological circumstances in the estuary and the adjacent tidal flats and salt marshes. A quantification of the salinity variability in this respect is highly desirable, too. The vertical distribution of salinity is strongly dependent on the rate of turbulent mixing which is particularly strong at times of flooding. Then a high vertical shear of velocities (surface velocities surpass the bottom ones) results in overturning of water. At ebb tide lower mixing rates occur due to the buoyancy effect (light water remains above denser water). Other mixing processes are naturally caused by wind forcing and (in the waterways) even by ship movements.

Field measurements (RIETHMÜLLER, DUWE and PFEIFFER, 1984) seem to indicate that the speed of horizontal velocity and the vertical velocity gradient have an overriding importance for the vertical eddy viscosity as against vertical stratification. There is a strong reduction of vertical eddy viscosity during slack water favouring a stronger influence of the horizontal density gradient at the bottom. One aspect of this is that the baroclinic pressure gradient is most obvious at high water causing the surface currents first to turn, whereby the bottom currents are further accelerated by the density driven forcing.

The strongest salinity gradients occur in the Lower Elbe normal to sharp depth gradients. Topographically induced current shear and advection at certain places generate strong but relatively shortlived frontal systems. As an example in Fig. 3 field measurements are shown made in a tidal channel near the edge of the tidal flats of the Neufelder Watt which is flooded by salty water originating from the German Bight (location see point A in Fig. 2). At high water this clashes with relatively fresh water of the Elbe river, plunging down the edge and flowing under the river water. This shows up in the measurements where the saltier water reaches the measurement position first at the bottom at 3 meters depth. This causes an instantenuous change in current direction pointing southward into the main channel. Therefore the source of the saltier water cannot be the main waterway. The surface currents remain going northeastward like one would expect during flood tide. Only about an hour later the surface front reaches the same horizontal position which was only 200m away from the edge of the tidal flats. This special feature of *frontogenesis* occurs once a tidal cycle and is confined to the southeastern edge of the tidal flats north of the main waterway.

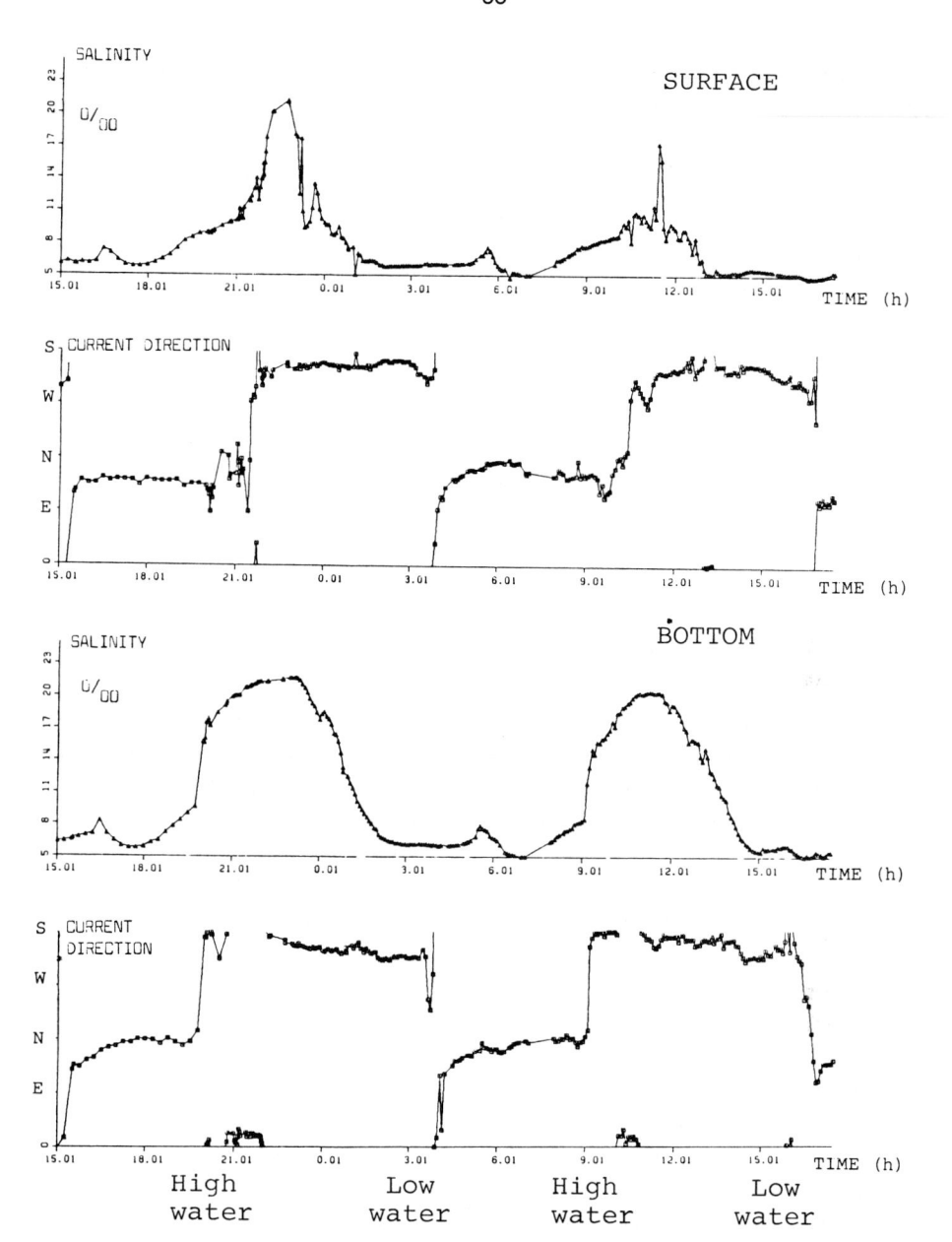

FIGURE 3: Measurements of salinity and current direction in Point A (Fig. 2) on 9th and 10th May 1984 near surface and bottom (3 m below msl). (RIETHMÜLLER, DUWE and PFEIFFER, 1984)

3. Modelling the currents and salinity transport in the Elbe

3.1. Goals of experiments and numerical modelling

In order to gain a more profound insight into the estuarine system of the Lower Elbe considerable field work and theoretical studies have been done in close cooperation between the Institut für Meereskunde (Universität Hamburg) and the GKSS Forschungszentrum Geesthacht, since 1981. Extensive measurements were being made between 1982 and 1984 in the area to obtain information about the temporal and vertical variations of velocities, salinity, temperature and suspended matter. These experiences were used in the establishment of a highly resolving numerical model designed to forecast waterlevels, currents and the transport of passive substances, of salt and temperature. The model is seen as a basic tool to get better information for answering questions about water quality and pollution problems, sedimentation and erosion, the distribution of suspended matter and the impact of dams or dredging. The following considerations and results are drawn from these aforementioned activities.

3.2. The semi-implicit circulation model

The circulation model used in the numerical investigations is based on a semi-implicit two-time level scheme introduced by BACKHAUS (1983). This was modified to cope with shallow water problems and falling-dry processes by DUWE and HEWER (1982). A series of verifications and the three-dimensional generalization used here were published by DUWE, HEWER and BACKHAUS (1983).

In the computations described below it was found necessary to treat the vertical eddy viscosity term implicitly, since the time step exceeded the value given by the explicit stability criterion. Because the term cancels out after vertical integration (the boundary conditions are taken semi-implicitly anyway), it does not effect the implicit scheme as such. The actual computations of the vertical velocity profiles are made after the water level and its corresponding barotropic pressure gradient have been derived from the divergence terms of the vertically integrated transports. The vertical eddy viscosity coefficient A_v was chosen according to results of PERRELS and KARELSE (1982) in tidal flume experiments, i.e. depending on vertical current shear and stratification:

$$A_v = L_m^2 \cdot |u_z| \cdot e^{-4 \cdot Ri}$$

with a mixing length L_m depending on the distance of the surface

$$L_m = k \cdot (z + z_0) \text{ for } 0 \leqslant z \leqslant 0.25 \cdot H$$

$$L_m = k \cdot (0.25 \cdot H + z_0) \text{ for } 0.25 \cdot H \leqslant z \leqslant H$$

where

H actual water depth

k von Kármán's constant

u horizontal velocity component

z Cartesian co-ordinate in upward direction

z_0 roughness length

The dependence on the local Richardson number Ri did not have a significant effect on the computed currents (apart from strong stratification during slack water), but the dependence on the speed and on the vertical shear of the horizontal velocities was found to be of crucial importance. This was especially the case during slack water when field measurements seem to indicate a strong de-coupling of bottom and surface currents. In this period the vertical velocity gradients were in the order of 0.02 s^{-1}, giving a significantly smaller eddy viscosity than during flood and ebb tide. During most of the tidal cycle however the velocity profile varies between nearly logarithmic and linear depending on topography if the wind effect can be ignored.

The quadratic bottom friction coefficient was held constant ($r = 0.0025$). It was found that a better computation of the near-bottom velocities did improve the results significantly more than a variation of r. No horizontal eddy viscosity was used in the computations.

3.3. The salt transport algorithm

For the computation of the advective salinity transport the finite difference algorithm used was first proposed by BOOK et al. (1975). This scheme was applied to the conventional upstream algorithm with a flux-corrected transport step giving a scheme with second order accuracy. It was relatively straightforward to implement even accounting for shallow water pecularities such as tidal flats and strong current shear. Moreover it proved to be computationally fast enough to enable a high spatial resolution which was necessary for the three-dimensional simulation of the salinity field variations. The turbulent mass exchange coefficient, $D_{V'}$, was also chosen according to PERRELS and KARELSE (1982) to enable a consistent modelling approach:

$$D_V = L_m^2 \cdot |u_z| \cdot e^{-18 \cdot Ri}$$

with u_z vertical gradient of horizontal velocity u.

Molecular diffusion processes were neglected as they are negligible compared with turbulent mixing in the area considered. If one would aim at resolving salinity variations with a spatial wavelength of less than say 10 times the grid size the scheme applied here would show its limitations. However in the investigations described below the finite difference approach proved to be sufficient to give satisfactory results on the larger scale tidally and density driven circulation.

3.4. The Lower Elbe model and numerical results

The following results are derived from a three-dimensional model of the estuary with a horizontal resolution of 250 m and a vertical resolution of 1.5 m (2 m at the surface). The depth distribution of the model is shown in Fig. 4. It has 22.787 *wet* grid points and is run with a time step of 150 s which is made possible by using the semi-implicit scheme. At the open boundaries water levels are prescribed taken from models of the German Bight and the Upper Elbe estuary, respectively. River discharges are taken into account at 5 positions. A *mean* salinity variation is prescribed at the western boundary, in the east fresh water is assumed.

In Fig. 5 results of model computations for the current speed in the main waterway (see point B in Fig. 2) are compared with measurements for a tidal cycle on the 5th October 1982. Clearly shown is the phase lag between surface and bottom current at 10 m depth. At low water the bottom friction causes the bottom current first to turn, whereas at high water the baroclinic pressure gradient more than compensates this phenomenon to the effect that the current turns first at the surface. The qualitatively

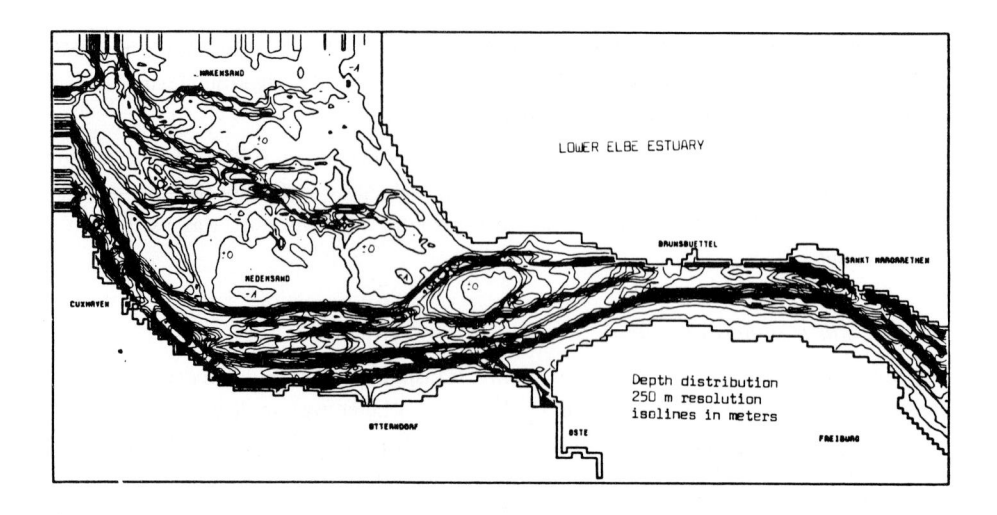

FIGURE 4: Topography of the three-dimensional Lower Elbe model; isolines are given in meters below msl.

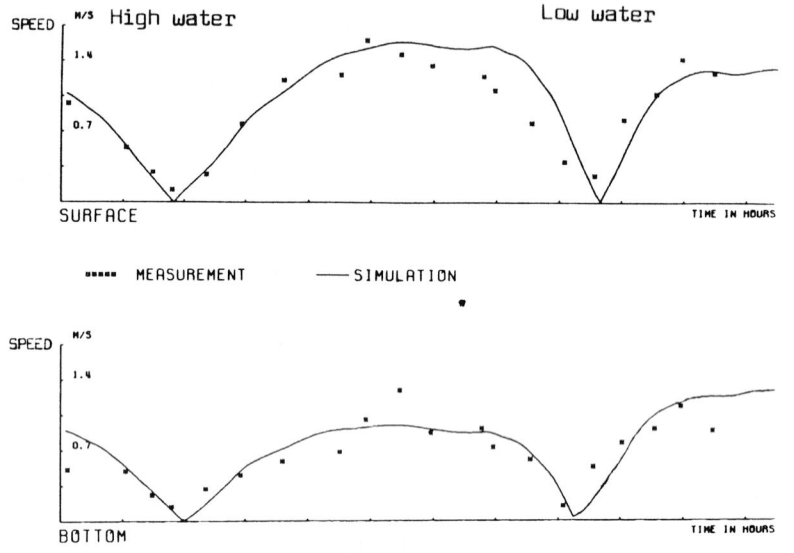

FIGURE 5: Bottom and surface current speed at position B (Fig. 2): measurement and computation for the 5th October 1982 (RIETMÜLLER, DUWE and PFEIFFER; 1984)

good agreement between observations and numerical results indicates that the circulation patterns describe the general current field in the estuary satisfactorily. As an example of the variation of the salinity field during a tidal cycle the surface distribution of density is given as computed by the model for a typical april situation with high fresh water discharge during neap tide (see Fig. 3, 6 and 7). Especially interesting is the formation of a very pronounced density front during high tide at the very position where measurements were taken for Fig. 3. The identations of the isolines in or near the waterway show the effect of strong horizontal current shear between the deep river bed (mean depth of 14 m) and the adjacent shallow areas. Here tendency to some sort of frontogenesis is shown very marked in those areas where observational experience would like to see them.

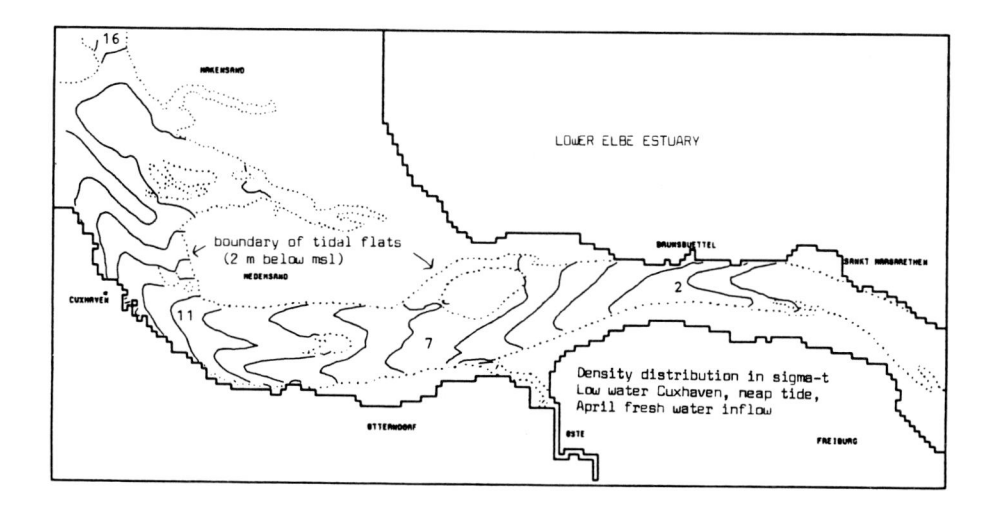

FIGURE 6: Density distribution in sigma-t during low water at Cuxhaven. Computation for neap tide and April (i.e. high) fresh water discharge

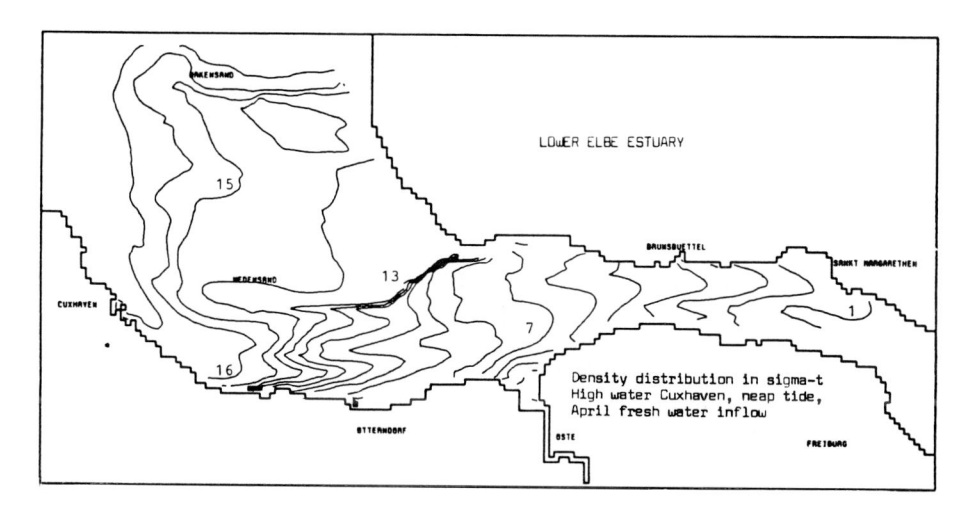

FIGURE 7: Density distribution for the same conditions described above during high water at Cuxhaven

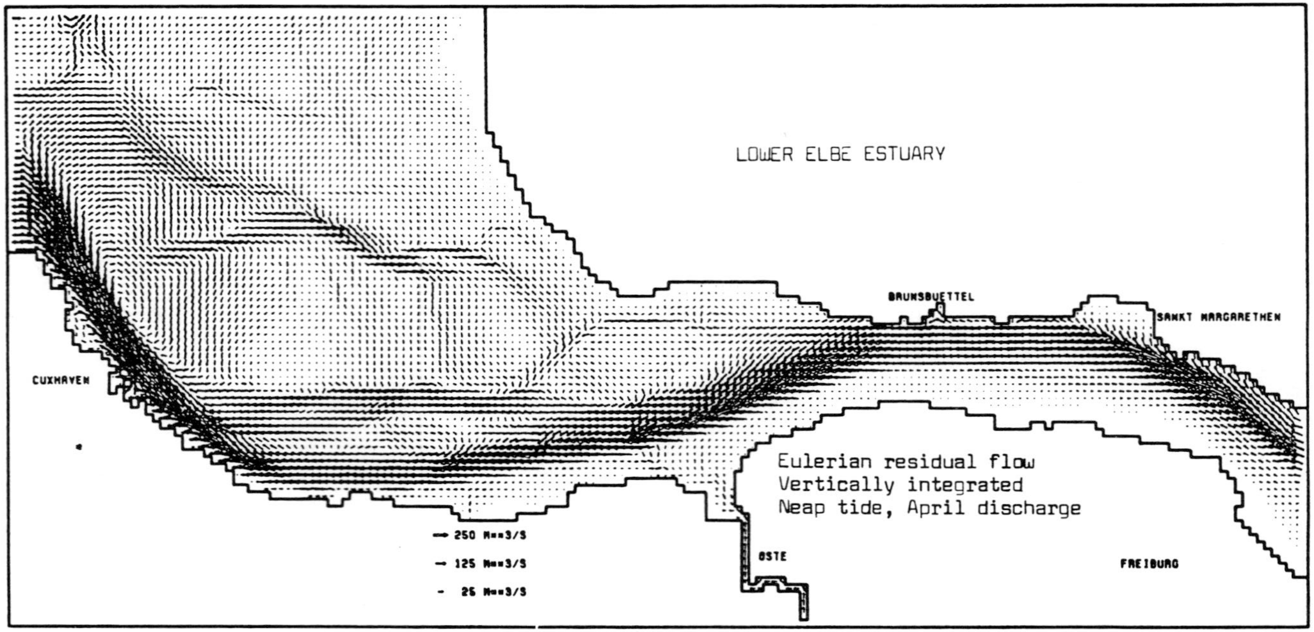

FIGURE 8: Vertically integrated Eulerian residual transport for neap tide conditions and April discharge

Fig. 8 presents the Eulerian residual transport for the same situation showing strong downstream transports in the deep river bed accompanied by upstream transports on the tidal flats. The amount of water transported back and forth during a tidal cycle is computed to be 10 times larger than the normal river discharge pointing to a longer residence time than might be expected at first sight. Comparable to the normal fresh water discharge is the inflow of seawater with high salinities during high water over the tidal flats north of the river making the edge of these sands one of the most interesting areas for investigations of salt intrusion in the Lower Elbe estuary.

4. Conclusion

The combination of the semi-implicit three-dimensional circulation model and the flux-corrected transport algorithm has proved to be a useful means for describing qualitatively the dynamics in a tidal estuary like the Lower Elbe with respect to the salinity distribution during a tidal cycle is simulated satisfactorily, as well, within the resolution limits of the model system. However the development of very sharp gradients or even fronts can only be simulated by the finite difference scheme to a certain extent, showing the need for further work on more sophisticated transport schemes like e.g. the Lagrangian tracer technique. These are, however, at present computationally not feasible to implement in such a large circulation model as presented here.

Acknowledgments

The authors would like to thank Messrs. R. RIETHMÜLLER and K. PFEIFFER for their cooperation in the evaluation of data material used in this paper. This work has been supported by the GKSS Forschungszentrum Geesthacht.

References

BOOK, D.L. BORIS, J.P. and HAIN, K., 1975. Flux-corrected transport II: Generalizations of the method. J. of Comput. Physics 18.

BACKHAUS, J.O., 1983. A semi-implicit scheme for the shallow water equations for application to shelf sea modelling. Continental Shelf Research, 2, pp. 243-254.

DUWE, K.C. and HEWER, R.R., 1982. Ein semi-implizites Gezeitenmodell für Wattgebiete. Deutsche Hydrographische Zeitschrift, 35, pp. 223-238.

DUWE, K.C. HEWER, R.R and BACKHAUS, J.O., 1983. Results of a semi-implicit two-step method for the simulation of markedly nonlinear flow in coastal seas. Continental Shelf Research, 2, pp. 255-274.

PERRELS, P.A.J. and KARELSE, M., 1982. A two-dimensional, laterally averaged model for salt intrusion in estuaries. Delft Hydraulics Laboratory. Publication No. 262.

RIETHMÜLLER, R. DUWE, K.C. and PFEIFFER, K., 1984. Personal communication. Measurements were made by GKSS Forschungszentrum Geesthacht.

ZALESAK, S.T., 1979. Fully multi-dimensional flux-corrected transport algorithm for fluids. J. of Comput. Physics 31.

III

RESIDUAL CURRENTS

Generalised Theory of Estuarine Dynamics

D. Prandle

Institute of Oceanographic Sciences, U.K.

ABSTRACT

The aim of this paper is to encourage interest in simplified theories relating to estuarine dynamics. Fundamental characteristics of estuarine dynamics are illustrated and explained using such simplified theory. These characteristics include tidal response (elevations, depth-mean currents and vertical current structure) and residual circulation associated with river flow, wind-forcing and density gradients (both well-mixed and fully stratified). Using throughout a consistent dynamical approach, this theoretical framework is extended to examine (1) the length of saline intrusion and (ii) the factors governing the levels of stratification in estuaries. At all stages, the theory developed is related to field observations from both major estuaries and flume studies.

1. Introduction

The purpose of this paper is to encourage interest in simplified theories relating to estuarine dynamics. Armed with a basic theoretical framework, researchers can more readily assimilate detailed findings from any specific system and make fuller use of their range of experience. In the initial stages of any estuarine study, such theoretical frameworks can assist in the planning of field work or in deciding the need for or extent of any modelling effort. In subsequent stages complex or even conflicting results from either observations or models can often be interpreted with the aid of simple theory.

Here we summarise some theoretical studies relevant to estuaries in which tidal forcing predominates. In section 2 we examine tidal response with an emphasis on surface elevations, in section 3 we consider tidal currents with particular emphasis on vertical structure while in section 4 we introduce residual circulation and the related phenomenon of salinity intrusion. The oscillatory tidal response is by far the most energetic component in estuaries of the type considered, yet most present research is concerned with the second order influence of residual circulation. This is a reflection of (i) achievement in modelling of tidal propagation, now routine and with little remaining research content and (ii) the importance of sediment and pollutant movements which are often directly related to residual circulation. While steady progress is being made in the development of sophisticated numerical models to simulate such secondary effects, it is interesting to note a revival of interest in the simple box modelling approach (COKELET (1984)).

2. Tidal response in estuaries — surface elevations

Tidal propagation in a narrow channel of rectangular cross-section can be described by the following linearized equations: motion along the X axis

$$\frac{\partial U}{\partial T} + g\frac{\partial Z}{\partial X} + SU = 0 \qquad (1)$$

continuity

$$\frac{\partial Z}{\partial T} + \frac{1}{B}\frac{\partial}{\partial X}(BHU) = 0 \tag{2}$$

where U is axial velocity, Z elevation, g gravitational acceleration, T time, B breadth, H depth and S friction coefficient.

2.1. Breadth and depth varying with some power of distance X

The geometry of many estuaries may be approximated by the functions

$$B(X) = B_L(X/\lambda)^n \tag{3}$$

and

$$H(X) = H_L(X/\lambda)^m \tag{4}$$

with X measured from the head of the estuary. The parameter λ is simply a unit of horizontal dimension, adopting H_L as a unit of vertical dimension and P, the tidal period, as a unit of time we adopt the following identity

$$\lambda = (gH_L)^{1/2}P \tag{5}$$

thus for the particular case of H_L constant, λ corresponds to the tidal wavelength.

Using these scaling parameters we transform to dimensionless parameters as follows:

$$x = X/\lambda, \quad t = T/P, \tag{6}$$
$$z = Z/H_L, \quad b = B/\lambda,$$
$$h = H/H_L,$$
$$u = UP/\lambda \text{ and } s = SP$$

PRANDLE and RAHMAN (1980) showed that the above equations yield the following solution for tidal elevation z at any location x at any time t for any tidal period P:

$$z = A\left(\frac{ky_x}{ky_M}\right)^{1-\nu}\frac{J_{\nu-1}(ky_x)}{J_{\nu-1}(ky_M)}e^{i2\pi t} \tag{7}$$

where $Ae^{i2\pi t}$ is the tidal elevation at the mouth x_M and

$$\nu = \frac{n+1}{2-m} \tag{8}$$

$$y_x = \frac{4\pi}{2-m}x^{\frac{2-m}{2}}, \quad y_M = \frac{4\pi}{2-m}x_M^{\frac{2-m}{2}} \tag{9}$$

$$k = \left(1 - \frac{is}{2\pi}\right)^{1/2} \tag{10}$$

and $J_{\nu-1}$ is a Bessel function of the first kind and of order $\nu-1$.

Fortunately the formidable expression (7) can be simply illustrated in diagrammatic form, namely Fig. 1. This figure constitutes a general response diagram showing the variation in tidal elevation along the length of an estuary. Thus, having determined the geometrical parameters ν and y_M for a particular estuary, the variation in amplitude and phase in the estuary can be read along the relevant vertical line. Moreover it can be easily shown that if the tidal period P is doubled the corresponding value of y_M is halved. Thus this general response diagram indicates the response at all positions for all tidal periods and for all estuaries (reasonably approximated by (3) and (4)). One qualification is necessary, the response is for a dimensionless friction factor $s = 0.2\pi$, whereas it can be shown that s may vary typically over a range $0.2\pi < S < 2\pi$. The

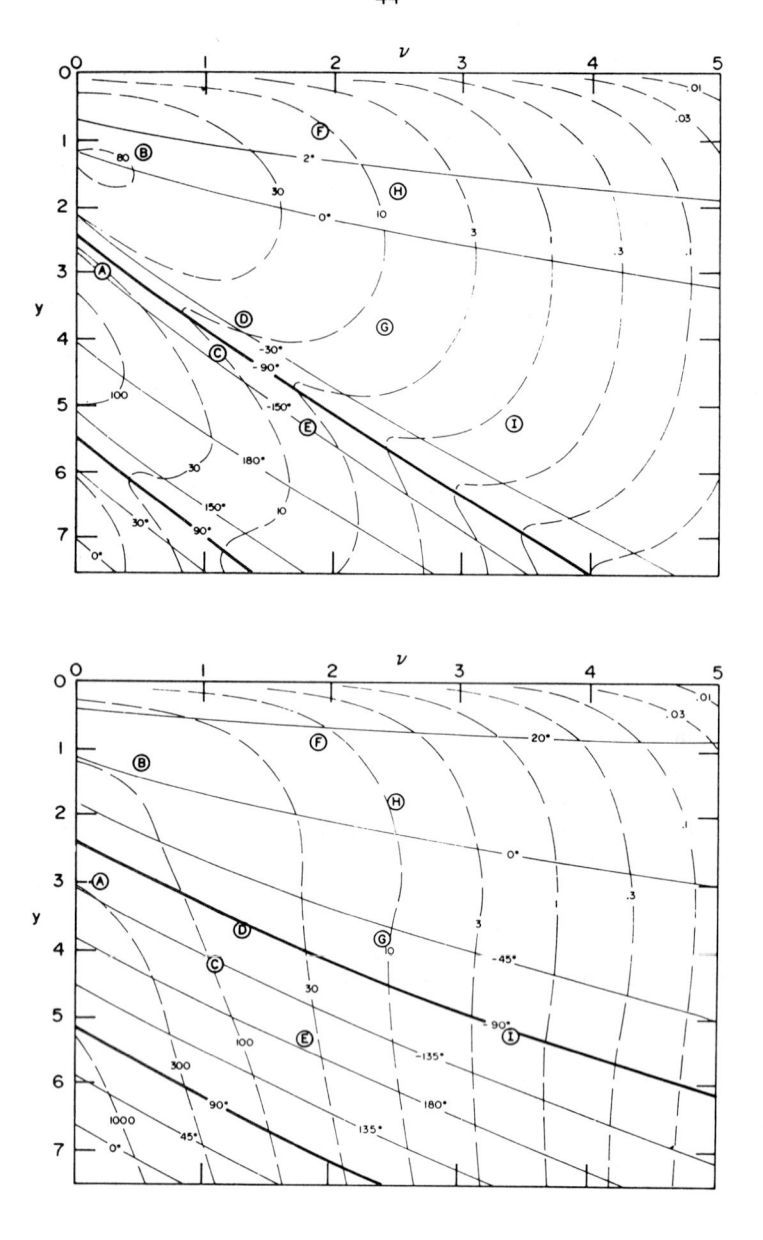

FIGURE 1 & 2   Response diagram for amplitude of tidal elevations. ν geometrical shape parameter, y propertional to distance from the head ($y=0$). Dashed contours show relative magnitude of tidal amplitude. Continuous contours show phase lag. Symbols A to I indicate ν and y values at the mouth of estuaries listed in Table 1. Fig. 1, low friction factor $s=0.2\pi$; Fig. 2, high friction factor $s=2\pi$.

n	m	ν		Length (Km)		α	β
-0.7	0.7	0.2	A Fraser	135	Agassiz	-2.8	2.8
0	0	0.5	B Rotterdam W.	99	Rotterdam	0	0
0.7	0.4	1.1	C Hudson	248	Troy	2.2	1.3
1.0	0.4	1.3	D Potomac	184	Chain Bridge	3.6	1.4
2.1	0.3	1.8	E Delaware	214	Trenton	5.3	0.8
2.7	0	1.9	F Miramichi	55	Newcastle	46.6	0
1.5	1.0	2.4	G Bay of Fundy	635	Minas Basin	3.9	2.6
2.3	0.7	2.5	H Thames	95	Richmond	14.1	4.3
1.7	1.2	3.4	I Bristol Channel	623	Sharpness	3.4	2.4
1.5	1.9	19.5	J St. Lawrence	418	Trois Tivieres	1.3	1.6

TABLE 1. Geometrical parameters for 10 estuaries α and β values apply to the M_2 constituent

response diagram for $s=2\pi$ is shown in Fig. 2, the overall pattern is similar but the magnitude and sharpness of the response are reduced.

An examination of these response diagrams explains a number of features commonly encountered in estuarine studies, namely:

(i) The *quarter-wavelength* resonance or primary mode common to most estuaries. Note that the positions indicated A to I designate the mouths of 9 major estuaries for the case of the M_2 semi-diurnal constituent. PRANDLE and RAHMAN (1980) show that the corresponding responses extracted from the general response diagram closely resemble the observed responses.

(ii) For a diurnal tidal constituent the y_M values A to I are halved, hence we expect a relatively small amplification of such constituents. For MSf, a 14-day constituent, the reduction in the y_M values would indicate little or no amplification or phase difference along the estuaries.

(iii) For quarter-diurnals or other higher harmonics we might expect high amplification, large phase differences and one or more nodal positions. However it is important to distinguish between the response to external forcing indicated by the present analysis and the generation of higher harmonics (and MSf) by non-linear processes within an estuary - the latter lying beyond the scope of the present study.

2.2 Breadth and depth varying exponentially with distance X

Replacing (3) and (4) with the following

$$B(X) = B_0 e^{nx} \tag{11}$$

and

$$H(X) = H_0 e^{mX} \tag{12}$$

we convert to dimensionless form as before but replacing H_L by H_0 so that

$$\lambda = \sqrt{gH_0}\,P \tag{13}$$

and obtain geometrical relationships

$$b(x) = b_0 e^{\alpha x} \tag{14}$$

and

$$h(x) = e^{\beta x} \tag{15}$$

In this case there is no straightforward analytical solutions and the separate influences of α and β cannot be accounted for by a single parameter as was the case with ν in the earlier study. However, numerical solutions to the relevant equations were obtained by PRANDLE (1984a). The equivalent *general response diagram* is shown in Fig. 3. In this case the orthogonal axes refer to the parameters α and β. The contours show the amplification between the amplitude of the tidal elevation at the head of the estuary relative to the value at the first nodal position. However for estuaries with values of $\alpha + 2\beta \geq 10$ no nodal position occurs and in this case the amplification shown is relative to the value for $x = 1$ where the latter value closely approximates the asymptote at $x = \infty$. This demarkation in the response of estuaries was not evident in the earlier study. While Fig. 3 shows the amplification for $s = 0.2\pi$, Fig. 4 shows the equivalent response for $s = 2\pi$. The symbols A to J again indicate the amplification for 10 major estuaries between the head and the first nodal position (not the mouth) for the semi-diurnal constituent M_2. To determine the maximum response for other tidal constituents from Fig. 3 and 4 we note that the values for α and β are directly proportional to period, thus for diurnal constituents α and β are doubled while for quarter-diurnal constituents α and β are halved.

In consequence we may deduce the following conclusions from the *general response diagrams* i.e., Fig. 3 and 4.

(i) The response of any estuary may be likened to the free vibrations of a damped simple harmonic oscillator thus: estuaries of type I with $\alpha + 2\beta \ll 10$ are under-damped and elevation amplitudes vary in an oscillatory manner along the x-axis. Estuaries of type II with $\alpha + 2\beta \approx 10$ are critically damped and produce maximum amplification. Estuaries of type III with $\alpha + 2\beta \gg 10$ are over-damped and elevations increase monotonically towards the head with little amplification and little sensitivity to frictional effects.

(ii) The large tidal ranges experienced in the Bay of Fundy and Bristol Channel are shown to be a consequence of their particular values of α and β (G and I in figures 3 and 4), i.e. $\alpha + 2\beta \approx 10$, and not simply due to their resonant lengths as commonly believed.

(iii) Construction of tidal barriers at a location x_1, effectively increases α and β by $e^{\beta x/2}$, thus for both Fundy and the Bristol Channel the sensitivity to barrier construction is evident while in the Thames (H) barrier construction has little influence.

3. Tidal Currents

3.1 Depth-averaged

The tidal current response in any estuary can be calculated in the same manner as described for tidal elevations in section 2. The equivalent *response diagrams* for current show that, in many estuaries for semi-diurnal tides, a position of maximum tidal current amplitude will be found landwards of the mouth. The significance of such positions appears to have received little attention to-date, perhaps some further consideration is warranted?

For the case of exponentially varying channel geometry (section 2.2), a position of maximum currents is found in all estuaries irrespective of the value from the demarkation parameter $\alpha + 2\beta$. In this latter case, it can be shown that the maximum current position lies closer than to the head (in terms of the dimensionless parameter x) as α and β increase.

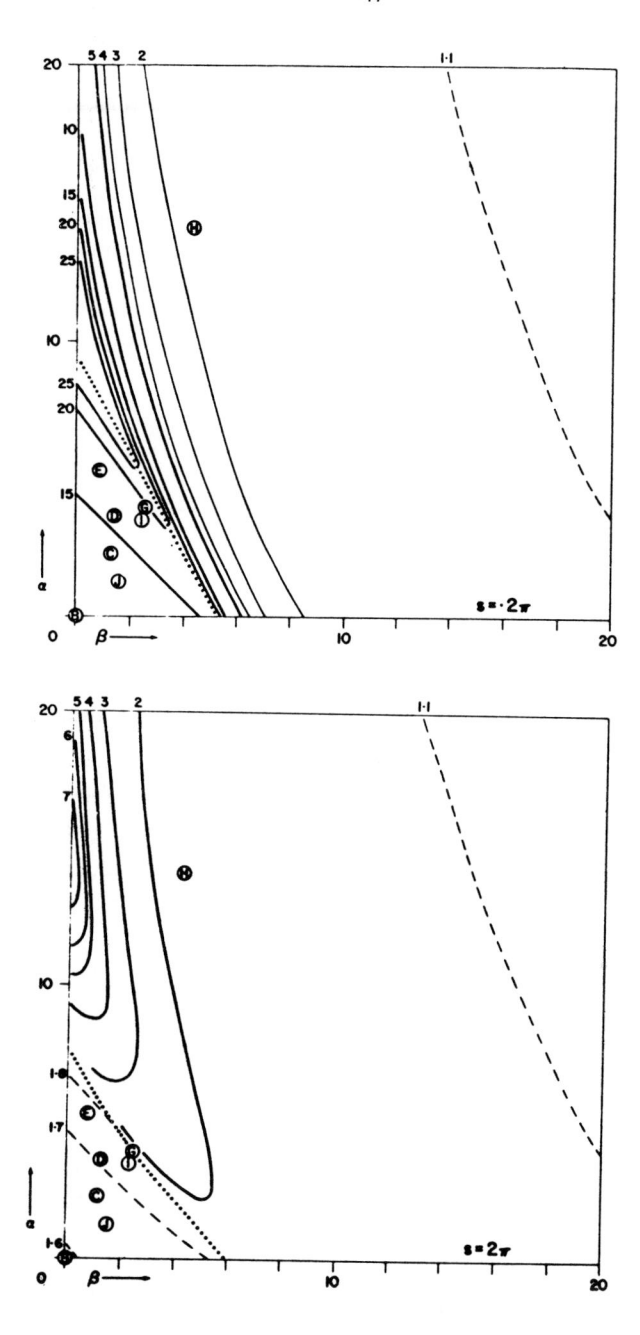

FIGURES 3 & 4 Response diagram for amplitude of tidal elevations. α and β rate of exponential increase in breadth and depth respectively. Contours show the ratio of tidal amplitude at the head to amplitude at the first nodal position. For $(\alpha+2\beta) \geqslant 10$ nodes of tidal amplitude do not occur and contours show the ratio of tidal amplitude at the head to the amplitude at $x = 1$ (in most cases the latter approximates the value at an infinite distance from the head). Symbols A to J indicate α and β values for estuaries listed in table 1 (M_2 constituent).

3.2 Current structure

While the simple theory introduced in earlier sections accounts for tidal elevations and depth-averaged currents, in many problems information regarding the vertical structure of tidal currents is required. Many studies have examined this problem by meshing boundary layer theory with the bulk-flow equations, see for example JOHNS and DYKE (1971). However, by confining our interest to flow outside of the immediate bed level, we can devise a simple theory for vertical structure assuming a quadratic friction law incorporating a slip velocity at the bed.

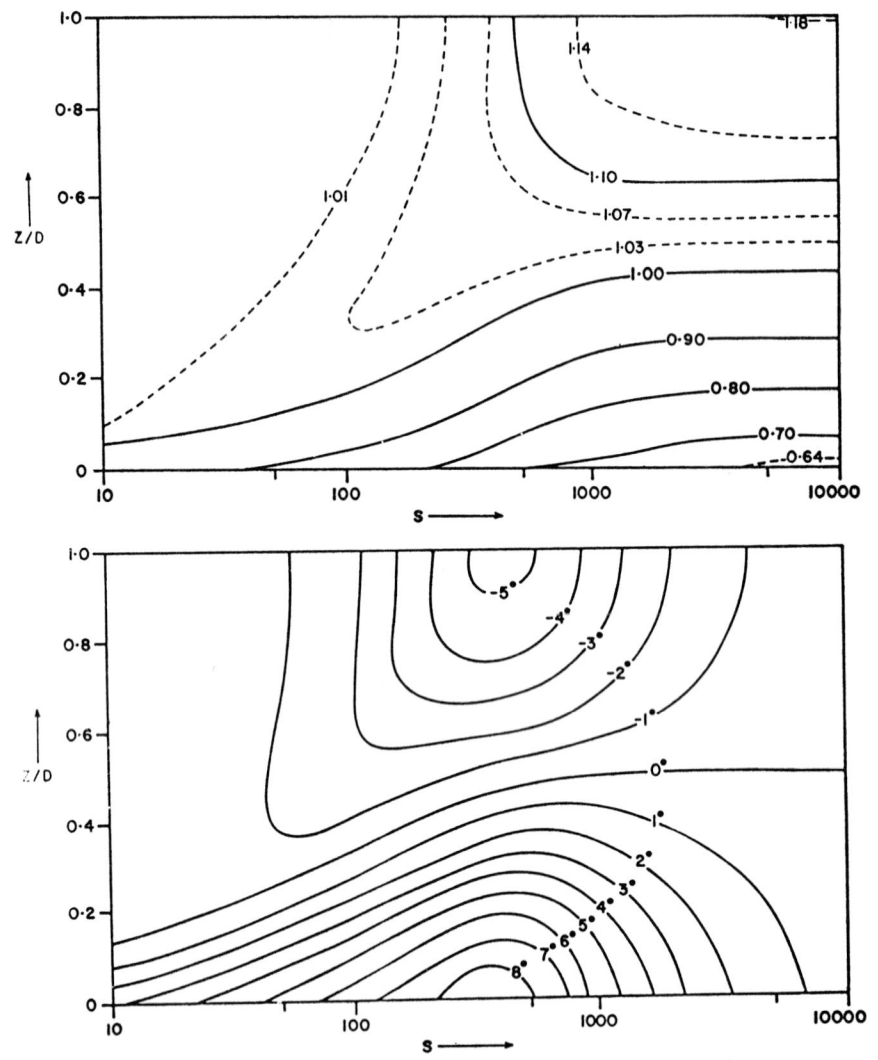

FIGURES 5 & 6 Tidal current profiles as a function of Strouhal Number ($S = \hat{U}P/D$). $\hat{U}$ tidal amplitude, P tidal period, D water depth. (Values for $k = 0.0025$). Figure 5 contours show amplitude in proportion to depth-mean value. Figure 6 contours show phase advance in relation to depth-mean value.

Thus, for tidal flow confined to one horizontal dimension, neglecting vertical acceleration, convective and density terms, at any height Z above the bed we obtain

$$\frac{\partial U}{\partial T} = -g\frac{\partial \zeta}{\partial X} + \frac{\partial}{\partial Z}E\frac{\partial U}{\partial Z} \tag{16}$$

where ζ is the surface elevation and E the vertical eddy viscosity. PRANDLE (1982a) provided a simple solution to (16) for a tidal constituent of period P:

$$\frac{U(Z)}{\overline{U}} = (e^{bz} + e^{-bZ+2y})/\mathbf{R} + \mathbf{Q} \tag{17}$$

where $\overline{U}$ is the depth averaged velocity amplitude

$$\mathbf{b} = \left(\frac{i\omega}{E}\right)^{1/2}, \quad (i=\sqrt{-1}, \omega=2\pi/P) \tag{18}$$

$$\mathbf{R} = (1-e^{2y})(\mathbf{j}-1/\mathbf{y}-1) - 2e^{2y} \tag{19}$$

$$\mathbf{Q} = \{\mathbf{j}(1-e^{2y}) - 1 - e^{2y}\}/\mathbf{R} \tag{20}$$

$$\mathbf{j} = 3\pi E\mathbf{b}/8k|\overline{U}| \tag{21}$$

$$\mathbf{y} = \mathbf{b}D \tag{22}$$

(D water depth)
The derivation of (17) includes the following assumptions
(i) E is not a function of time or depth
(ii)   zero surface stress
(iii)   at $Z=0$, a bed stress $\frac{8}{3\pi}k|\overline{U}|U_{Z=0}$

Comparison of the velocity profiles described by (17) with observations suggested the following approximation for the vertical eddy viscosity

$$E = k\overline{U}D \tag{23}$$

The expression (17) is then a function of the single parameter kS, where S is the Strouhal No. given by $|\overline{U}|P/D$. Figs. (5) and (6) show the resulting vertical variations in velocity amplitude and phase respectively for $k=.0025$. These diagrams illustrate that the depth-averaged velocity almost always occurs at $z=0.4D$, as traditionally assumed by engineers. Moreover in most strongly tidal estuaries $S>10000$ and hence the velocity distribution will tend toward the asymptotic value shown for large S.

This theoretical approach can be readily extended to account for an eddy viscosity varying linearly with depth (PRANDLE 1982b). In addition, by resolving fully three-dimensional tidal motion into separate clockwise and anti-clockwise rotations, this simple recti-linear theory can be used to explain the vertical structure of tidal ellipses.

4. Residual currents and salinity intrusion

4.1 Well-mixed estuaries

Using the same approach and similar assumptions as outlined in section 3.2 we can derive the steady-state residual current structure associated with forcing by (a) freshwater flow Q (b) surface wind stress τ and (c) well-mixed longitudinal density gradient $\partial\rho/\partial X$. Thus we obtain (PRANDLE 1984b) a residual velocity structure associated with:
(a) fresh water flow (per unit breadth) Q

$$U_0 = -0.89\frac{Q}{D}\left\{-\frac{\eta^2}{2} + \eta + \frac{\pi}{4}\right\} \tag{24}$$

where $\eta = Z/D$
(b) surface wind stress τ

$$U_w = \frac{\tau}{\rho k \hat{U}} \{0.574\eta^2 - 0.149\eta - 0.117\} \tag{25}$$

($\hat{U}$-amplitude of tidal velocity)
(c) density gradient $\partial \rho / \partial x$

$$U_M = \frac{g}{\rho} \frac{\partial \rho}{\partial x} \frac{D^2}{k\hat{U}} \{\frac{-\eta^3}{6} + 0.269\eta^2 - 0.037\eta - 0.029\} \tag{26}$$

		U	S/F	W/F	$\bar{U}/\hat{U}$	S_+	$\delta s/\bar{s}$
Vellar		0.8	0.1	0.1	2.0	3	2
Columbia		0.06	40	0.7	0.06	6	0.5
James		0.005	90	23	0.014	47	0.25
Tees		0.02	10	1.3	0.013	476	0.2
Southampton W.		0.002	100	33	0.003	760	0.1
Tay		0.02	6	1.0	0.010	1354	0.1
Thames		0.015	5	3	0.017	634	0.05
Mersey		0.002	83	10	0.001	9268	0.05
Bristol		0.007	4	3	0.003	18338	0.02
Flume	T10	1.99	7	1.6	0.170	4	2
test:	T20	1.99	10	1.1	0.118	6	1.5
	T30	1.49	7	2.1	0.125	8	1
WES	11	1.71	1.8	1.3	0.131	28	1.30
tests	10	1.71	2.0	1.0	0.095	47	0.79
	16	0.61	4.3	3.8	0.047	91	0.74
	14	0.61	2.6	2.3	0.029	403	0.24
		(m s −1)					

TABLE 2: Estuary and flume date relating to salinity intrusion

We can interpret the magnitudes of these residual flow components by introducing the depth-averaged equation of motion for steady state residual flow:

$$\frac{\partial \zeta}{\partial x} + \frac{H}{2\rho} \frac{\partial \rho}{\partial x} - \frac{\tau}{\rho g D} + \frac{4}{\pi} \frac{k\hat{U}Q}{gD^2} = 0 \tag{27}$$

By defining the dimensionless parameters:

$$S = \frac{H}{\rho} \frac{\partial \rho}{\partial x}, \quad W = \frac{\tau}{\rho g D}, \quad F = \frac{k\hat{U}Q}{gD^2} \tag{28}$$

We note the depth-averaged forcing terms associated respectively with density, wind and bed friction are in the ratio

$$\frac{S}{2} : W : \frac{4F}{\pi} \tag{29}$$

Hence in terms of these parameters from (24), (25) and (26) the residual velocities at the surface and bed are

	Surface	Bed	
(a) freshwater flow Q	$1.14\, Q/D$	$0.70\, Q/D$	(30)
(b) wind stress τ	$0.31\, \dfrac{W}{F}\dfrac{Q}{D}$	$-0.12\, \dfrac{W}{F}\dfrac{Q}{D}$	(31)
(c) density gradient $\partial\rho/\partial x$	$0.036\, \dfrac{S}{F}\dfrac{Q}{D}$	$-0.029\, \dfrac{S}{F}\dfrac{Q}{D}$	(32)

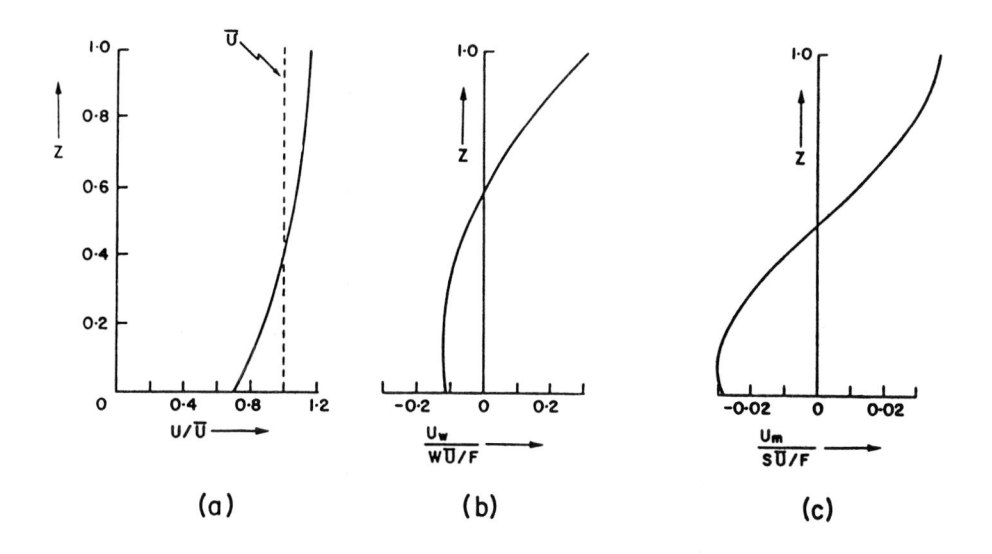

FIGURE 7. Vertical structure for (steady-state) residual components
(a) freshwater flow $\bar{U}=Q/D$
(b) wind stress τ
(c) longitudinal density gradient $\partial\rho/\partial x$ W,S and F defined by (28).

The corresponding continuous profiles are shown in Fig. 7.

Thus by calculating representative values of S, W and F for any estuary (or possibly a section from an estuary) we can immediately see how sensitive the residual circulation is to changes in freshwater flow or density gradient or to the onset wind forcing. Table 2 shows typical values for 9 major estuaries and indicates the range of sensitivity encountered. (In some cases, the inter-relationships between density gradient and fresh water flow may obscure their separate impacts).

4.2 Stratified estuary

Again following the same general approach, it is possible to compute the residual velocity profile for a wedge-type intrusion. Assuming no mixing of salt and freshwater but continuity of both stress and velocity at the interface we obtain: for the top freshwater layer of density ρ:

$$U_T = -\frac{Q}{D}\frac{\epsilon}{d}\left\{\frac{1}{\gamma}\left(\frac{\eta^2}{2}-\eta+d-\frac{d^2}{2}\right)-0.308d(1-d)\right\} \tag{33}$$

for the lower salt layer of density $\rho + \Delta\rho$:

$$U_L = \frac{-Q}{D} \frac{\epsilon(1-d)}{d^2} \{-0.574\eta^2 + 0.149\eta d + 0.117d^2\} \quad (34)$$

$$\epsilon = d / \{(1-d)^2 (\frac{(1-d)}{3\gamma} + 0.308d)\}$$

d is the fractional height of the interface. γ is the ratio of eddy viscosity in the top layer to that in the bottom.

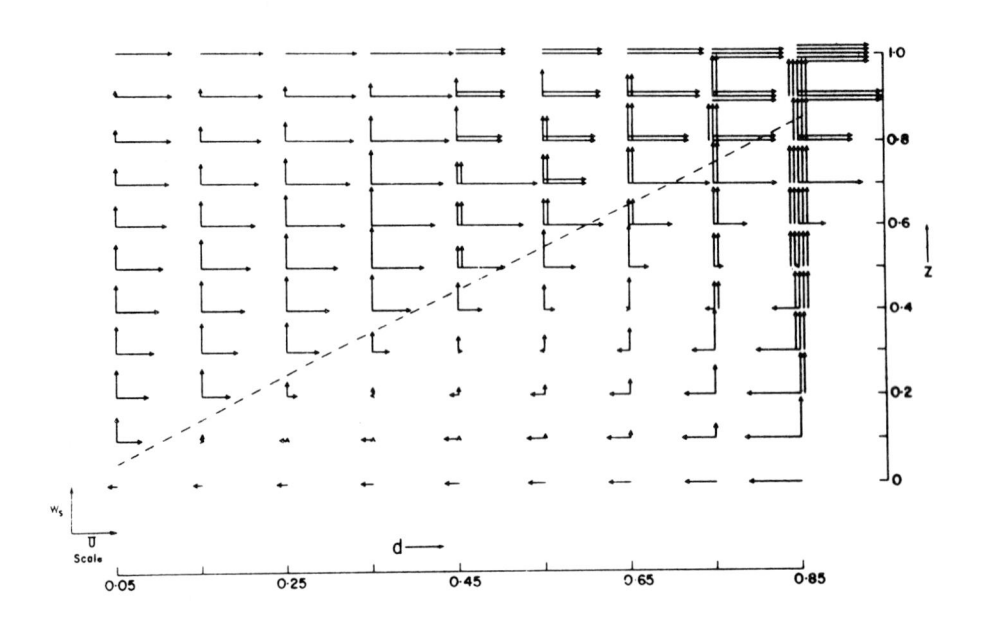

FIGURE 8. Residual velocities in a stratified wedge. z is fractional height (0 to 1, bed to surface). d is fractional height of the interface (also shown by dashed line).

Equations (33) and (34) may be further simplified by assuming that the eddy viscosity in the two layers are given by a product of $k\hat{U}$ and the layer thickness, then:

$$U_T = -1.56 \frac{Q}{D} \frac{1}{(1-d)^3} \{\frac{\eta^3}{2} - \eta - 0.808d^2 + 1.616d - 0.308\} \quad (36)$$

$$U_B = -1.56 \frac{Q}{D} \frac{1}{d^2(1-d)} \{-0.574\eta^2 + 0.149\eta d + 0.117d^2\} \quad (37)$$

or at the surface

$$U_T = \frac{1.26}{(1-d)} \frac{Q}{D} \quad (38)$$

and at the bed

$$U_B = -\frac{0.18}{(1-d)} \frac{Q}{D} \quad (39)$$

Fig. 8 shows the above velocity profiles over the length of a wedge intrusion. These simple solutions reproduce all of the observed features of such intrusions.

4.3 Salinity intrusion

As part of the analysis used in deriving the velocities for a wedge intrusion (section 4.2), PRANDLE (1984b) found that for a horizontal channel of constant cross section, the intrusion length L is given by

$$L = 0.26 \frac{gH^3}{k\hat{U}Q} \frac{\Delta\rho}{\rho} \tag{40}$$

Prandle tested the applicability of the above formulation by comparison with two sets of available flume data. For the first set (RIGTER 1973), a correlation faction of 0.97 was obtained between observed intrusion lengths and values calculated using

$$L = 0.19 \frac{gH^3}{k\hat{U}Q} \frac{\Delta\rho}{\rho} + L_1 \tag{41}$$

(L_1 is a constant introduced to allow for vagaries in mixing conditions at the mouth involved in experimental flumes).

For the second set (IPPEN and HARLEMAN 1961), a correlation factor of 0.84 was obtained between observed intrusion lengths and values calculated from

$$L = 0.13 \frac{gH^3}{k\hat{U}Q} \frac{\Delta\rho}{\rho} + L_2 \tag{42}$$

In both flume experiments, flow conditions were generally *partially mixed* yet the intrusion lengths can be predicted from a relationship derived for wedge-type intrusion. Thus the formulation (40) may be of use to provide qualitative estimates of changes in salinity intrusion in estuaries arising from small modifications to: (i) the depths or roughness of a channel, (ii) river flow or (iii) tidal range.

4.4 Stratified or mixed?

This success in applying the intrusion length formula led to a further examination of just when and where the dynamics pertaining to (i) well-mixed conditions or (ii) wedge conditions apply. One useful guide in answering this question is the Hansen and Rattray Stratification Diagram shown in Fig. 9 (HANSEN and RATTRAY (1966). The diagram was derived from a theoretical modelling study, here we explain the salient features from the simpler dynamical derivations outlined in section 4.

The two basic parameters which form the orthogonal axes in the Stratification Diagram are

(i) $\delta s / \bar{s}$ salinity difference between bed and surface divided by depth-mean value (tidal averages)
(ii) $U_s / \bar{U}$ residual velocity at surface divided by depth-mean value.

Four estuarine types are identified with sub-divisions into (a) mixed or (b) stratified according to whether $\delta s / \bar{s} <> 0.1$.

In *type 1 estuaries* residual flow is landwards at the bed. From the results shown in section 4.1, landward flow requires $S/F > 24$ and hence $U_s / \bar{U} > 2$. Thus, this latter result explains the demarkation line between estuaries of types 1 and 2 in the Stratification Diagram. Moreover, the results from 4.1 can be used to provide the alternative axis parameter S/F to replace $U_s / \bar{U}$ as shown in Fig. 9.

In a similar fashion, the upper demarkation line for estuaries of *types 3 and 4* may be explained directly from the results obtained in section 4.2. Thus from (38)

$$\frac{U_s}{\overline{U}} = \frac{1.26}{(1-d)} \tag{43}$$

while by assuming $\rho = 1 + as$, (a constant) we have simply

$$\frac{\delta s}{S} = \frac{\Delta \rho}{\Delta \rho d} = \frac{1}{d} \tag{44}$$

and hence from (43) and (44)

$$\frac{\delta s}{S} = \frac{U_s / \overline{U}}{U_s / \overline{U} - 1.26} \tag{45}$$

The latter expression (45) agrees precisely with the upper demarkation line on the Stratification Diagram. The relationships (43) and (44) provide alternative axes parameters in terms of d as shown in Fig. (9). From these alternate axes we note that in *type 4* estuaries the interface must be at $d < 0.75$ while in a stratified estuary of *type 3* the interface must be close to the surface i.e. $d \to 1$.

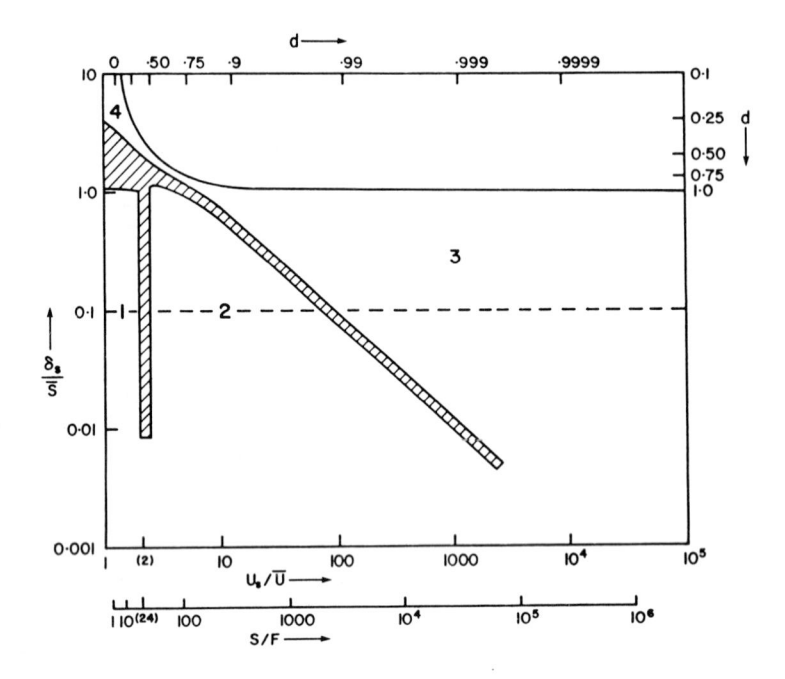

FIGURE 9. Hansen and Rattray Stratification Diagram. $\delta s / \overline{s}$ salinity difference between bed and surface: depth-mean salinity. $U_s / \overline{U}$ residual velocity at the surface: depth-mean value. d fractional height of the interface for a salt wedge S and F defined by (28).

Clearly, the dynamics pertaining to residual current structure can explain the basis of the Stratification Diagram. However, when data from separate estuaries are plotted on this diagram they show a broad spread proving there is no one-to-one relationship between the degree of mixing and the residual current dynamics. To provide such a direct relationship, it is necessary to consider energy balances, effectively this may be seen as introducing the equation of continuity to supplement the original equation of motion.

Following the earlier studies of IPPEN and HARLEMAN (1961) and SIMPSON and

HUNTER (1974), we evaluate the potential energy gained in mixing a stratified wedge, J, in comparison with the available energy derived from the tidal dissipation G. The rate of gain of P.E. is $\frac{1}{2}\Delta\rho g D^2 \overline{U}$ while tidal energy is dissipated over a mixing length L at a rate $\frac{4}{3\pi}k\rho \hat{U}^3 L$. Hence we can define a Stratification Number, St, given by

$$St = \frac{G}{J} = 0.85 \frac{k\hat{U}^3 L}{\Delta\rho/\rho g D^2 \overline{U}} \tag{46}$$

or from (28)

$$St = 0.85 / \frac{S}{F}(\frac{\overline{U}}{\hat{U}})^2 \tag{47}$$

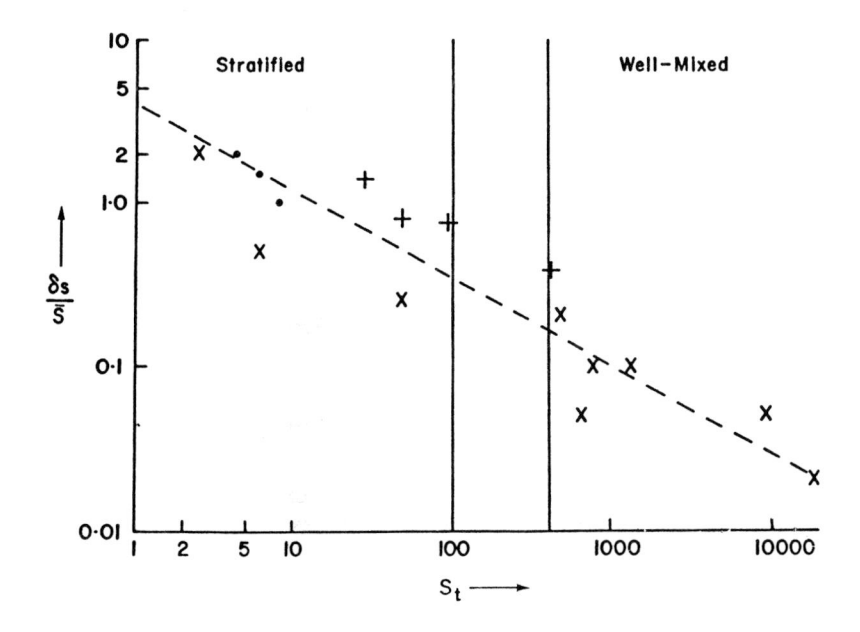

FIGURE 10. Level of stratification ($\delta s / \bar{s}$) versus Stratification No. St. (log-log scale).
$\delta s / \bar{s}$ salinity difference between bed and surface: depth mean salinity
St defined by (46). Data listed in Table 2.

Fig. 10 shows a plot of St versus $\delta s / \bar{s}$ using data from 9 major estuaries and 6 data points from the flume tests described above (see Table 1). The points all lie close to the line

$$\delta s / \bar{s} = 4St^{-0.55} \tag{48}$$

and thus confirm the conclusion of Ippen and Harleman that the Stratification Number is a direct indicator of the level of stratification. The demarkation values of $S = 100$ and 400 or $\delta s / \bar{s} = 0.15$ and 0.32 may be considered as separating three zones of mixing, namely well-mixed, partly mixed and stratified.

Fig. 11 represents an alternative method of representing the data shown in Fig. 10. In this latter figure, for any estuary the level of stratification $\delta s / \bar{s}$ can now be read directly from the gross estuarine parameters S / F and $\bar{U} / \hat{U}$. Thus this diagram may be used in a similar fashion to the Hansen-Rattray Stratification Diagram with the additional advantage that the sensitivity of the existing level of stratification to changes in parameters such as freshwater flow $\bar{U}$ or tidal velocity $\hat{U}$ may be readily understood.

5. Conclusions

A number of salient features of estuarine dynamics have been illustrated and explained using simple analytical solutions to basic flow equations.

(i) General tidal response diagrams have been derived which predict the changes in amplitude and phase of both elevations and currents within any estuary for all tidal constituents. Estuarine shape is shown to be responsible both for the large amplification and the sensitivity to barrier construction in the Bay of Fundy and Bristol Channel. Moreover, estuarine shape can be used to delineate a basic classification system for tidal response.

(ii) The vertical structure of tidal currents is related to the Strouhal Number $S = \hat{U}P / D$ where $\hat{U}$ is the amplitude of the tidal velocity, P the tidal period and D the depth.,

(iii) Steady-state residual velocity structures are determined for (a) freshwater flow Q, (b) wind stress τ and (c) well-mixed longitudinal density gradient $\partial \rho / \partial x$. The relative magnitude of these flow components is related to dimensionless numbers S, W and F where the latter refer respectively to forcing terms associated with density gradient, wind stress and bed-friction. By determining the values of S, W and F for any specific estuary, the sensitivity of the residual circulation in the estuary to wind stress, density gradients or changes in river flow is immediately evident.

(iv) The residual velocity structure for a stratified wedge type salinity intrusion is indicated. In addition a formula is derived to predict the length of intrusion in a horizontal channel of constant cross-section. By adjustment of one coefficient, this formula accurately predicts intrusion lengths in flumes over a range of mixing conditions. Thus the formula may be used to provide a first indication of the effect on intrusion lengths of changes in (a) channel depth or roughness, (b) river flow or (c) tidal conditions.

(v) These studies of residual current structure can be used to explain the principle features of the Hansen-Rattray Stratification Diagram. However, the level of stratification cannot be directly associated with the dynamics of current structure. Energy considerations must be used to explain stratification levels, in this way stratification may be linked to both current structure and the flow ratio (i.e. the ratio of residual velocity to tidal velocity). Thus a simple diagram can be constructed relating these three parameters and directly indicating the sensitivity of the level of stratification to estuarine conditions.

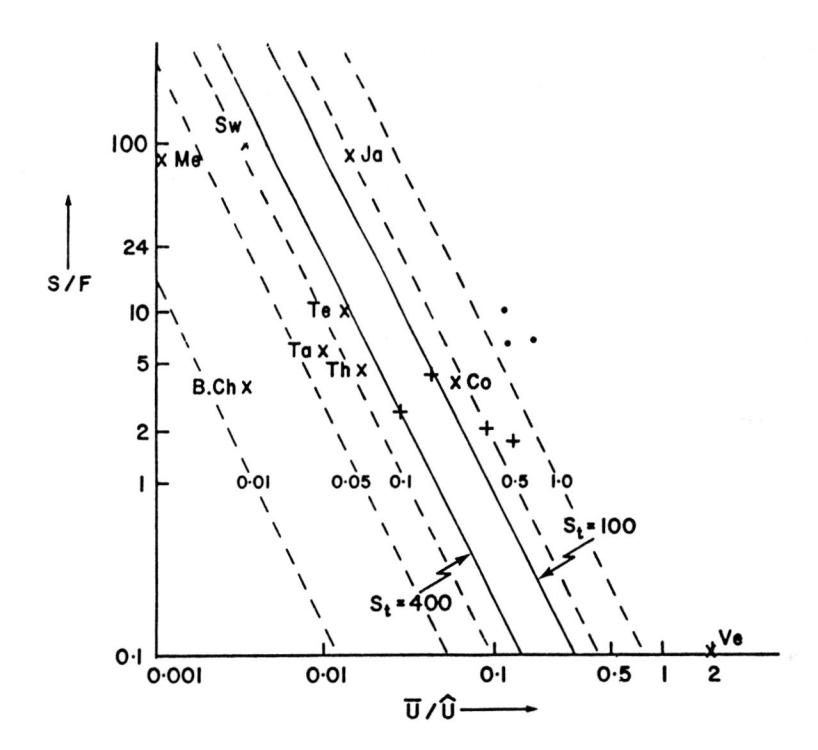

FIGURE 11. S/F versus *flow ratio* ($\overline{U}/\hat{U}$), (log-log scales). Dashed lines represent contours of $\delta s/\overline{s}$. See captions for Fig. 9 and 10 for further definitions.

References

COKELET E.D., STEWART R.J. & EBBESMEYER C.C. 1984. The exchange of water in fjords: a simple model of two-layer advective reaches separated by mixing zones. To appear in: Proc. Am. Soc. Civ. Engrs. 19th Coastal Eng. Conf., Houston.

HANSEN D.V. & RATTRAY M.J. 1966. New dimensions in estuary classification. Limnol. Oceanog., 11, 319-326.

IPPEN A.T. & HARLEMAN D.R.F. 1961. One-dimensional analysis of salinity intrusion in estuaries. Technical Bulletin No. 5, Committee on Tidal Hydraulics, Waterways Experiment Station, Vicksburg, Mississippi.

JOHNSON B. & DYKE P. 1971. On the determination of the structure of an offshore tidal stream. Geophys. J.R. Astron. Soc. 23, 287-297.

PRANDLE D. & RAHMAN M. 1980. Tidal response in estuaries. J. Phys. Ocean., 10, 1552-1573.

PRANDLE D. 1982a. The vertical structure of tidal currents and other oscillatory flows. Continental Shelf Research 1, 191-207.

PRANDLE D. 1982b. The vertical structure of tidal currents. Geophys. & Astrophys. Fluid Dynamics, 22, 29-49.

PRANDLE D. 1984a. Classification of tidal response in estuaries from channel geometry. to appear in : Geophys. J.R. Astron. Soc.

PRANDLE D. 1984b. On salinity regimes and the vertical structure of residual flows in narrow tidal estuaries. to appear in: Est. Coast. & Shelf. Sciences.

RIGTER B.P. 1973. Minimum length of salt intrusion in estuaries. Proc. Am. Soc. Civ. Engnrs., J. Hyd. Div. 99, 1475-1496.

SIMPSON J.H. & HUNTER J.R. 1974. Fronts in the Irish Sea. Nature 250, 404-406.

Synoptic Observations of Salinity, Suspended Sediment and Vertical Current Structure in a Partly Mixed Estuary

R.J. Uncles, R.C.A. Elliott, S.A. Weston

Natural Environment Research Council,
Institute for Marine Environmental Research, U.K.

D.A. Pilgrim, D.R. Ackroyd

Department of Marine Science,
Plymouth Polytechnic, U.K.

D.J. McMillan and N.M. Lynn

Department of Nuclear Science and Technology,
Royal Naval College, U.K.

ABSTRACT

Synchronized observations along the axis of a partly mixed estuary during spring and neap tidal cycles are presented. Significant salinity stratification occurred near the fresh water-brackish water interface, which was also the location of the turbidity maximum. The axial salt flux during the spring tide was largely controlled by tidal pumping, whereas transport due to vertical shear was dominant during the neap tide. Transport of suspended sediment due to tidal pumping greatly exceeded that due to vertical shear during both surveys, and probably contributed to the formation of the turbidity maximum.

The depth-averaged, mass transport Stokes drift was directed up-estuary, although speeds were very small in the lower reaches. The observed vertical structure of the mass transport Stokes drift differed in form between the upper and lower reaches. Near the mouth, speeds in the lower half of the column were very small; near the head, speeds maximized in the lower half of the column.

1. Introduction

Analyses are presented of synoptic observations of residual fluxes of water, salt and suspended sediment in the Tamar, which is a partly mixed estuary in the southwest of England (Fig. 1). Attention is given to the vertical structure of residual currents in the upper and lower reaches of the estuary.

The primary objectives of this work were:
(1) to obtain a synoptic set of data on residual fluxes which could be used to generalize conclusions based on earlier, non-synoptic data (UNCLES et al., 1985); and
(2) to obtain a simultaneous picture of vertical profiles of residual current along the estuary for subsequent modelling studies. Vertical residual circulations, especially the gravitational circulation, are known to have an important influence on salt and contaminant transport in partly mixed estuaries; this arises because of their associated vertical shear dispersion (e.g. HANSEN and RATTRAY, 1966; DYER, 1973; OFFICER, 1976; FISCHER et al., 1979; RATTRAY and DWORSKI, 1980). However, tidal pumping can also act as an important transport mechanism for salt in mesotidal and macrotidal, partly mixed and well mixed estuaries (UNCLES and JORDAN, 1979; HUGHES and RATTRAY, 1980; LEWIS

and LEWIS, 1983).

Tidal pumping, in conjunction with resuspension of bed sediment, also appears to be an important mechanism for transport of suspended sediment in these estuaries (ALLEN et al., 1980; UNCLES et al., 1985). This process may contribute to the formation of the turbidity maximum — a phenomenon which occurs in the Tamar and many other estuaries.

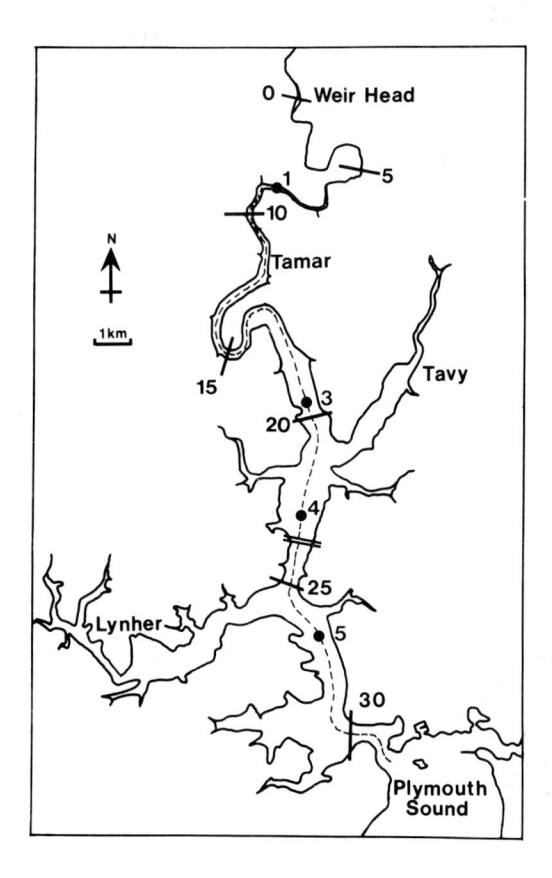

FIGURE 1. Sketch chart of the Tamar estuary, showing its sub-division into 5 km intervals, and station positions (●).

Distributions of salinity and suspended sediment along the axis of the Tamar are shown in Fig. 2. Measurements were made at approximately high water on the spring tide preceeding the synoptic surveys described here. These profiles show significant stratification in salinity near the fresh water-brackish water interface, and the existence of a pronounced turbidity maximum in this region of the estuary.

Four anchor stations were positioned along the axis of the estuary, and were worked simultaneously over one neap and one spring tidal cycle. Residual fluxes of salt and suspended sediment are interpreted in terms of the transport due to the residual flows of water, tidal pumping and vertical shear.

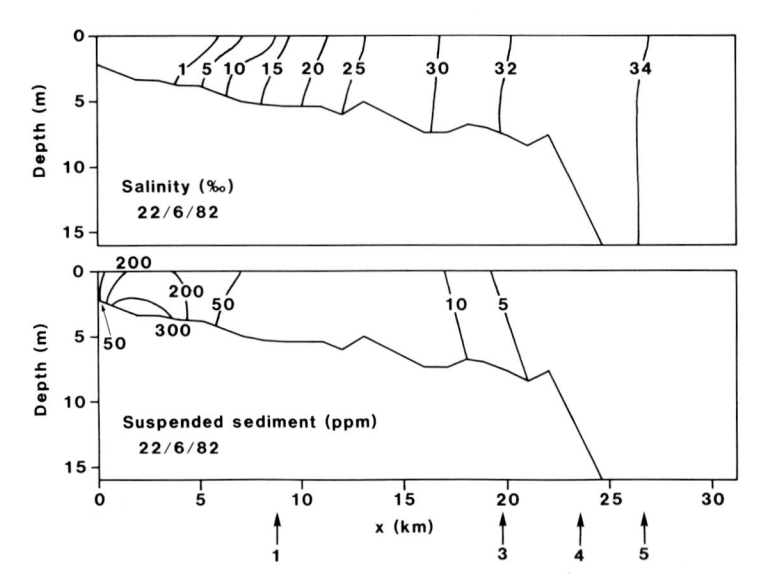

FIGURE 2. Salinity and suspended sediment concentrations at high water along the axis of the Tamar during 22/6/1982. The tidal range was 4.8 m (typical spring tide) and the freshwater run-off across Weir Head into the upper reaches was 8.2 m^3 s^{-1} (low, typical summer value).

2. Observations and treatment of data

Simultaneous observations were made along the central channel at stations 1, 3, 4 and 5 (Fig. 1) during a neap (N) tide on 30/6/82, and during a spring (S) tide on 7/7/82. The tidal ranges in the lower reaches were 2.7 m and 3.9 m; mean neap and spring tidal ranges in the lower reaches are 2.2 m and 4.7 m, respectively. Freshwater run-off was similar and low (typical summer values) for the two surveys.

For both surveys, synchronized readings were taken at all four stations every half hour for a 12.5 hour period. Total water depth and vertical profiles of current velocity, salinity, temperature and suspended sediment were measured. With the exception of the current meters, which were factory calibrated, all equipment was intercalibrated before and after each survey.

It is convenient to measure (or interpolate) data at fixed fractions of the instantaneous non-dimensional depth, σ ($\sigma=0$ and 1 at the seabed and surface, respectively). The procedure for interpolating such data to standard fractional depths has been discussed previously (KJERFVE, 1975 and 1979). In this study, standard fractional depths were taken to be $\sigma = 0.1, 0.3, 0.5, 0.7$ and 0.9. The axial, x, component of motion along the estuary is considered here, with H the total depth, U the velocity at non-dimensional height σ above the bed, and q a *stream function*, such that:

$$dq/d\sigma = HU \quad \text{with} \quad q(\sigma=0) = 0 \quad \text{and} \quad q(\sigma=1) = Q.$$

Q is the rate of volume transport of water per unit width, $Q = H\overline{U}$, where the overbar denotes a depth average. The residual rate of transport per unit width is given by:

$$\frac{d<q>}{d\sigma} = <HU> = h(u_E + u_S) = hu_L, \tag{1}$$

in which diamond brackets denote a tidal average, and where:

$$h = <H> \qquad (2)$$

$$u_E = <U> \qquad (3)$$

$$u_S = <\tilde{H}\tilde{U}>/h \qquad (4)$$

with

$$\tilde{U} = U - <U> \quad \text{and} \quad \tilde{H} = H - <H> \qquad (5)$$

and

$$u_L = d<q>/hd\sigma = u_E + u_S. \qquad (6)$$

Here, u_E is referred to as the Eulerian residual current, although it relates to values at fixed positions in the (x, σ) coordinate system, rather than at fixed positions in space. The mass transport residual current is denoted by u_L. The current u_S is referred to as the mass transport Stokes drift, to distinguish it from the related Stokes drift defined by LONGUETT-HIGGINS (1969). A suitable analytical form for the Stokes drift in highly non-linear tidal flows (such as those considered here) does not appear to exist. The subscripts on u_E, u_S and u_L serve to indicate that these currents correpond to the familiar Eulerian, Stokes and Lagrangian residual currents for one-dimensional flows. Depth averaged values, $\bar{u}_E$, $\bar{u}_S$ and $\bar{u}_L$ are derived by integrating over σ from 0 to 1. In particular:

$$\bar{u}_L = <Q>/h \qquad (7)$$

A freshwater-induced residual current, $\bar{u}_F$, can be defined. If $<A>$ is the tidally-averaged area of a cross-section, and $<Q_F>$ the rate of input of freshwater volume up-estuary of this section (also averaged over a tidal cycle), then:

$$\bar{u}_F = <Q_F>/<A> \qquad (8)$$

The residual rate of transport of salt per unit width of column is (in ppt cm s^{-1}, see UNCLES and JORDAN, 1979):

$$F = F_L + F_{TP} + F_V \qquad (9)$$

where F_L is due to the residual flow of water, $\bar{u}_L$, and F_{TP} and F_V are due to tidal pumping and vertical shear, respectively. If S denotes the instantaneous salinity, and $s=<S>$, $\tilde{S}=S-\bar{s}$, then :

$$F = <\overline{HUS}>/h \qquad (10)$$

$$F_L = \bar{u}_L \bar{s} \qquad (11)$$

$$F_{TP} = <\tilde{Q}\tilde{S}>/h \qquad (12)$$

and

$$F_V = <\overline{HU'S'}>/h \qquad (13)$$

where

$$S' = S - \bar{S} \quad \text{and} \quad U' = U - \bar{U} \qquad (14)$$

The residual rate of transport of suspended sediment per unit width of column is (in ppm cm s^{-1}, where ppm = parts per million by weight of water):

$$G = G_L + G_{TP} + G_V \qquad (15)$$

Subscripts have the same meaning as those for the salt flux (equation (9)). If P denotes the instantaneous suspended sediment concentration, and $p=<P>$, $\tilde{P}=P-\bar{p}$, then

$$G = <H\overline{UP}>/h \tag{16}$$

$$G_L = \bar{u}_L\bar{p} \tag{17}$$

$$G_{TP} = <\tilde{Q}\tilde{P}>/h \tag{18}$$

and

$$G_V = <H\overline{U'P'}>/h \tag{19}$$

where

$$P' = P - \bar{P}$$

The accuracy with which tidal averages were derived was estimated by generating synthetic data for each station. Essentially, the observed data at a station were randomly perturbed by an amount which depended on instrument errors (UNCLES et al., 1985). All observed residual currents and fluxes are plotted with 95% confidence intervals where these are larger than the plotting symbols. Plotted errors must be considered minimum values because they do not take into account errors due to possible small movements of the boat during observations.

Station	U_T	h	$<Q_F>$	$\bar{u}_F$	$\bar{s}$	$\bar{p}$	$\Delta s/\bar{s}$
	(cm s^{-1})	(m)	(m^{3}s^{-1})	(cm s^{-1})	(ppt)	(ppm)	
1(S)	26	2.9	3.5	2.10	4.1	183	0.43
3(S)	20	4.4	3.5	0.26	26.6	18	0.04
4(S)	20	6.6	4.4	0.12	31.1	8	0.05
5(S)	16	15.2	5.7	0.07	32.1	4	0.02
1(N)	17	2.9	5.4	3.30	2.4	69	0.92
3(N)	13	5.7	5.4	0.40	26.3	10	0.24
4(N)	12	6.6	6.7	0.18	---	7	---
5(N)	9	16.0	8.4	0.10	31.0	3	0.06

TABLE Basic properties of anchor stations 1, 3, 4 and 5. The spring tide (S) was worked on 7/7/82 with a tidal range of 3.9 m and the neap tide (N) on 30/6/82 with a tidal range of 2.7 m.

A summary of background data for each station is given in the Table. This shows:

(1) the mean, depth-averaged current speed, $\bar{U}_T$;

(2) the mean depth, h;

(3) the freshwater inputs, $<Q_F>$, which vary along the estuary because of inputs from the LYNHER and TAVY (see Fig. 1);

(4) the cross-sectionally averaged residual currents generated by freshwater inputs, $\bar{u}_F$;

(5) the depth and tidally-averaged salinity, $\bar{s}$;

(6) the depth and tidally-averaged suspended sediment concentration, $\bar{p}$; and

(7) the bottom to surface salinity stratification, $\Delta s/\bar{s}$.

3. Results

General transport processes are considered first, beginning with the depth-averaged residual fluxes of salt and suspended sediment. This is followed by results for the depth-averaged residual transport of water, and the vertical structure of currents.

3.1 Residual fluxes of salt

The residual flux of salt per unit width of water column is given by equation (9). If conditions were uniform over a cross-section, and if a steady-state existed, then from equation (11):

$$F_L = \overline{u_L s} = \overline{u_F}\overline{s} = F_F. \tag{20}$$

so that F_L is the flux of salt which is carried down-estuary by cross-sectionally averaged freshwater induced currents. In steady-state, the residual flux of salt over any cross-section would be zero ($F=0$ in equations (9) and (10)), and equation (9) would become:

$$-1 = F_{TP}/F_F + F_V/F_F. \tag{21}$$

F_F is defined by equation (20), and can be computed from data in the Table. F_{TP} and F_V have been estimated from the observations (using equations (12) and (13)).

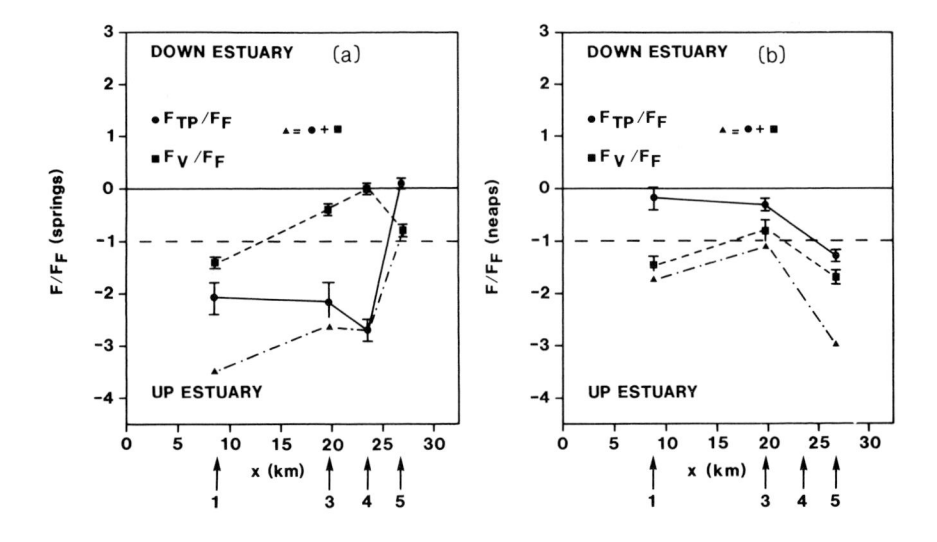

FIGURE 3 Observed depth-averaged values of the non-dimensional salt fluxes due to tidal pumping (●), vertical shear (□), and their sum (△).
(a) Spring tide data.
(b) Neap tide data.
95% confidence intervals are shown for the tidal pumping and vertical shear.

Data for F_{TP}/F_F (the non-dimensional salt flux due to tidal pumping) and F_V/F_F (that due to vertical shear) are given in Fig. 3(a) for the spring tide observations. Except for F_{TP} at station 5 (which was not significantly different from zero), salt fluxes F_{TP} and F_V were always directed up-estuary (the negative direction), and tidal pumping was dominant. The sum of these fluxes is also shown in Fig. 3(a); this sum was comparable with or greater than that required to balance the down-estuary advection of salt due to freshwater inputs (F_F).

Components of the residual fluxes of salt for the neap tide observations are shown in Fig. 3(b). Fluxes due to tidal pumping and vertical shear were directed up-estuary. Transport due to vertical shear dominated that due to tidal pumping throughout the estuary. The sum of these fluxes is also shown in Fig. 3(b), and was again comparable with or greater than that required to balance the down-estuary salt flux due to freshwater flows. Because this excess up-estuary flux of salt occurred for both surveys, it was unlikely to have been a consequence of unsteady conditions, but was probably due to the existence of strong cross-estuary variations in the fluxes.

The observations determined F_L rather than F_F, and these are equal only in the unlikely event that $\bar{u}_F = \bar{u}_L$ (equation (20)). However, the residual flow at one station in a section would be almost balanced by opposing flows elsewhere on the section; a net contribution to the sectionally-averaged salt flux would exist only if these lateral circulations correlated with salinity (FISCHER, 1972; UNCLES et al., 1983).

3.2 Residual fluxes of suspended sediment

The residual flux of suspended sediment is given by equation (15). If conditions were uniform over a cross-section, and if a steady-state existed, then from equation (17):

$$G_L = \bar{u}_L \bar{p} = \bar{u}_F \bar{p} = G_F \qquad (22)$$

and equation (15) would become

$$G / G_F - 1 = G_{TP} / G_F + G_V / G_F \qquad (23)$$

Data for G_{TP} / G_F (dimensionless tidal pumping) and G_V / G_F (dimensionless vertical shear) during the spring tide observations are drawn in Fig. 4(a). Tidal pumping of sediment was much larger than that due to vertical shear and freshwater-induced currents. The pumping was directed up-estuary at station 1 and down-estuary elsewhere. At this station the tidal currents were strongly asymmetrical, with flood currents exceeding ebb currents. This is a characteristic feature of estuarine tidal flows in shallow water (KREISS, 1957; UNCLES, 1981). At station 1 the stronger flood currents produced enhanced resuspension of bottom sediment. This was transported into the estuary. Ebb currents produced less resuspension and less transport of sediment, so that the residual transport was directed up-estuary. The dimensional fluxes can be computed from data in the Table, using equation (22). Although not obvious from Fig. 4(a), the flux at station 1 was of order 100 times larger than that at station 5.

Similar data for the neap tide are shown in Fig. 4(b). Tidal pumping was again dominant, and was directed up-estuary at stations 1, 3 and 5. The up-estuary transport at station 5 was apparently due to the advection of a near-surface patch of higher turbidity water through this station. The patch was observed on the early flood, but was either dispersed before the ebb, or else returned seawards via another path. The dimensional flux at station 5 was less than 5% of that at station 1. The mechanisms producing the up-estuary transport at stations 1 and 3 were identical to those operating during the spring tide.

A pronounced turbidity maximum was observed in the upper reaches during both surveys. The spring tide maximum was much larger, as shown in the Table. As shown in Figs 4(a) and (b), transport of suspended sediment due to vertical shear was either directed down-estuary, or was not significantly different from zero. Transport due to tidal pumping was much larger than that due to vertical shear, and was directed up-estuary near the head for both surveys. This effect may have contributed to the formulation of the turbidity maximum.

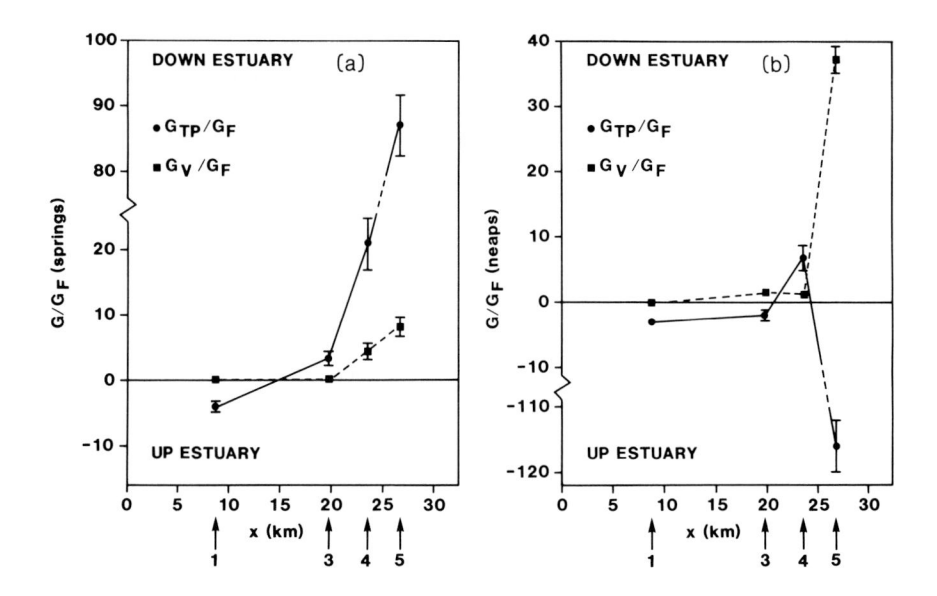

FIGURE 4. Observed depth-averaged values for the non-dimensional suspended sediment fluxes due to tidal pumping (●) and vertical shear (■), at:
(a) Spring tides; and
(b) Neap tides.
95% confidence intervals are shown for the tidal pumping and vertical shear.

3.3 Residual fluxes of water

Data for $\bar{u}_S$, $\bar{u}_L$ and $\bar{u}_F$ are given in Fig. 5(a) for the spring tide survey. The mass transport Stokes drift was directed up-estuary in the deep channel. Values decreased from roughly 5 cm s^{-1} near the head to less than 0.5 cm s^{-1} in the lower reaches. The mass transport residual current, $\bar{u}_L$, was very different from $\bar{u}_F$ and was generally directed up-estuary. This implies the existence of lateral circulation patterns.

Data for the neap tide are given in Fig. 5(b). The mass transport Stokes drift was again directed up-estuary, although speeds were roughly halved compared with the spring tide. The disparate values of $\bar{u}_F$ and $\bar{u}_L$ in the middle and lower reaches again imply the existence of significant lateral circulation patterns.

3.4 Vertical structure of currents

Vertical profiles of the Eulerian residual current along the estuary are shown for both surveys in Fig. 6. The mass transport residual currents were similar and, with the exception of station 4, exhibited a gravitational circulation throughout the estuary. The near-bed, up-estuary flow did not occur in the Eulerian current at station 1 during the neap tide survey. This was a result of the greater freshwater run-off at that time (see the Table), which displaced $\bar{u}_E$ down-estuary.

Profiles of u_E, u_L and u_S at stations 1 and 5 for both surveys are shown in Fig. 7. Error bars (95% confidence intervals) are given for the spring tide data. These are of similar magnitude for the neap tide data (not shown), and for data at other stations. The gravitational circulation in u_L at station 1 was much more pronounced than that in

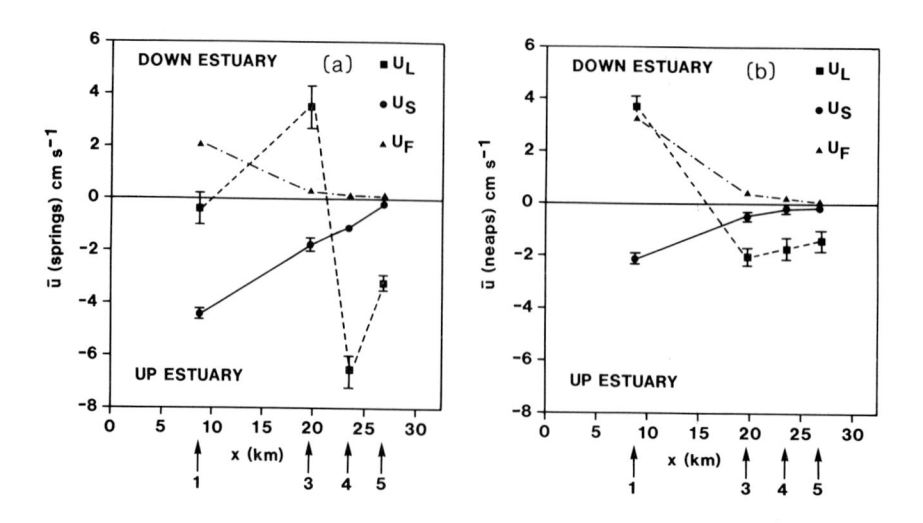

FIGURE 5. Observed depth-averaged values of u_S (●), u_F (Δ), and u_L (■).
(a) Spring tide data;
(b) Neap tide data.
95% confidence intervals are shown for u_S and u_L.

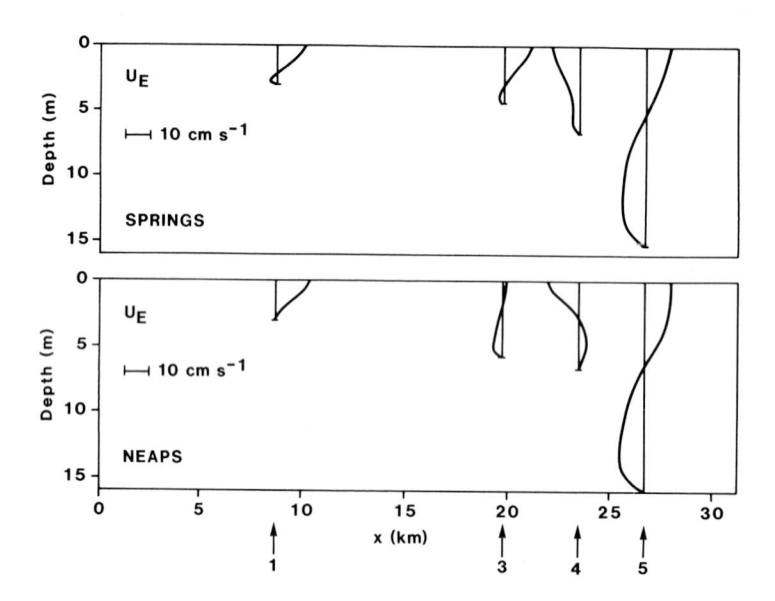

FIGURE 6. Vertical profiles of Eulerian residual current along the axis of the estuary during the spring (upper) and neap (lower) tide surveys.

u_E. This is because $\bar{u}_E$ was displaced down-estuary in order to compensate for the up-estuary directed mass transport Stokes drift. In steady-state, with no freshwater flows and with uniformity over a cross-section, $\bar{u}_L = 0$ and $\bar{u}_E = -\bar{u}_S$; the directional sense of u_L through the column would depend on the subtle balance of u_E and u_S with depth.

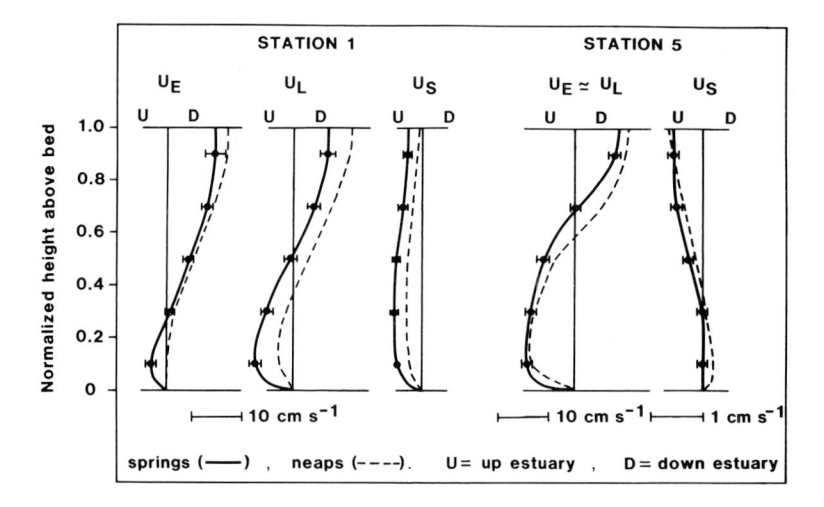

FIGURE 7. Vertical profiles of u_E, u_L and u_S at stations 1 and 5 during the spring (—) and neap (- - -) surveys. 95% confidence intervals are shown for the spring tide survey.

Vertical profiles of u_S had a different form in the upper and lower reaches (Fig. 7). Near the mouth, u_S (although very small) decreased with depth, and, for the spring tide, was not significantly different from zero near the bed. The near-bed, down-estuary mass transport Stokes drift for the neap tide was significantly different from zero; therefore, this was a real effect. Near the head, u_S was directed up-estuary throughout the depth, and had its maximum speed in the lower half of the water column.

The vertical structure of the mass transport Stokes drift in the lower reaches (station 5) can be understood from equation (4):

$$u_S = <\tilde{H}\tilde{U}>/h \approx \frac{1}{2}H_2 U_2 \cos(\gamma_2 - \epsilon_2)/h \qquad (24)$$

where

$$\tilde{U} = U_2 \cos(\omega_2 t - \gamma_2) + \text{overtides} \qquad (25)$$

and

$$\tilde{H} = H_2 \cos(\omega_2 t - \epsilon_2) + \text{overtides} \qquad (26)$$

The semi-diurnal angular frequency is denoted by ω_2. Overtides are small compared with the semi-diurnal tides, and have been ignored. The amplitude of the tidal current, U_2, decreases near the bed, owing to the greater influence of friction. This will tend to reduce u_S near the bed. The mass transport Stokes drift is zero when $(\gamma_2 - \epsilon_2) = 90°$, and is directed up-estuary when $(\gamma_2 - \epsilon_2)$ exceeds 90° — becoming larger with increasing γ_2. Ignoring the effects of density gradients, which can be justified (see later), the greater influence of friction means that phases of the tidal currents are advanced (eg. PROUDMAN, 1953), so that γ_2 decreases with increasing depth. This effect will reduce $(\gamma_2 - \epsilon_2)$ such that $(\gamma_2 - \epsilon_2)$ approaches 90°, which, in addition to the reduction in U_2, will further reduce u_S. When $(\gamma_2 - \epsilon_2)$ falls below 90°, the mass transport Stokes drift will be directed down-estuary, which is the case near the bed at station 5 (Fig. 7). The depth-averaged mass transport Stokes drift must be directed up-estuary in the deep-channel because it supplies the energy required to compensate frictional dissipation of the tide (eg. UNCLES and JORDAN, 1980).

The decrease with increasing depth of the tidal current amplitude and phase (U_2 and γ_2) due to seabed friction must have also occurred at station 1 in the upper reaches. However, at this station u_S reached a maximum (up-estuary) speed in the lower half of the water column, in contrast to that at station 5. This may have been a consequence of the increased importance of density gradients in the upper estuary. The relative magnitudes of density and surface slope forcing can be estimated from the simplest version of the 'x' component of acceleration (eg. see equation (23) of UNCLES and JORDAN (1979)):

$$\partial_t U = -g\partial_x \zeta - gH(\rho^{-1}\partial_x \rho)(1-\sigma) + H^{-2}\partial_\sigma (N\partial_\sigma U), \tag{27}$$

using the notation $\partial_t = \partial/\partial t$. Here, ζ is surface elevation, ρ density, g acceleration due to gravity and N the coefficient of vertical eddy viscosity. Equation (27) ignores Coriolis force, advective accelerations, and vertical variations in density.

The density forcing term in equation (27) is proportional to total depth, and thus varies cyclically in time with the semi-diurnal period. This forcing will modify the semi-diurnal currents. The tidal component of depth-averaged density gradient forcing is approximately:

$$K = -\frac{1}{2}g\tilde{H}(\rho^{-1}\partial_x \rho).$$

the surface slope forcing is

$$L = -g\partial_x \tilde{\zeta}.$$

The semi-diurnal components of these terms, L_2 and K_2, can be estimated from the observations using the depth-averaged version of equation (27). At station 5 (lower reaches):

$$K_2/L_2 \approx 0.03.$$

Whilst at station 1 (upper reaches):

$$K_2/L_2 \approx 0.10.$$

Therefore, the effects of density gradients on U_2 and γ_2 (and thus u_S, see equations (24)-(26)) were much more pronounced in the upper reaches, and near the seabed (see equation (27) with $\sigma \ll 1$). It is likely that some significant modification of U_2 and γ_2 resulted from $K_2/L_2 \approx 0.1$. This may have been the cause of the differing behaviour of u_s in the upper and lower reaches.

4. Conclusions

The data presented in this paper represent spring and neap tide observations in the deep channel of the estuary during periods of low freshwater run-off. The measurements complement those discussed by UNCLES et al. (1985), which were non-synoptic, and which were generally applicable to medium or high run-off conditions.

Salt fluxes due to tidal pumping and vertical shear were either directed up-estuary for both surveys, or else were not significantly different from zero. The axial salt flux during the spring tide was largely controlled by tidal pumping, whereas transport due to vertical shear was dominant during the neap tide. These results are essentially the same as those deduced from observations made under high run-off conditions, except that no systematic picture emerged for the transport near the head during neap tides, owing to stronger winds, the non-synoptic nature of the data, and the pronounced stability of the water column; it appeared that wind-stress had a significant influence on the transport of near-surface water for those observations. An inspection of data given by DYER (1978) shows that tidal pumping was also the dominant transport mechanism for salt in the

Thames and Gironde Estuaries during spring tides.

The suspended sediment concentrations in the upper estuary during the spring tide survey were much higher than during the neap tide. This was because of enhanced resuspension of bed sediments in the faster tidal currents, and the presence of a stronger turbidity maximum, which occurred in the low salinity water near the head of the estuary. Concentrations in the lower estuary were small. These results are similar to those found for the high run-off, non-synoptic observations, except that for those data the turbidity maximum was absent during the small neap tides sampled.

Transport of suspended sediment due to tidal pumping greatly exceeded that due to vertical shear during both surveys. This transport was directed up-estuary near the head, and probably contributed to the turbidity maximum. Similar results were obtained during spring tides under high run-off for the non-synoptic data (UNCLES et al., 1985). However, for the small neap tides, and during periods of high run-off, the observed fluxes were either negligible, or else directed down-estuary. In these cases a turbidity maximum did not form. Data in this paper indicate that the turbidity maximum increases in strength with increasing tidal range and decreasing freshwater run-off. The importance of tidal pumping to the spring tide sediment budgets of the Gironde and Thames Estuaries can be deduced from data given in DYER (1978).

Generally, the observed depth-averaged mass transport residual currents were very different from the cross-sectionally-averaged currents resulting from freshwater inputs, and were directed up-estuary at some stations. This implies the existence of strong lateral circulation patterns.

The depth-averaged, mass transport Stokes drift was directed up-estuary, and was very small in the lower reaches. Therefore, the depth-averaged Eulerian and mass transport residual currents were very similar in the lower reaches. In the upper reaches, the Eulerian residual currents had a significantly stronger down-estuary component than did the mass transport residual currents; this was in response to the stronger mass transport Stokes drift.

The observed vertical structure of u_S differed in form between the upper and lower reaches. Near the mouth, the mass transport Stokes drift in the lower half of the column was much smaller than in the upper half, and was not significantly different from zero. Near the head, the mass transport Stokes drift maximized in the lower half of the column. The different behaviour of u_S in the upper and lower reaches of the estuary may be a consequence of the greater importance of density gradients in the upper reaches. This possibility requires further investigation.

Acknowledgements

We are grateful to Mr. J.A. Stephens and Ms. T. Woodrow for assistance with data analyses, illustrations and word-processing.

This work, which was based on a multi-institute field sampling exercise, forms part of the Physical Processes programme of the Institute for Marine Environment Research, a component of the Natural Environment Research Council (NERC). It was partly supported by the Department of the Environment on Contract No. DGR 480/48.

References

ALLEN, G.P., SALOMON, J.C., BASSOULLET, P., DU PERIHOAT, Y and DE GRANDPRE, C., 1980. Effects of tides on mixing and suspended sediment transport in macrotidal estuaries. Sedimentary Geology, 26, pp.69-90.

DYER, K.R. 1973. Estuaries: A physical introduction. John Wiley and Sons, 140pp.

DYER, K.R. 1978. The balance of suspended sediment in the Gironde and Thames Estuaries. In: Estuarine Transport Processes, B. KJERFVE (ed.). The Belle W. Baruch library in marine science, No.7, 331pp.

FISCHER, H.B. 1972. Mass transport mechanisms in partially stratified estuaries. J. Fluid Mech., 53, pp. 671-687.

FISCHER, H.B., LIST E.J., KOH, R.C.Y., IMBERGER, J. and BROOKS, N.H., 1979. Mixing in inland and coastal waters. Academic Press, New York, 484pp.

HANSEN, D.V. and RATTRAY, M. 1966. New dimensions in estuary classification. Limnol. Oceanogr., 11, pp. 319-326.

HUGHES, F.W. and RATTRAY, M. 1980. Salt flux and mixing in the Columbia River Estuary. Estuar. Coastal Mar. Sci., 10, pp. 479-493.

KJERFVE, B. 1975. Velocity averaging in estuaries characterized by a large tidal range to depth ratio. Estuar. Coastal Mar. Sci., 3, pp. 311-323.

KJERFVE, B. 1979. Measurement and analysis of water current, temperature, salinity and density. In: Estuarine Hydrography and Sedimentation, K.R. DYER (ed.), C.U.P., 230pp.

KREISS, H. 1957. Some remarks about non-linear oscillations in tidal channels. Tellus, 9, pp. 53-68.

LEWIS, R.E. and LEWIS, J.O. 1983. The principal factors contributing to the flux of salt in a narrow, partially stratified estuary. Estuar. Coastal Shelf Sci., 16, pp. 599-626.

LONGUETT-HIGGINS, M.S. 1969. On the transport of mass by time-varying ocean currents. Deep-Sea Res., 16, pp. 431-447.

OFFICER, C.B. 1976. Physical oceanography of estuaries (and associated coastal waters). John Wiley and Sons, 465pp.

PROUDMAN, J. 1953. Dynamical Oceanography. Methuen, London.

RATTRAY, M. and DWORSKI, J.G. 1980. Comparison of methods for analysis of the transverse and vertical circulation contributions to the longitudinal advective salt fluxes in estuaries. Estuar. Coastal Mar. Sci., 11, pp. 515-536.

UNCLES, R.J. 1981. A note on tidal asymmetry in the Severn Estuary. Estuar. Coastal Shelf Sci., 13, pp. 419-432.

UNCLES, R.J. and JORDAN, M.B. 1979. Residual fluxes of water and salt at two stations in the Severn Estuary. Estuar. Coastal Mar. Sci., 9, pp. 287-302.

UNCLES, R.J. and JORDAN, M.B. 1980. A one-dimensional representation of residual currents in the Severn Estuary and associated observations. Estuar. Coastal Mar. Sci., 10, pp. 39-60.

UNCLES, R.J., BALE A.J., HOWLAND, R.J.M., MORRIS, A.W. and ELLIOTT, R.C.A. 1983. Salinity of surface water in a partially-mixed estuary, and its dispersion at low run-off. Oceanologica Acta, 6, pp. 289-296.

UNCLES, R.J., ELLIOTT, R.C.A. and WESTON, S.A. (1985). Observed fluxes of water, salt and suspended sediment in a partly mixed estuary. Estuar. Coastal Shelf Sci., 20, pp. 147-168.

Analysis of Residual Currents Using a Two-Dimensional Model

Po-Shu Huang,

Dong-Ping Wang,

Tavit O. Najarian

Najarian and Associates, Inc.
USA

ABSTRACT

Use is made of a transient, two-dimensional (in a vertical plane) model to analyse residual currents in oscillating flow environments. LONGUET-HIGGINS' (1969) approximation is employed to compute Stokes velocities for small amplitude tides. The influence of geometry, bottom friction, density gradient, vertical eddy viscosity, and phase and amplitude differences in tidal forcing on residual currents and estuaries and sea-level canals are revealed through a series of model simulations.

1. Introduction

Estuaries and waterways serve the useful purpose of transporting effluents discharged from urban areas to the open oceans. The net transport of discharged pollutants and nutrients in tidal waters is related to residual currents. The amplitude of residual currents is a function of the freshwater inflows, the density gradients and the tidal amplitudes which drive the net circulation. The judicious allocation of wasteload discharges into an estuary and the optimization of estuarine water resources require an accurate description of the residual currents. Ordinarily, such descriptions are derived from the analysis of current measurements at a few fixed stations in the domain of interest.

Current meter measurements describe the velocity at a particular point in the flow field. The averaging of these real time velocities over the dominant time scale of one tidal period provides the residual Eulerian currents at fixed locations. The mass transport velocity describes the rate of displacement of pollutants in a flow field. This rate of displacement is the mean Lagrangian velocity of a particular mass discharged into a coastal environment. Therefore, the mean Lagrangian velocity profiles provide a measure of pollutant displacement or transport over a tidal cycle.

The two-dimensional, laterally-integrated equations describing circulation in oscillating flow environments are quite complex. Analytical solutions to these equations are attainable through simplifying assumptions regarding the geometry of the solution domain and the non-linear terms describing bottom friction and vertical eddy viscosities. On the other hand, numerical models provide the means to alleviate these limitations in describing the net circulation in tidal waters.

LONGUET-HIGGINS (1969) showed that the mean Lagrangian velocity can be approximated, to the second order accuracy, as the sum of the mean Eulerian and Stokes velocities. ZIMMERMAN (1979) discussed the error in the mean Lagrangian velocity computations obtained using LONGUETT-HIGGINS approximation for small amplitude tides. To eliminate such truncation errors, CHENG (1982) resorted to modelling the mean Lagrangian velocity by following the path of particles released in the domain of interest. However, the modelling approach taken by CHENG can be computationally prohibitive for the extensive sensitivity analysis presented below.

2. Model equations and assumptions

The mean Lagrangian velocity computations obtained using LONGUETT-HIGGINS' approximation remain valid for cases when the spatial gradients in the velocity field are small and the tidal excursion is small compared to the tidal wave length (ZIMMERMAN, 1979). The mean Lagrangian velocity can be expressed by

$$U_L = \bar{u} + U_s \tag{1}$$

where U_L = mean Lagrangian velocity; u = Eulerian velocity; and U_s = Stokes velocity. The overbar denotes time integration over a period of one tidal cycle. The second term in equation 1 is the Stokes velocity. LONGUETT-HIGGINS (1969) showed that the Stokes velocity for small displacement tide is equal to

$$U_s = \overline{\int u dt \frac{\partial u}{\partial x}} + \overline{\int w dt \frac{\partial u}{\partial z}} \tag{2}$$

where w = vertical component of tidal velocities.

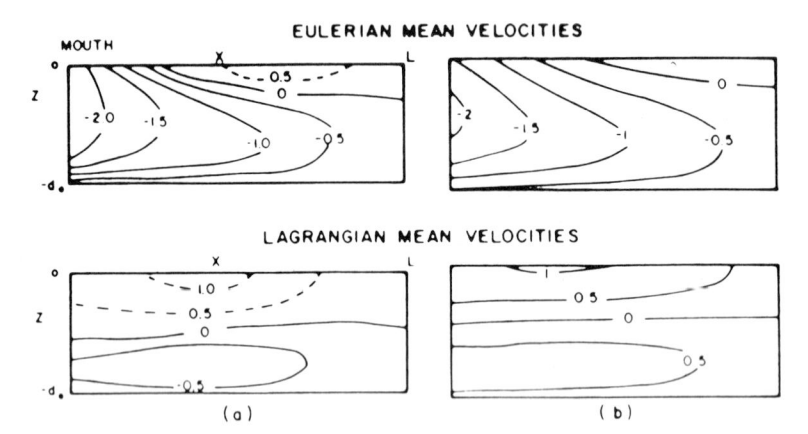

FIGURE 1. Nondimensional results of
(a) analytical solution; and
(b) numerical solution for residual velocities in a rectangular, homogeneous channel (seaward velocities are negative)

The mean Lagrangian velocities in oscillating flows are derived here from the accurate description of tidal velocities and water surface variations. The exercise of the transient two-dimensional model developed by WANG and KRAVITZ (1980) resulted in the time series solutions of u, w, ρ and η where ρ is the density and η is the water surface elevation above mean sea level. The mean Eulerian and Stokes velocities are computed from equations 1 and 2 by time-averaging over a tidal period the transient Eulerian velocities and their spatial derivatives. The governing equations of the two-dimensional (in the vertical plane) model are shown elsewhere (see WANG and KRAVITZ, 1980 and NAJARIAN et al. 1984).

The existing version of the model provides the option of analyzing residual currents with either linear or quadratic boundary friction approximations. Similarly, the model can be exercised with either constant values of vertical eddy viscosities or with temporally and spatially varying values. The latter option is based on the MUNK-ANDERSON (1948) approximation of vertical viscosities.

3. Sensitivity of residual currents to different flow parameters

The two-dimensional numerical model was first verified against published analytical solutions for residual currents in homogeneous estuaries. The model exercises were carried out under the assumption of constant vertical eddy viscosities and linear bottom and side friction. Fig. 1 shows the comparison of model-generated and analytical solution of mean Eulerian, Stokes and mean Lagrangian velocities. The analytical solutions are well documented by IANNIELLO (1977, 1979).

A series of numerical model simulations were performed to evaluate the influence on residual currents in an estuary of different flow parameters. A single harmonic tide was imposed at the mouth of a prismatic channel. Table 1 lists the simulations and their associated parameters.

Simulation	Cross Section	Friction	Density Structure
1 (Fig. 2)	rectangular	linear	homogeneous
2 (Fig. 3)	rectangular	quadratic	homogeneous
3 (Fig. 4)	trapezoidal	quadratic	homogeneous
4 (Fig. 5)	trapezoidal	quadratic	with gradient

TABLE 1. Simulation parameters for sensitivity analysis

To eliminate the influence of the freshwater inflows on residual currents, the discharges into the idealized estuary were all set to zero. The first two of the experiments were conducted to analyse the sensitivity of residual currents to the description of bottom friction. In the first case linear bottom friction was assumed such that

$$\tau_b = \gamma u \tag{3}$$

where γ is the friction coefficient, and u is the velocity at the bottom layer. In the second case the quadratic bottom friction was assumed in the model simulation. The latter stress was approximated by

$$\tau_b = C_D \rho |u| u \tag{4}$$

Fig. 2 shows the residual velocity profiles for a rectangular homogeneous estuary with linear bottom friction. The channel is 70 km long and 10 m deep. The flow is driven by a single harmonic tides at the mouth of 30 cm amplitude. The maximum tidal velocity computed was 73 cm/sec which is less than 10% the celerity of the tidal wave. The mean Eulerian velocities are basically seaward and the Stokes velocities are landward. The mean Langrangian velocities are landward at the surface and seaward at the bottom. The total flow obtained by vertically integrating the mean velocity profiles results in a net zero mass transport in the estuary.

Fig. 3 shows the residual currents obtained for the same idealized channel with the quadratic bottom friction assumption. It is important to note here that the coefficients of the two friction parameters, γ and C_D were adjusted to ensure a fair comparison of results. The comparison of results shown in Figs 2 and 3 reveals two important differences. First, the assumption of quadratic friction causes a reduction in the

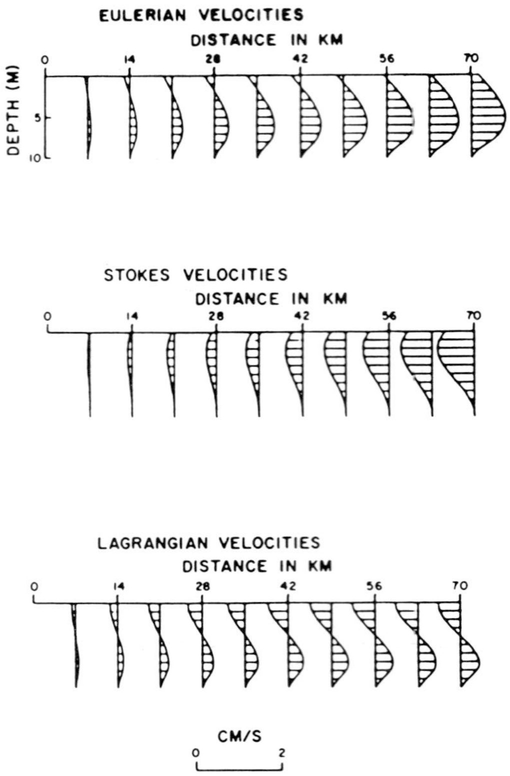

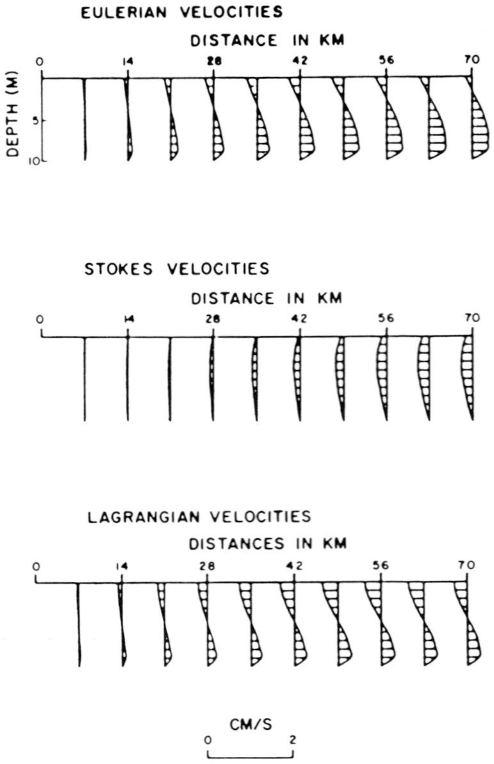

FIGURE 3. Residual velocity profiles for a rectangular homogeneous channel with quadratic boundary friction.

FIGURE 2. Residual velocity profiles for a rectangular homogeneous channel with linear boundary friction.

maximum residual velocities shown in the mean Eulerian and Lagrangian profiles. Second, the mean Eulerian velocities with quadratic bottom friction are landward at the surface and seaward at the bottom. Such is not the case for mean Eulerian currents obtained with linear bottom friction.

To analyze the sensitivity of residual currents to density gradients two new simulations were performed. For both simulations, a trapezoidal channel, 70 km long was assumed with side slopes of 2:1 and a depth of 10 m below mean sea level. A quadratic bottom friction was used in both simulations. The tidal amplitude was set at 1 m at the ocean boundary of the channel. In the first simulation, a homogeneous density was assumed. Fig. 4 shows the residual velocity profiles for the case described. In the second simulation a density gradient in the channel was assumed. Fig. 5 shows the residual velocity profiles for the same trapezoidal channel with a density gradient. In the latter simulation the two-dimensional model was exercised with ocean salinity varying from 19 ppt at the surface to 21 ppt at the bottom. These salinities were linearly interpolated to zero at the head of tide to describe the initial conditions. Finally, the eddy viscosity and diffusivity coefficients were allowed to vary with the local Richardson number (see MUNK and ANDERSON, 1948).

Whereas the shape of the mean Lagrangian velocity profiles in the homogeneous case are similar to those obtained in the rectangular channel, the mean Lagrangian velocity profiles shown in Fig. 5 exhibit a classical estuarine circulation with seaward surface currents and landward bottom currents. The comparison of the mean Lagrangian velocity profiles shown in Figs 4 and 5 reveal that the density-induced current is stronger than but opposite to tide-induced residual current.

4. Mass transport

The residual transport in Eulerian terms is given by

$$\overline{M} = \overline{\int_{-H}^{\eta} u\,dz} \tag{5}$$

where η = water surface elevation with respect to mean sea level; and H = water depth below mean sea level. Equation 5 can be used to derive the net mass transport only if Eulerian velocity profiles and the water surface elevations are known simultaneously. The net transport can also be derived from the following expression.

$$\overline{M} = \int_{-H}^{0} \overline{u}\,dz + \overline{u\eta} = \int_{-H}^{0} U_L\,dz \tag{6}$$

The latter equation shows that the net transport can be derived without the knowledge of the surface elevation variations if Lagrangian velocity profiles, instead of Eulerian velocity profiles, are used. In an estuary with harmonic tides and no freshwater inflow the net residual mass transport is zero although residual flows at different depths are present.

On the other hand, in a sea-level canal with two ocean boundaries a net residual transport is always present when the tides at the extremities of the canal are not in phase. This net transport results from the phase and amplitude differences in the two tides at the boundaries. To analyze the effect of the phase and amplitude differences between the tides on the net transport in a sea-level canal, two sets of simulations were conducted. The geometric parameters of the sea-level canal were based on a simplified (trapezoidal) schematization of the CHESAPEAKE and DELAWARE (C&D) canal.

To analyze the influence of the phase differences between the boundary tides, a set of simulations were conducted assuming that the tidal amplitudes at both ends of the canal

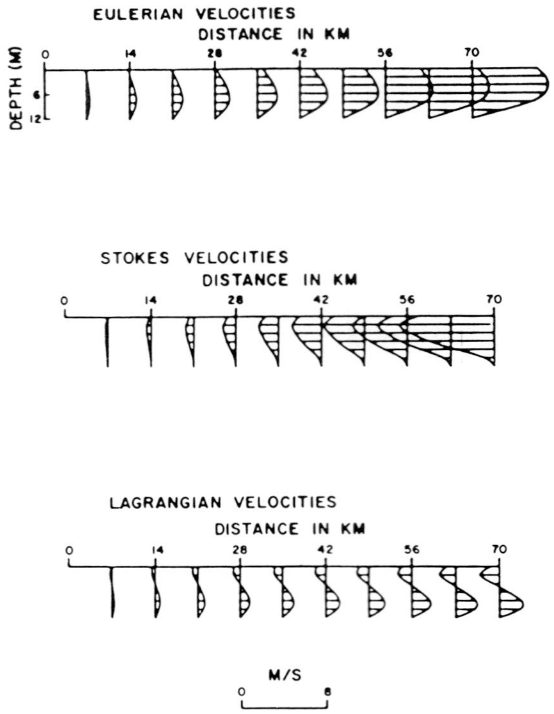

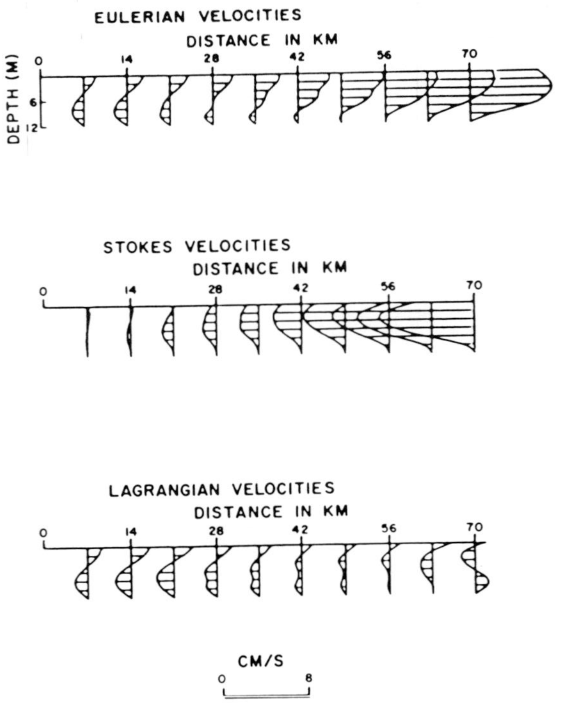

FIGURE 5. Residual velocity profiles for a trapezoidal channel with density gradient.

FIGURE 4. Residual velocity profiles for a trapezoidal homogeneous channel.

are 50 cm. A second set of simulations were conducted using the realistic mean tidal amplitudes at the extremities of the C&D canal. In this case, the tide was assumed to have an amplitude of 78.6 cm at the eastern end and 31.4 cm at the western end. Both sets of simulations were conducted by setting the phase difference in tides in increments of 30 degrees. Quadratic bottom friction and an eddy viscosity of 60 cm^2/sec were used in these simulations. The results of these simulations are shown in Fig. 6. For equal tidal amplitudes the net transport in the canal is maximum when the phase difference between the two tides is at 70°. With different tidal amplitudes the net transport is forced from the higher amplitude end towards the lower amplitude end over a wide range of phase differences.

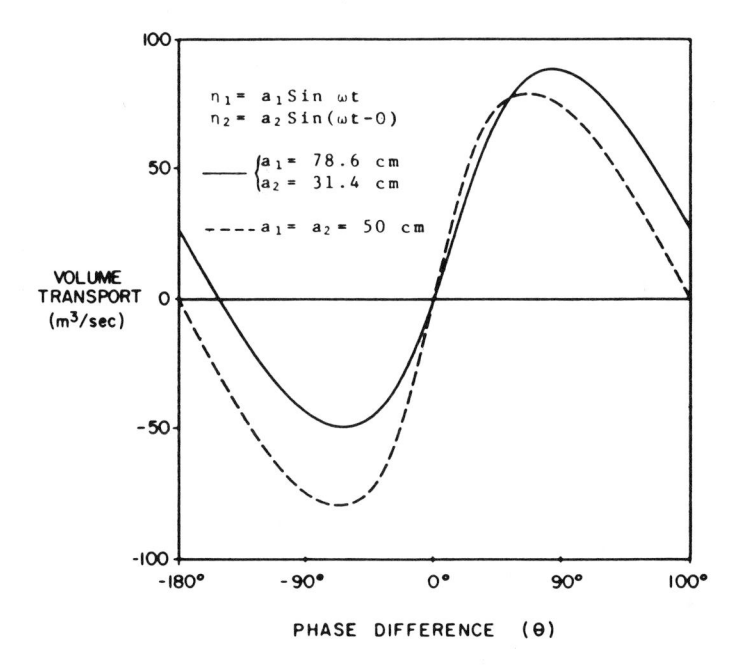

FIGURE 6   Net transport in a sea-level canal (— different amplitude and, --- equal amplitude tides)

The tides at the Chesapeake Bay end of the canal leads the tides at the Delaware end by an average of 1.2 hours. This translates to a phase difference of -35° for semi-diurnal tide. Fig. 6 shows that for such a phase difference in tides the net transport is on the order of 40 m^3/sec eastward. This result agrees with the current meter measurement of PRITCHARD et al (1974) and the physical model result by BOYD et al. (1973).

The net transport in the canal is also sensitive to the specification of eddy viscosity. Another set of simulations on the C&D canal were conducted to assess this sensitivity over a range of values from 5 cm^2/sec to 100 cm^2/sec. In addition to the constant eddy viscosity assumptions a simulation was also conducted with variable eddy viscosities whose coefficient is related to the local vertical gradient of the horizontal velocities. Here again, the bottom friction was assumed quadratic.

$$N_z = 100 \cdot \left| \frac{\partial u}{\partial z} \right|. \tag{7}$$

The results of the sensitivity analysis of residual current profiles to eddy viscosity is

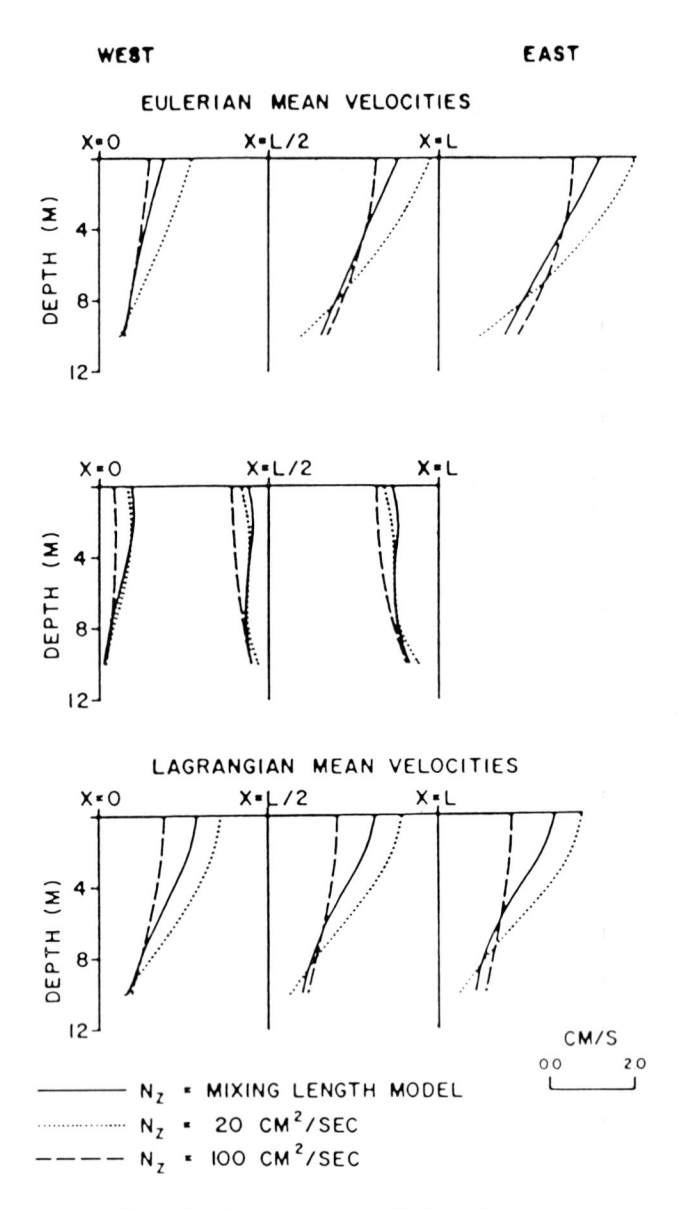

FIGURE 7   Residual velocity profiles in the Canal.

shown in Fig. 7. It is apparent that the mean Eulerian velocities are, in general, twice as large on the eastern boundary than on the western boundary of the canal. The Stokes velocities are directed landward from the boundaries in the direction of the propagation of tidal waves generated at the boundaries of the canal. The mean Lagrangian velocities are quite homogeneous, that is, they are all directed eastward and of the same magnitude everywhere along the length of the canal. The sensitivity analysis shows that with higher values of eddy viscosity coefficient the residual current profiles become more vertically uniform. The results of the residual current profiles obtained using the variable eddy viscosity coefficient of equation 7 were very similar to those obtained using a constant eddy viscosity coefficient of 60 cm^2/sec.

The sensitivity of the net transport to the value of the eddy viscosity coefficient is shown in Fig. 8. The net transport remains eastward in the canal. However, the amplitude of the net transport decreases with increasing eddy viscosity coefficient. This decrease in net transport with eddy viscosities is asymptotic such that beyond a value of $N_z > 100$ cm^2/sec the influence of increasing viscosities is minimal.

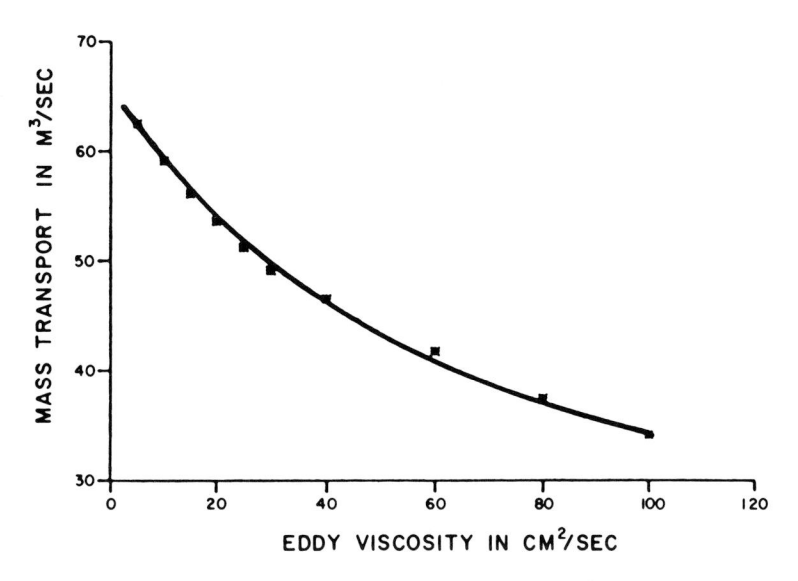

FIGURE 8   Mass transport in a sea-level canal as a function of Eddy viscosity.

5. Conclusions

A two-dimensional model developed to analyze the residual currents in estuaries and sea-level canals was used to analyze the sensitivity of such currents to important physical parameters. Although the model is limited in its present application to analysing residual currents in real world problems; nevertheless, it provides the means of investigating the intricacies of the issue concern. The model's limitation is due to the fact that it is based on the simplifying assumption derived by LONGUETT-HIGGINS (1969) related to the calculation of Lagrangian velocities in a small amplitude oscillating flow field. Thus, the model is inadequate for computation of residual flows in estuaries of rapid geometry changes.

The results presented here reveal the paradoxical nature of the residual currents. The mean Eulerian current profiles in a closed-end estuary with no freshwater inflow show a seaward flow with no compensating landward flow. Therefore, the simple integration of the mean Eulerian current profiles will not provide an accurate assessment of net transport in tidal flows. Such an assessment can only be obtained by integrating mean Lagrangian profiles.

The results of the sensitivity analysis show that the residual currents are influenced by the description of the bottom friction and the vertical eddy viscosities. The quadratic bottom friction and the large eddy viscosity coefficients tend to reduce the vertical gradients in the horizontal velocity profiles. Temporal and spatial variations in the density of estuary result in residual currents that can be stronger than tide-induced currents and they generate a net circulation which opposes tide-induced circulation.

The net transport in a sea-level canal is strongly dependent on the phase difference of tides at the two-ends. This net transport is also a function of the amplitude differences between the two tides. The net transport decreases with increasing eddy viscosity coefficient such that in the case of the C&D canal the net flow varies from $60 \, m^3/sec$ to $30 \, m^3/sec$ over the realistic ranges of the eddy viscosity coefficient.

References

BOYD, M.B., BOBB, W.H., HUVAL, C.J. and HILL, T.C. 1973. Enlargement of the Chesapeake and Delaware Canal — Hydraulic Mathematical Model Investigation, Technical Report H-73-16, U.S. Army Waterways Experiment Station, Oct. 1973.

CHENG, R.T. and CASULLI, V., On Lagrangian Residual Currents with Applications in South San Francisco Bay, California, Water Resources Research, Vol. 18, No.6, pp. 1652-1662.

IANNIELLO, J.P. 1977. Tidally-Induced Residual Currents in Estuaries of Constant Breadth and Depth, Journal of Marine Research, Vol. 35, No.4, pp. 755-786.

IANNIELLO, J.P. 1979. Tidally-Induced Currents in Estuaries of Variable Breadth and Depth, Journal of Physical Oceanography, Vol. 9, No.5, pp. 962-974.

LONGUETT-HIGGINS, M.S. 1969. On the Transport of Mass by Time-Varying Ocean Currents, Deep Sea Research, Vol. 16, pp. 431-447.

MUNK, W.H. and ANDERSON, E.R. 1948. A Note on the Theory of the Thermocline, Journal of Marins Research, Vol. 26, pp. 24-33.

NAJARIAN, T.O. WANG, D-P. and HUANG, P-S. 1984. Lagrangian Transport Model for Estuaries, Journal of Waterway, Port, Coastal and Ocean Engineering, Vol. 110, No.3, August 1984, ASCE.

PRITCHARD, D.W. and GARDNER, G.B. 1974. Hydrography of the C&D Canal, Technical Report No. 85, Chesapeake Bay Institute, The John Hopkins University, Baltimore, Md., Feb., 1974.

WANG, D.P. and KRAVITZ, D.W. 1980. A Semi-Implicit, Two-Dimensional Model of Estuarine Circulation, Journal of Physical Oceanography, Vol. 10, No.3, pp. 441-454.

ZIMMERMAN, J.T.F. 1979. On the Euler-Lagrange Transformation and the Stokes Drift in the Presence of Oscillatory and Residual Currents, Deep Sea Research, Vol. 26A, pp. 505-520.

Residual Currents: A Comparison of Two Modelling Approaches

H. Gerritsen

Delft Hydraulics Laboratory,
The Netherlands

ABSTRACT

Two methods for the computation of residual flow are investigated in applications to a limited area schematization of the southern half of the North Sea. The first method involves a tidal computation and filtering of the results to obtain the residual flow. In the second method the residual flow is obtained directly by solving a linear steady one-equation model in terms of the stream function of the residual transports, with non-linearities accounted for by known stress terms. It is shown that both methods yield comparable residual flow results provided that boundary and body forcings are known and are properly specified. In an investigation into the sensitivity of the models it is shown that the forcing terms are sensitive to variation in the tidal computation. The accuracy of the results of the direct model is thus subject to the accuracy of the tidal computation. It is found that linear superposition of tidal and wind contributions to the forcing in the direct model leads to substantial errors in the computed flow field since nonlinear interactions between the two are not accounted for. This implies that simulations of the tidally and wind-induced residual flow by means of the direct model always have to be preceded by a tidal computation including this wind field to determine the proper non-linear stress terms. Consequently, for Southern North Sea circumstances the direct model does not present a practical alternative to the approach of a tidal model followed by the application of a filtering procedure to the results in order to eliminate tidal contributions of periods of a day or less.

1. Introduction

The last two decades have shown a growing awareness of the possible hazards of increased navigation in open sea, discharges of pollutants and waste dumping activities. For policy and management studies one must be able to estimate and quantify the long term behaviour of pollutants entering open surface water from rivers or sewage and other industrial outlets. The intensified study of long term water motion and mass transports in estuaries, embayments and coastal seas is a direct result of this. Due to costs and other problems related to the large time and spatial scales involved, synoptic measurements are relatively scarce. In recent years, however, numerical simulation techniques have become available, and are permanently being extended and improved, which make it possible to simulate the large scale flow and transport phenomena, and to study the various aspects under controllable laboratory-like circumstances.

In well mixed shallow areas of seas and coastal waters, tide and meteorology (wind and pressure distributions) are the main mechanisms in the generation of flow. To study such phenomena various numerical tidal flow models can be used. Techniques for filtering of the so obtained results over a suitable time period, say a day, are applied to obtain the mean or longer term water motion. This motion is generally denoted by residual flow, or residual currents. The efficiency of the use of a tidal model is restricted by the requirement to resolve the tidal motion properly. In cases when the interest does not lie in the tidal flow itself, but in the residual flow on time scales of weeks or larger, the use of a tidal model is neither logical nor efficient, making it worthwhile to look for alternative ways.

In a literature review Alfrink (1980) concludes that there exist two important approaches to the simulation of depth-averaged residual flow. The first is based on the averaging of tidal flow results, and was followed by FLATHER (1976), TEE (1976), RAMMING (1976), MAIER-REIMER (1977), PINGREE and MADDOCK (1977), PRANDLE (1978), BACKHAUS (1980), LEENDERTSE et al (1981) and many others. In the second method the equations are time-averaged. The effect of the time-averaged non-linear terms in the tidal equations is accounted for by stress terms, to be determined once from the results of a tidal simulation. With the further assumption that the residual flow varies only slowly in time, the time derivatives are neglected. Differentiation of the momentum equations to y and x respectively, followed by their combination and the introduction of a stream function then yields a linear steady one-equation model in terms of the stream function of the residual transports. Models along these lines, which can be solved very efficiently, are described by NIHOUL and RONDAY (1975), and NIHOUL (1980). A similar approach was also investigated by VAN DE KREEKE and CHIU (1981).

Thus far there has been no agreement on the accuracy and feasibility of both model types. The accuracy argument is based on the order of magnitude difference between the first order (tidal) flow and the second order (residual) flow. For the Southern Bight of the North Sea, where representative values are $1\,m.s^{-1}$ and $5\,cm.s^{-1}$, respectively, a systematic 5% error in the tidal flow computation could lead to a 100% error in the time-averaged (= residual) flow. A model based on the time-averaged equations supposedly did not suffer from this error source (NIHOUL and RONDAY (1976), NIHOUL and RUNFOLA (1981)). Moreover, the linearity of the latter model would enable to simply superpose separately computed tidally and meteorologically induced residual flow, making this approach far more efficient and cost effective. A drawback, on the other hand, is formed by the need to determine the forcing terms representing the tide-averaged effect of the non-linear terms in the tidal equations.

A critical analysis of the linearization process in the development of a linear residual flow model is presented by HEAPS (1978). For example the friction term, which importance is shown among others by HUTHNANCE (1981), is linearized according to perturbation theory, yielding several extra terms on the residual level. The final set of equations, which as yet has not been translated into an operational numerical model, unfortunately is strongly coupled, and an efficient solution in terms of a stream function is no longer possible. Since the aim of the present study is the investigation of existing numerical models in view of an efficient determination of the residual flow in the North Sea, attention is restricted to the tidal model and to Nihoul's approach.

In this paper both numerical modelling approaches are compared. The paper addresses the formulations, the solution methods, the accuracy and sensitivity to parameter variations and the feasibility for practical residual flow computation in the Southern North Sea. For this purpose a series of simulations with both models were performed, a summary of which is given in Table 1. In sections 2 and 3 the methods are described. It is shown that their ability to compute the residual flow is similar, provided that in the linear direct model the full forcing is determined from the results of a simulation with the nonlinear time dependent model. In section 4 the influences of variation in bathymetry and grid size, and error transport are discussed. It is shown that the forcing terms in the linear direct model (model B) are sensitive to the resolution of the tidal model (model A) from which results they are determined. Section 5 addresses the feasibility of linear superposition of tide and wind in the linear direct model, since only if linear superposition can reasonably be applied, the direct model may be more practical than the approach solving the tidal equations. It appears that due to nonlinear interaction between tide plus meteorology in the forcing terms, linear superposition is not permitted, implying that for every tide plus wind simulation with the direct linear model a full computation with the tidal model must be performed. Finally, in section 6 some concluding remarks are made.

2. The tidal model: averaging of the results

In well-mixed shallow areas the large scale horizontal motion can be described by the shallow water equations:

$$\frac{\partial u}{\partial t} + u\frac{\partial u}{\partial x} + v\frac{\partial u}{\partial y} - fv + g\frac{\partial \zeta}{\partial x} + \frac{gu\sqrt{u^2+v^2}}{C^2(H+\zeta)} = A\left(\frac{\partial^2 u}{\partial x^2} + \frac{\partial^2 u}{\partial y^2}\right) + \frac{F_x}{(H+\zeta)} \quad (1)$$

$$\frac{\partial v}{\partial t} + u\frac{\partial v}{\partial x} + v\frac{\partial v}{\partial y} + fu + g\frac{\partial \zeta}{\partial y} + \frac{gv\sqrt{u^2+v^2}}{C^2(H+\zeta)} = A\left(\frac{\partial^2 v}{\partial x^2} + \frac{\partial^2 v}{\partial y^2}\right) + \frac{F_y}{(H+\zeta)} \quad (2)$$

$$\frac{\partial \zeta}{\partial t} + \frac{\partial}{\partial x}(u(H+\zeta)) + \frac{\partial}{\partial y}(v(H+\zeta)) = 0 \quad (3)$$

A1 (2)	Tide only	B1 (3)	Tide only; forcing determined from the results of A1
A2 (2)	Tide + western wind	B2 (3)	Tide and western wind. Forcing determined from the results of A2
A3 (4)	Smoothed bathymetry; tide only	B3 (4)	As A3. Forcing determined from the results of A3
A4 (4)	Refined grid; tide only	B4 (4)	As A4. Forcing determined from the results of A4.
		B5 (5)	Tide + western wind. Tidal forcing from A1; wind forcing added separately
		B6 (5)	Tide + western wind. Tidal body forcing from exp. A1. Wind forcing added separately. Boundary forcing from experiment A2.

TABLE 1. Numerical simulations with the tidal model (A) and the direct model (B). The number between brackets indicates the paragraph in which the simulation is described

where:

f	:coefficient of Coriolis acceleration	$[s^{-1}]$
g	:gravitational acceleration	$[m.s^{-2}]$
t	:time-coordinate	$[s]$
$u(x,y,t)$	:depth-integrated velocity (Eastward)	$[m.s^{-1}]$
$v(x,y,t)$	:depth-integrated velocity (Northward)	$[m.s^{-1}]$
x	:horizontal coordinate (East)	$[m]$
y	:horizontal coordinate (North)	$[m]$
A	:horizontal eddy diffusion coefficient	$[m^2.s^{-1}]$
$C(x,y)$	:bottom friction coefficient	$[m^{\frac{1}{2}}.s^{-1}]$
$F_x(x,y), F_y(x,y)$	:external body forces per unit of mass (wind)	$[m^2.s^{-2}]$
$H(x,y)$	:water depth below plane $z=0$	$[m]$

$\zeta(x,y,t)$:water elevation above $z=0$ [m]

For the derivation of this depth-integrated equation see e.g. DRONKERS (1964). Solving the Eqs. 1-2 in time and averaging the results of the computations over a suitable time period T so that the tidal signal is removed (generally once or twice the tidal period) yields the residual flow. In the definition given by ALFRINK (1980)

$$\zeta_0(x,y,t) = <\zeta(x,y,t)> = \frac{1}{T}\int_{t-T}^{t}\zeta(x,y,\tau)d\tau \tag{4}$$

$$u_0(x,y,t) = <u(x,y,t)> = \frac{1}{T}\int_{t-T}^{t}u(x,y,\tau)d\tau \tag{5}$$

$$U_0(x,y,t) = <U(x,y,t)> = \frac{1}{T}\int_{t-T}^{t}u(x,y,\tau)\{H(x,y)+\zeta(x,y,\tau)\}d\tau \tag{6}$$

and

$$u_{0T}(x,y,t) = \frac{<U(x,y,t)>}{H(x,y)+<\zeta(x,y,t)>} \tag{7}$$

denote the residual (mean) elevation, the (Eulerian) residual velocity, the (Eulerian) residual transport, and the (Eulerian) residual transport velocity, respectively. Note that these definitions allow for long term fluctuations in tide and wind.

The area of interest is the Southern North Sea between Dover Straits and the 56th parallel, see Fig. 1. For this region a North South oriented finite difference schematization was made with grid sizes $\Delta x = \Delta y = 10,000m$. The bottom topography is shown in Fig. 2. Assuming $A=0$, and setting $C=65m^{\frac{1}{2}}\cdot s^{-1}$, Eqs. 1-3 are solved with a second order accurate explicit finite difference scheme based on a staggered grid (GERRITSEN, 1983). The CFL condition for stability restricted the time step to $\Delta t=225s$. At closed boundaries vanishing normal velocity is prescribed:

$$(\vec{u},\vec{n}) = 0, \tag{8}$$

while at the open boundaries (Fourier type) tidal boundary conditions are obtained from results with a larger, encompassing model. Along the Northern boundary the conditions are prescribed in terms of water levels of the form:

$$\zeta(x,y,t) = \zeta(A_0,M_2,M_4,M_6) \tag{9}$$

At the Southern boundary a total flux

$$Q(x,y,t) = 11.10^4 + 183.10^4\cos(\omega_{M_2}t-\phi) \tag{10}$$

is prescribed, which is translated into normal velocities along the boundary using the local depth profile. The tidal frequency ω_{M_2} corresponds to a tidal period $T=12\frac{1}{2}$ hours. In order that the analytical mathematical problem has a solution, which moreover is unique, at both the open boundaries a second boundary condition should be prescribed when locally the direction of the flow corresponds to inflow. In the staggered grid formulation of the finite difference equations in the numerical model at hand, with finite grid size, however, the set of equations can be solved uniquely while one boundary condition is prescribed during all phases of the flow.

Given the comparative nature of the study the model was not calibrated in detail, but only in the general sense of assessing the overall representation of the flow motion during a tidal cycle on the basis of amplitude and phase plots.

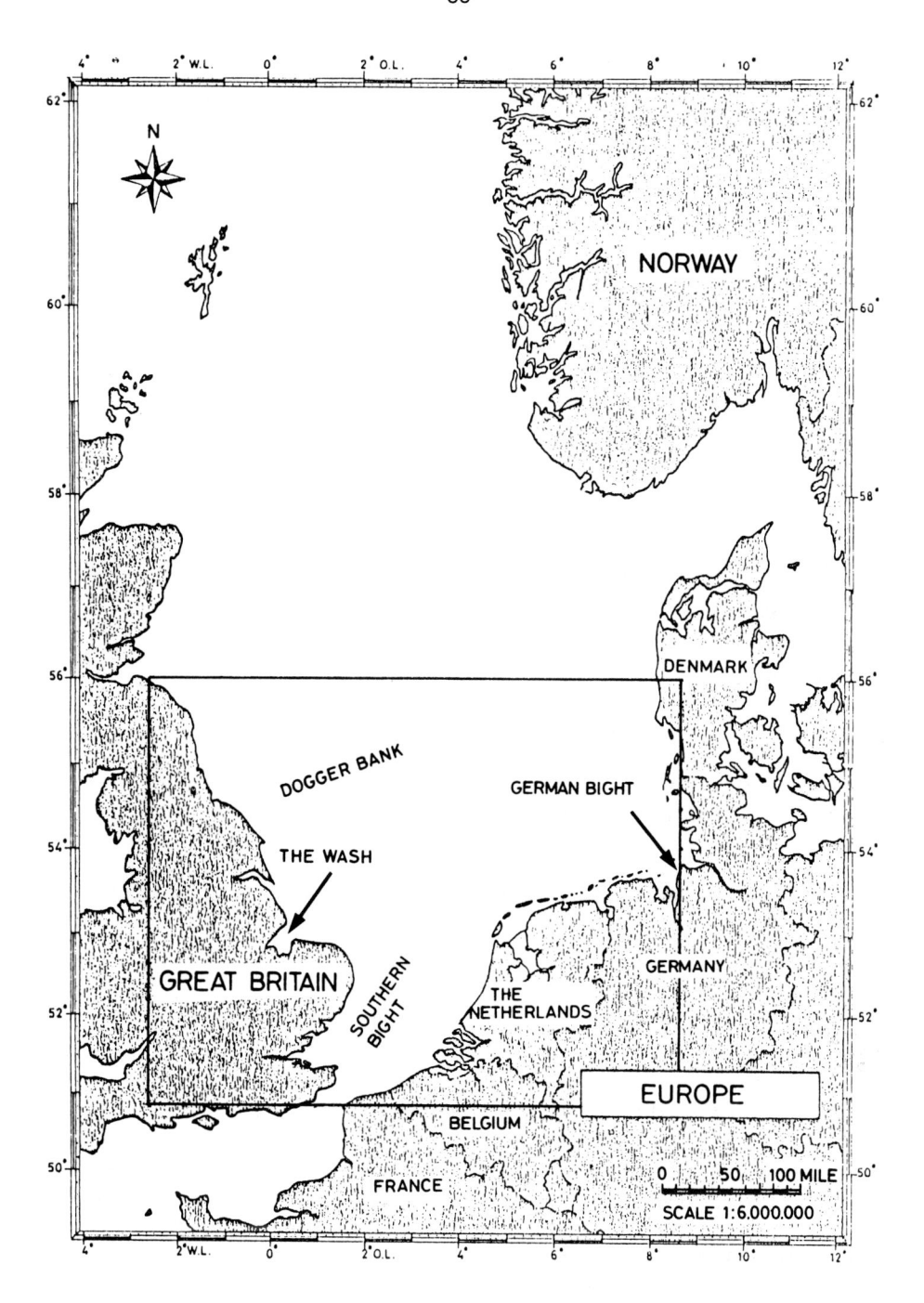

FIGURE 1. Map of the North Sea showing the region covered by the models

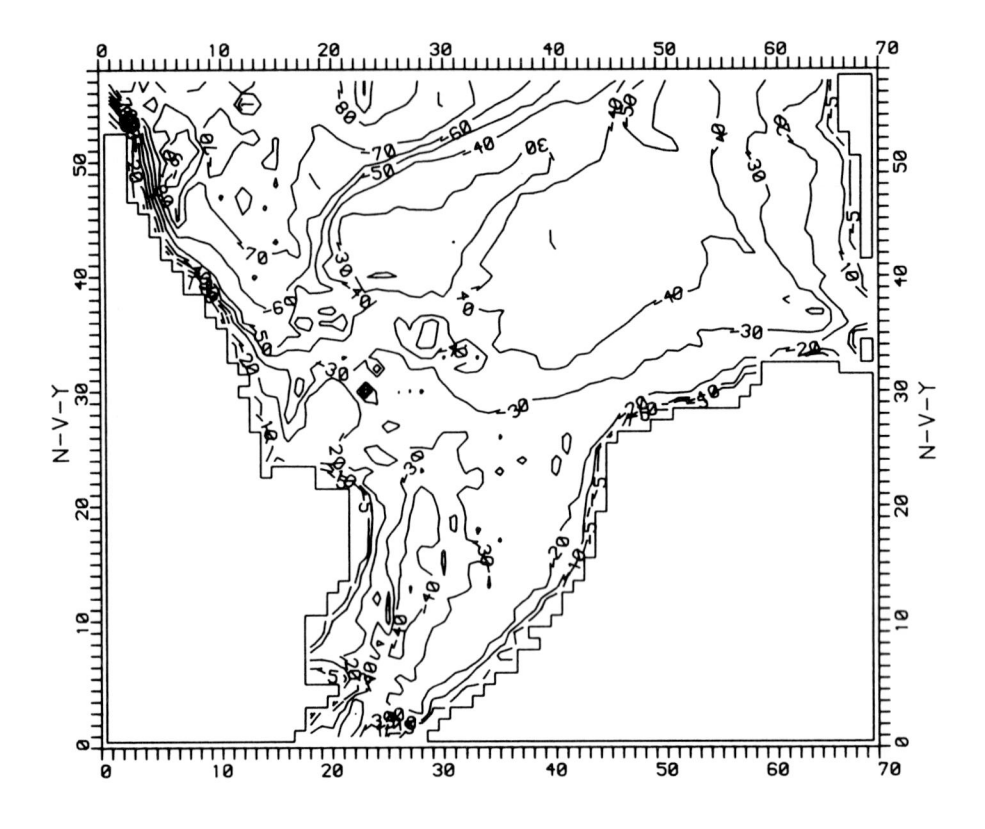

FIGURE 2. Bottom topography; lines of equal depths in meters

Starting from a situation of rest, the tidal flow computation simulation A1 is considered cyclic after three periods, using the criterion that everywhere in the field:

$$\frac{|\lambda(t+T)-\lambda(t)|}{\max_{\tau\in(t,t+T)}|\lambda(\tau)|} < 0.01 \tag{11}$$

in which λ stands for ζ, u, v, respectively. Applying a moving average over one tidal period to the tidal velocities, it was found that a similar condition:

$$\max_{\tau,\tau_1,\tau_2\in(t,t+T)}\left[\frac{|(u_0(\tau_1)-u_0(\tau_2))|}{\max|u_0(\tau)|}, \frac{|(v_0(\tau_1)-v_0(\tau_2))|}{\max|v_0(\tau)|}\right] < 0.01 \tag{12}$$

was only satisfied after six or seven tidal cycles. Here τ, τ_1 and τ_2 denote arbitrary points in time in the interval $(t, t+T)$. The response time, i.e. the time it takes until the results are periodic when the simulation is started from an initial state of rest, is

$$\begin{array}{l} 3-4 \text{ cycles } (37-50 \text{ hours}) \text{ for tidal flow} \\ 6-7 \text{ cycles } (75-87 \text{ hours}) \text{ for residual flow} \end{array} \qquad (13)$$

Figs. 3, 4 and 5 present the vector plots of the residual velocities, transports and transport velocities determined on the basis of the eighth tidal cycle. The global flow pattern is smooth. The clearly unphysical residual flow distribution in the Northern boundary region is certainly due to inaccuracies in the distribution of the prescribed waterlevels Eq. 9, and should therefore not be given further attention. At the closed boundary local deviations of the general pattern can be seen, resulting from insufficient accuracy in the local treatment of the advective terms. Although relatively unimportant in the tidal flow, they do show up in the residual flow field due to the systematic nature of such errors. Several eddies can be detected; one along the Dutch, and two or three off the English coast. Near the southern boundary the residual transports show a much more regular pattern than the residual velocities. The residual transport field shows no eddy along the Dutch coast, and the southbound flow area along the English coast also differs significantly. These differences are a direct result of the nonlinear relation Eq. 6 between transports and velocities, which may be expected to differ most in areas, where the amplitude depth ratio is large.

In a second simulation, (A2), a constant uniform western wind with magnitude $|W| = 5\sqrt{2} \, m.s^{-1}$ is added to the tidal forcing. Assuming a drag coefficient $c_D = 3.2 \ast 10^{-6}$ in the quadratic wind stress formulation (DRONKERS, 1964) yields:

$$F_x = c_D W_x |W| = 1.6 \; 10^{-4} \, m^2.s^{-2} \qquad (14)$$
$$F_y = c_D W_y |W| = 0.0$$

The addition of wind did not influence the response times Eq. 13.

The total tide plus wind induced residual transport field is shown in Fig. 6. The net wind-induced residual transports under these tidal circumstances were obtained by subtracting corresponding tidal flow fields with and without wind, and averaging the results over T, see Fig. 7. The zero net wind-induced residual transport through the Southern boundary reflects the fact that the wind was only prescribed in the body force while the tidal boundary conditions Eqs. 9 and 10 were applied unchanged, as estimates for the effect due to wind were not available. The prescription of water levels along the Northern boundary allows for the large net zero circulation through the boundary, with inflow in the Western, and increased outflow in the continental section. Not shown here is the corresponding large wind-induced pile up of water along the continental coast, which amounts to more than 25 cm in the German Bight.

3. The direct model: solution of the tide averaged equations

The formulation of the direct model starts from the shallow water equations in terms of transports U and V. After integration of the equations over the tidal period, the assumption that the residual quantities vary only slowly in time makes way for the introduction of a stream function ψ for the residual transports:

$$U_0 = -\frac{\partial \psi}{\partial y}, \quad V_0 = \frac{\partial \psi}{\partial x} \qquad (15)$$

Differentiation then immediately yields the equation for ψ:

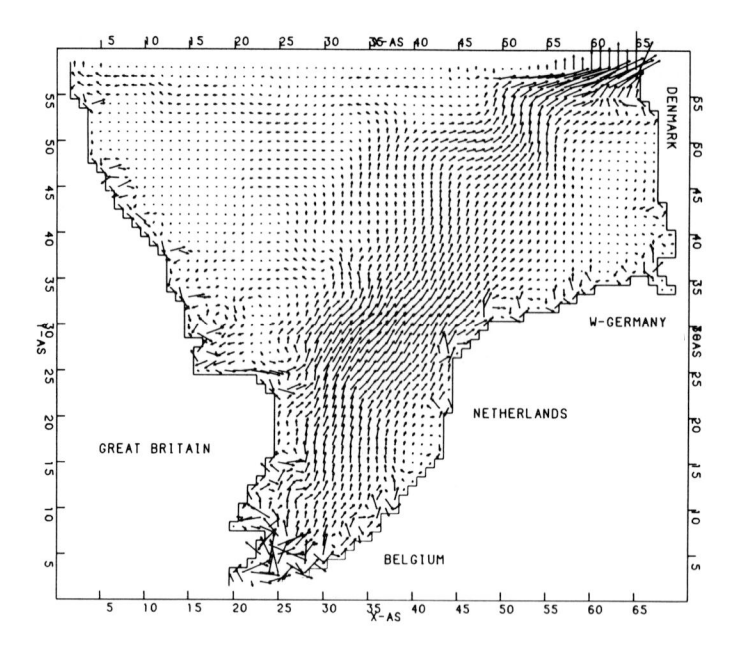

FIGURE 3. Simulation A1. Flow field of the tidally induced residual velocities

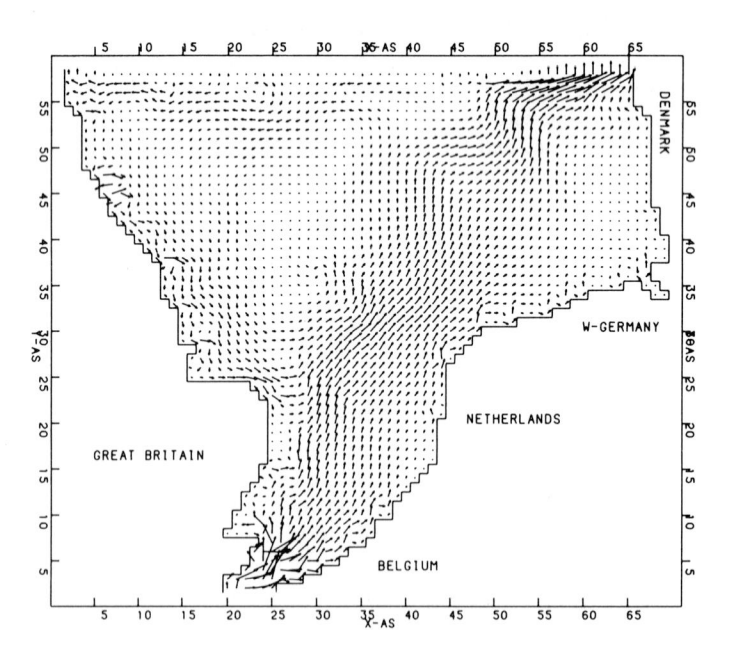

FIGURE 4. Simulation A1. Flow field of the tidally induced residual transports

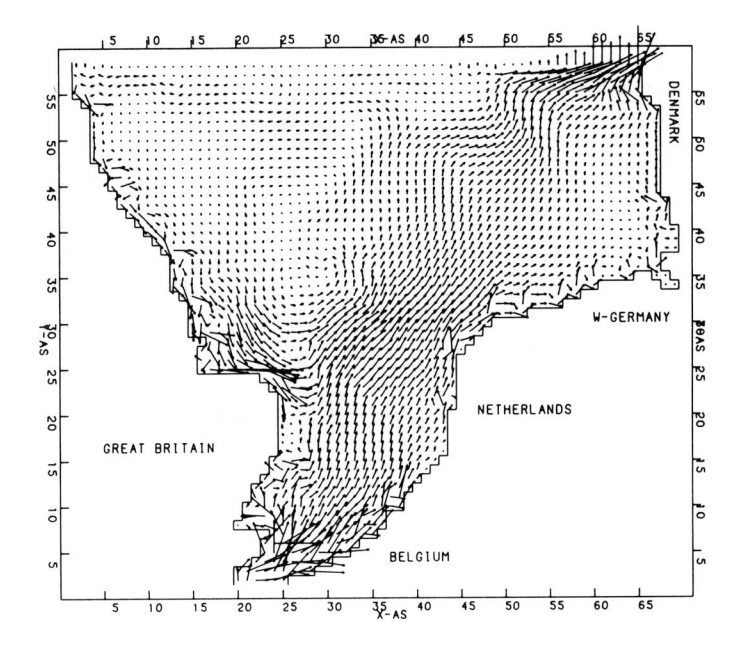

FIGURE 5.   Simulation A1. Flow field of the tidally induced transport velocities

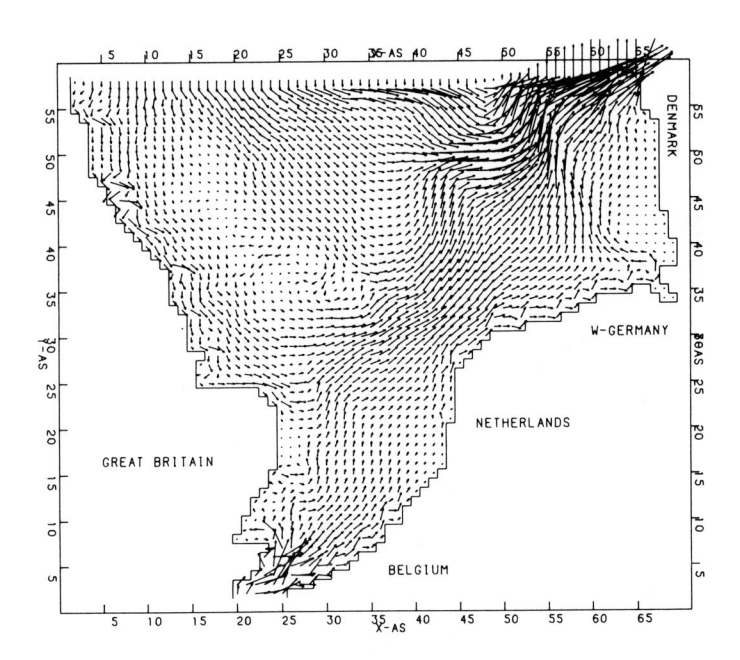

FIGURE 6.   Simulation A2. Residual transports due to tide and wind

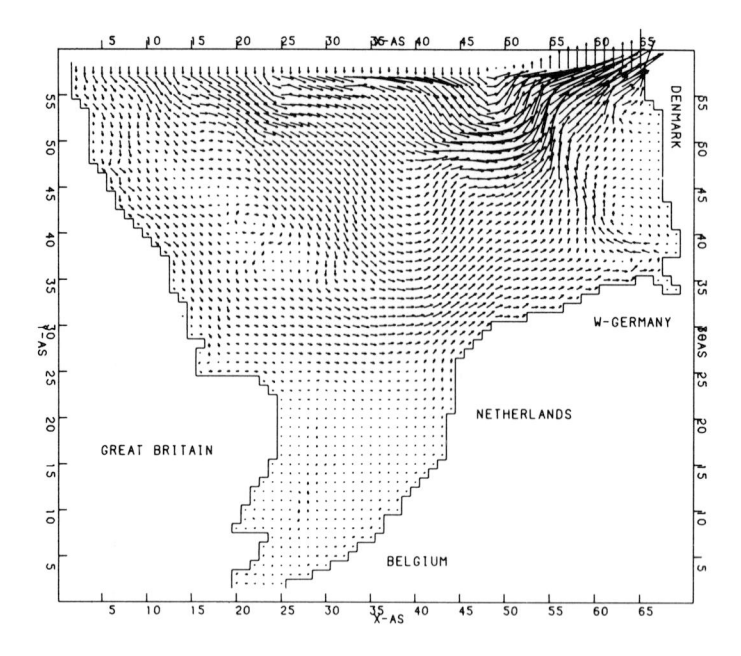

FIGURE 7. Simulation A2. Net wind-induced residual transports

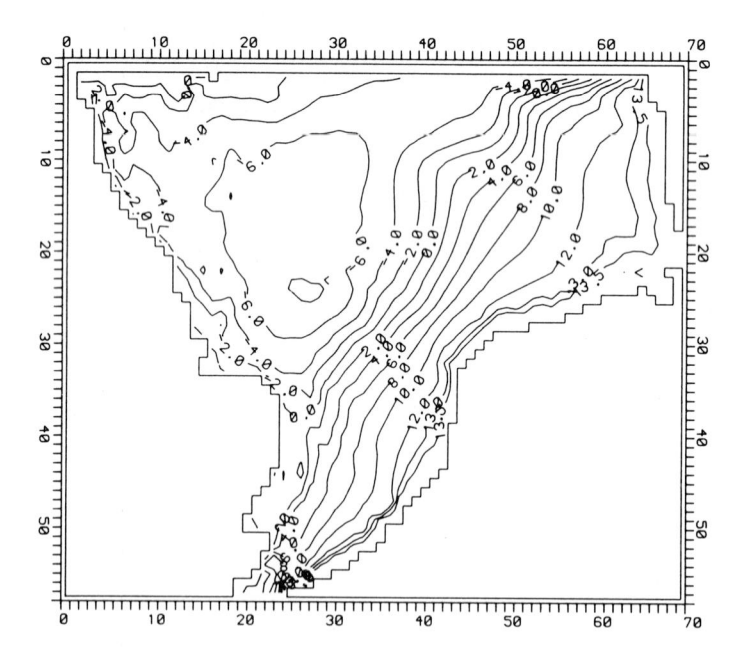

FIGURE 8. Simulation B1. Stream function ψ of tidally induced residual transports

$$\frac{\partial}{\partial x}\frac{k}{(H+\zeta_0)^2}\frac{\partial \psi}{\partial x}+\frac{\partial}{\partial y}\frac{k}{(H+\zeta_0)^2}\frac{\partial \psi}{\partial y}-\frac{\partial}{\partial x}\frac{f\frac{\partial \psi}{\partial y}}{(H+\zeta_0)}+\frac{\partial}{\partial y}\frac{f\frac{\partial \psi}{\partial x}}{(H+\zeta_0)} \quad (16)$$
$$=\frac{\partial}{\partial x}\left(\frac{\theta_y}{(H+\zeta_0)}\right)-\frac{\partial}{\partial y}\left(\frac{\theta_x}{(H+\zeta_0)}\right)$$

Here

$$k = \frac{g(H+\zeta_0)}{C^2} < \frac{\sqrt{U^2+V^2}}{(H+\zeta)^2} > \quad (17)$$

is a friction coefficient, while

$$\theta_x = <F_x> + \tau_x - <G_x> \quad (18)$$
$$\theta_y = <F_y> + \tau_y - <G_y>$$

are the total stress terms.

The tidal, or mesoscale Reynolds stress terms τ_x and τ_y, and the mesoscale friction stress terms G_x and G_y are given by:

$$\tau_x = -<g(\zeta-\zeta_0)\frac{\partial}{\partial x}(\zeta-\zeta_0)> - <\frac{\partial}{\partial x}(U\frac{U}{H+\zeta})> - <\frac{\partial}{\partial y}(U\frac{V}{H+\zeta})> \quad (19)$$

$$\tau_y = -<g(\zeta-\zeta_0)\frac{\partial}{\partial y}(\zeta-\zeta_0)> - <\frac{\partial}{\partial x}(U\frac{V}{H+\zeta})> - <\frac{\partial}{\partial y}(V\frac{V}{H+\zeta})>$$

$$<G_x> = \frac{g}{C^2} < \frac{(U-U_0)\sqrt{U^2+V^2}}{(H+\zeta)^2} > \quad (20)$$

$$<G_y> = \frac{g}{C^2} < \frac{(V-V_0)\sqrt{U^2+V^2}}{(H+\zeta)^2} >$$

It is stressed that the only assumption in the derivation of Eq. 16 concerns the negligibility of the time variations of the residual flow quantities. Assume furthermore:

$$\frac{|\zeta_0|}{H} \ll 1, \quad \frac{|\nabla \zeta_0|}{|\nabla H|} \ll 1 \quad (21)$$

This condition is easily satisfied in practice. Eq. 16 now constitutes a one equation model for the residual stream function ψ. The model is linear in ψ. Consequently it also constitutes a linear model for the residual transports U_0 and V_0. The body forcing terms τ_x, τ_y, G_x, G_y, and the coefficient k can be supplied by the results of a one time simulation with a tidal model. Note that when linear superposition does not hold due to substantial nonlinear interaction between tide and meteorology, a one time determination of the forcing terms is not satisfactory. Boundary conditions are prescribed in terms of ψ along the boundary which, given Eq. 15, implies the normal residual transport to be known.

The elliptic Eq. 16 is discretized to second order accuracy on the rectangular grid described in section 2. It is solved for ψ by means of a standard overrelaxation technique, applying the following criterion for convergence:

$$\max_{i,j}|\psi^{(s+1)}-\psi^{(s)}| < 10^{-5} \quad (22)$$

Numerical differentiation then yields the residual transports U_0 and V_0.

First a simulation of tide only was considered (B1). The terms k, τ_x, τ_y, G_x and G_y were determined from the results of the corresponding simulation A1 with the tidal model, while $<F_x>=<F_y>=0$. The set

$\psi=0$ (no through flow) along the English coast

$$\psi(x)=\int_{x_0}^{x}V_0(x)dx \text{ along the open boundaries} \tag{23}$$

$\psi=$ constant (no through flow) along the continental coast,

forms a matching set of boundary conditions. Here $V_0(x)$ is the residual transport through the boundary determined in simulation A1. The resulting distributions of the stream function ψ and the residual transports are shown in Figs. 8 and 9 respectively. The representation of the global flow patterns in Figs. 9 and 4 shows a good correspondence. This is not surprising since the equations are the same, and only the assumption Eq. 21 and the solution method can give rise to differences. Looking in more detail, two differences can be noted. Since in the direct method the condition at closed boundaries not only prescribes the direction of the flow but fully fixes the flow, systematic errors such as shown in the tidal model residuals do not appear. Furthermore, in places where the residual transport of simulation A1 (Fig. 4) is less smooth, this effect seems to be enhanced in the present results shown in Fig. 9. The whole English coastal area, including the gyre touching the main South North stream, is affected by this feature. Since the elliptic solution method has an integrating and smoothing nature rather then a tendency to emphasize discontinuities, transport or, when appropriate, enhancement of inaccuracies through the stress terms seems to be a likely origin of these irregularities.

For a second simulation with wind included, (B2), the terms τ_x, τ_y, G_x, G_y, k and the boundary conditions (Eq. 23) were extracted from the results of the simulation A2 with the tidal model, in which wind (Eq. 14) was also included. The stream function distribution in Fig. 10 gives an explicit quantification of the strong wind induced incoming transport through the Western part of the Northern boundary. The global distribution of the transports, Fig. 11, again compares well with the residual transport field obtained with the tidal model simulation A2 (Fig. 6). Once more the results of the direct model are less smooth in areas with large and irregular gradients. Due to the rectifying nature of the applied homogeneous and constant wind forcing this feature is suppressed somewhat in comparison with the case of tide only. The global flow representation is similar in both model types, though. In this general sense the direct model is capable of giving results comparable to those obtained by averaging the results of a tidal model. That is to say, at least for the cases considered, where the full forcing in the direct model is obtained from the corresponding simulation with the tidal model.

To assess the usefulness of the numerical model in practical applications and problem solving the obtained model results must be compared with measurements. It was found that as far as the main flow features are concerned the tidal residual flow field shows a good correspondence with the measured residual water movements with negligible wind as presented by RIEPMA (1980). A similar comparison can be made for the modelled residual flow field due to tide and a steady western wind. For this case Riepma notes a large West-East residual flow and a piling up of water in the German Bight, which is also shown by the models. This comparison gives good confidence in the ability to represent the global characteristics of the residual flow by the present models.

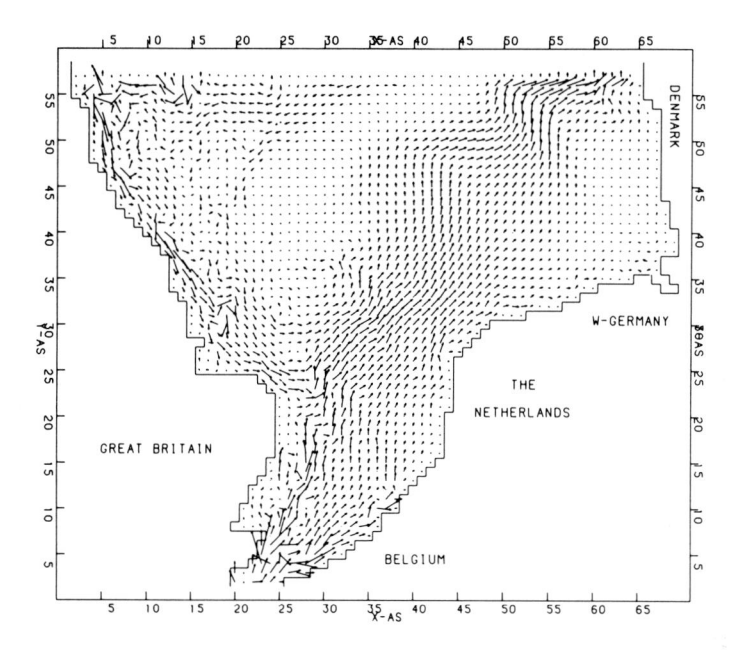

FIGURE 9. Simulation B1. Flow field of the tidally induced residual transports

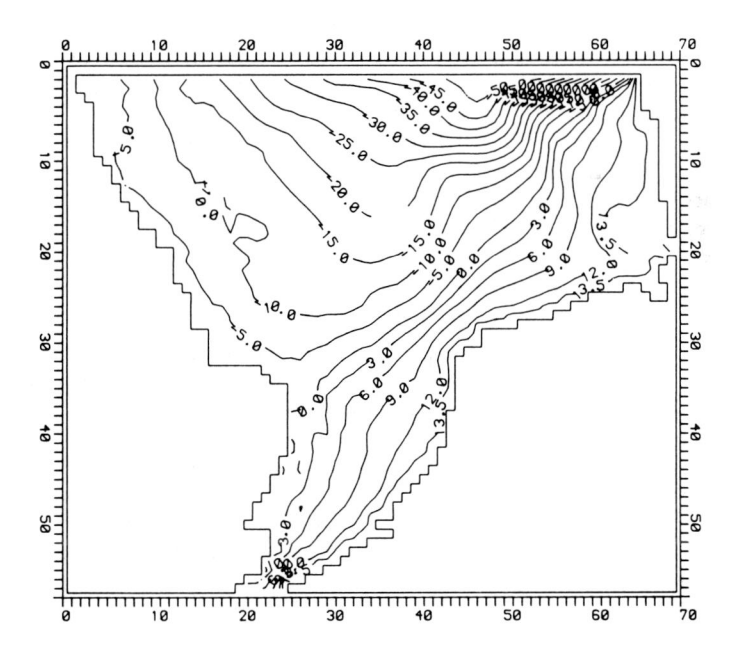

FIGURE 10. Simulation B2. Stream function ψ of residual transports due to tide and wind

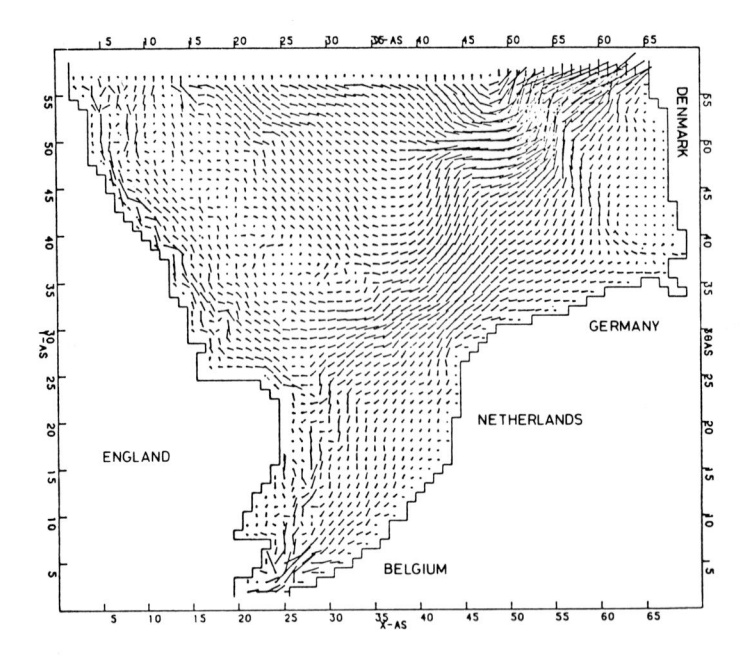

FIGURE 11. Simulation B2. Residual transports due to tide and wind

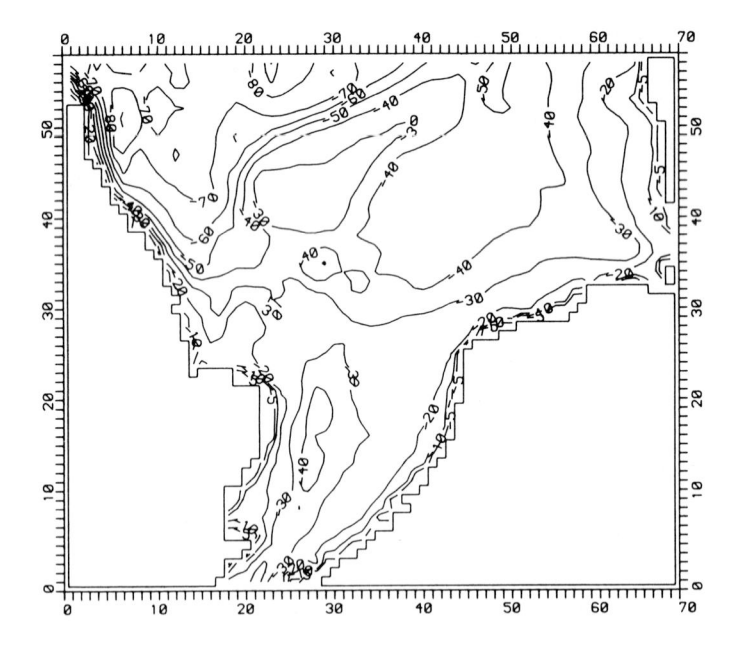

FIGURE 12. Bottom topography after smoothing; lines of equal depths in meters

4. Error transport, bathymetry smoothing and grid size variation

4.1 Error transport

In this section the sensitivity of the flow representation in both models is studied further. Of particular interest here is the sensitivity or insensitivity of the results of the direct model to variations or inaccuracies in the results obtained in the tidal computation. This sensitivity investigation directly addresses the question whether or not the accuracy argument about the independence of the direct model on inaccuracies in the tidal model computation holds. The introduction of simplifications in the derivation of the direct model and variations in the numerical discretization used in its solution are therefore considered to be of minor interest. They generally lead to a decrease in accuracy, and inherently have bearing only on the behaviour of the direct model. Variations in bottom topography and numerical gridsize on the other hand influence the representation of residuals in both models.

4.2 Bathymetry smoothing

For simulation A3, the bathymetry was smoothed twice applying the operator defined by:

$$H'_{ij} = [(1-0.5)I + 0.125(E_{01} + E_{10} + E_{-10} + E_{0-1})]^2 H_{ij} \tag{24}$$

in which I is the identity operator and E_{nm} denotes a shift operator over n and m grid points in positive x and y-direction, respectively. The resulting bathymetry (Fig. 12) is considerably more regular than the original one, cf. Fig. 2. Repeating the first simulation (A1; tide only) with the smoothed bathymetry resulted in an only marginally smoother residual flow field (not shown here). The stress terms, however, are visibly more smooth. The related simulation B3 with the direct model, all forcing based on the results of simulation A3, led to a much smoothed flow pattern, see Fig. 13. Detailed comparison shows it to be not only more smooth than that of Fig. 9, simulation B1, but, due to the combined effect of bathymetry in B3 and a smoother stress pattern, also more smooth than in simulation A3. The direct model improves significantly if the original topography is smoothed to eliminate short wave-length contributions, which apparently have a large influence on the terms of Eqs. 17-20. This is a further confirmation of the earlier observed phenomenon (cf. Fig. 9 and 4, simulations B1 and A1) that inaccuracies in the tidal model (results) may effect the direct model results. The stress terms constitute the medium through which such influence takes place.

4.3 Grid size refinement

In order to isolate the influence of grid refinement the grid size in both directions was halved: $\Delta x' = \Delta y' = 5000m$. The boundary schematization was not altered, while the extra bathymetry values were determined via linear interpolation, leading to a dynamically equivalent topography for the new grid.

The residual transports in the fine grid tidal computation (A4) are not essentially different from those of the orignal computation (A1): The spatial behaviour is somewhat more smooth locally, but no overall changes in the representation can be noted, see Fig. 14. This also holds for the corresponding fine grid computation B4 (Fig. 15). The irregularities noted in the original simulation B1 are again reduced significantly, though not to the extent that was seen in the experiment with smoothed bottom topography. The similar effect of grid size refinement and bathymetry smoothing is easily explained when one realizes that linear interpolation in the bathymetry effectively increases its resolution, as does the smoothing. Apparently the stress terms in the direct model are very sensitive to the resolution, especially of bathymetry, in the tidal computation from which results they

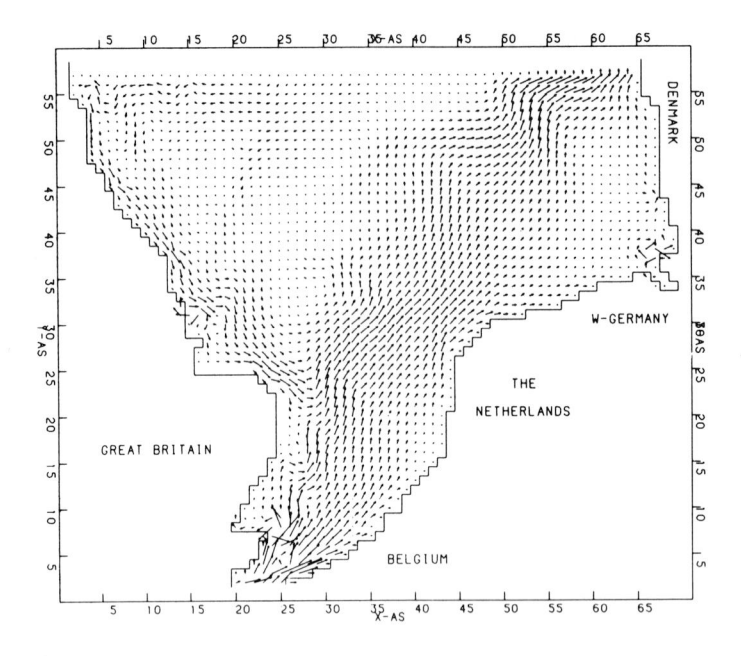

FIGURE 13.   Simulation B3. Tidally induced residual transports (smoothed topography)

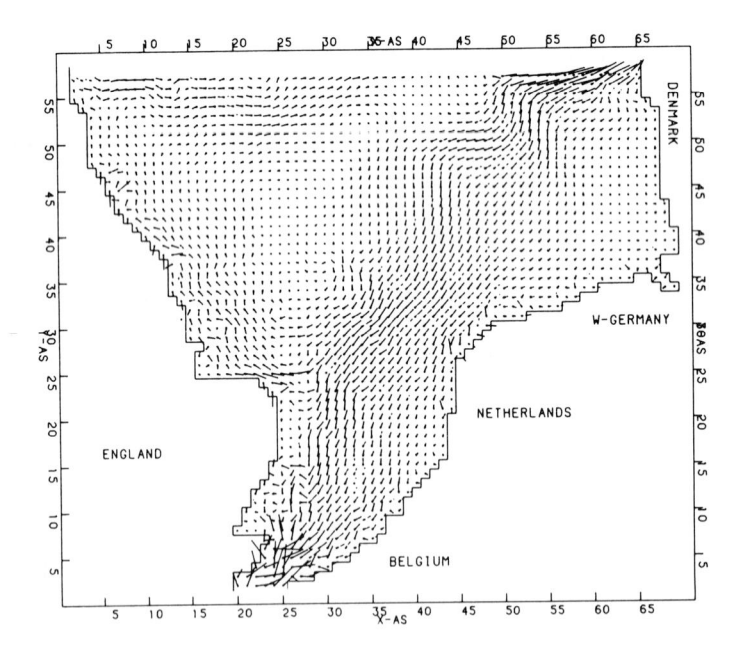

FIGURE 14.   Simulation A4. Tidally induced residual transports in the refined grid computation

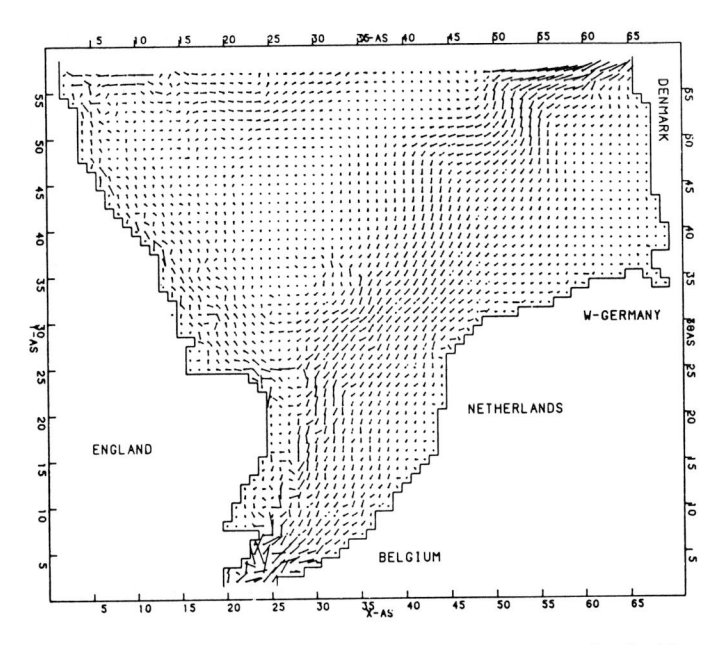

FIGURE 15. Simulation B4. Tidally induced residual transports; refined grid computation

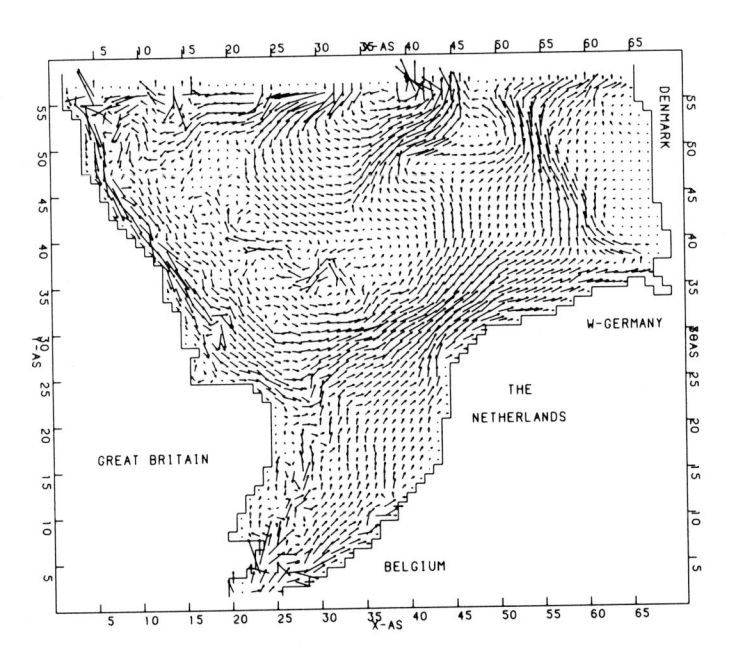

FIGURE 16. Simulation B5. Residual transports in case of linear superposition of tide and wind

are determined.

5. Linear superposition of tidal and meteorological forcings

In one of the above sections the linearity of the direct model was already noted. From a mathematical point of view this allows for a one time determination of the residuals due to tide, after which the residuals generated by various additional wind fields can be computed separately. For a given tide and wind situation the total residual flow field is then simply the sum of the two corresponding residual flow fields. In principle this property gives the direct model a large advantage in efficiency and simplicity of application over the time dependent tidal model approach. In practice however, the situation may be less straightforward if the net effect on the residuals of non-linear interaction between tide and meteorological state is not negligible.

Figure 16 shows the residual flow field of a simulation with the direct model in which the wind stress Eq. 14 was combined with the stress and boundary forcing applied in simulation B1, the case of tide only. Comparing Fig. 16 with Figs. 11 and 4, and comparing Figs. 17 and 10 for the corresponding stream function, it is obvious that linear superposition gives rise to a largely different flow pattern. It is also obvious however, that the wind field drastically changes the flow pattern through the large open boundary, which is not accounted for. In a subsequent simulation (B6) boundary conditions were prescribed that accounted for full nonlinear interaction between tide and wind (obtained from the results of simulation A2), while the concept of linear superposition was retained in the body forcing. There are still large differences when results are compared with those obtained with a fully nonlinear treatment. Fig. 18 illustrates these differences in terms of the stream function. Although the differences at the boundary vanish per definition, errors are still substantial in the interior. These are due to the absence of wind influence in the stress terms τ_x, τ_y, G_x, G_y, and in k, Eqs.17-20. Since these influences depend both on the wind strength and its direction, they cannot be accounted for in a simple way. Recomputation of the residual flow after the stress terms and k have been updated with residual contributions known from a first computation would seem to present a viable option. The iteration process could be repeated several times until convergence is reached. For the North Sea with its unique geometry, tidal and meteorological characteristics this does not present a practical approach primarily because of stability and convergence.

In an attempt to update the stress terms iteratively with newly computed residual transport information, DIJKSTRA (1980) in his computation for the Southern North Sea found that the process was both unstable and divergent. This indicates that for every simulation of tidally and wind-induced residuals with the direct model (B), a full nonlinear computation with a tidal model, including this windfield, must be made to provide the correct nonlinear forcing terms for the direct model.

6. Conclusions

In the preceding paragraph two methods for the computation of residual flow were compared. A tidal model, with averaging of the results, and a direct, linear, model containing stress terms which represent the time-averaged effect of nonlinearities.

From the comparison of both models with respect to residual flow modelling in the Southern North Sea the following conclusions can be drawn:
- *response time*. The response time of the residual flow to a zero initial state is roughly a factor two larger than that of the tidal flow; the response times found were 75-87, and

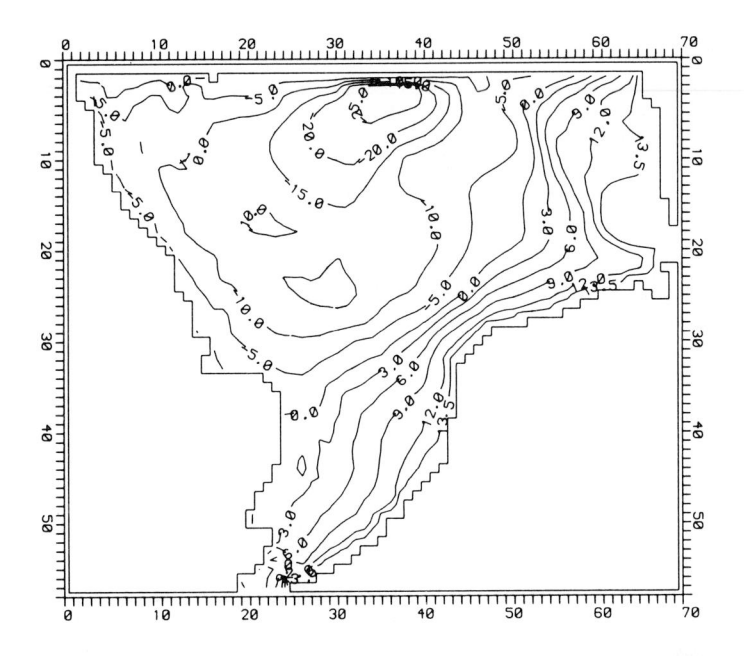

FIGURE 17   Simulation B5. Stream function in case of linear superposition of tide and wind

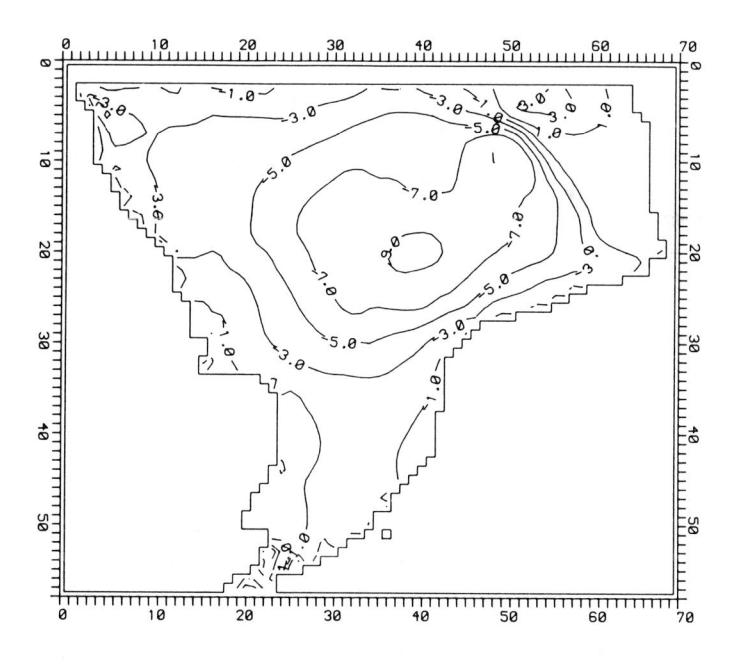

FIGURE 18   Simulation B6. Effect on the stream function of linear superposition in the body forcing τ_x, τ_y, G_x, G_y and in k only.

37-50 hours, respectively.

- *basic flow representation.* With the assumption that the boundary forcings and body forcings are known, both methods can yield an accurate and comparable representation of the residual flow. This holds both for the case of tide only, and for the case when wind is included, provided that in the latter case all forcing terms are determined from the results of the tidal computation with wind, thus including the effects of nonlinear interactions.
- *error transports.* The forcing terms in the direct model are sensitive to variations in the tidal computation, especially the resolution of the bathymetry. This leads to a dependence of the direct model results on the results of the tidal computation. In other words: the accuracy of the results of the direct model is subject to the accuracy of the tidal computation.
- *linear superposition.* The linear superposition of tidal and wind contributions in the body forcing and boundary forcing of the direct model results in substantial errors in the computed flow field since the nonlinear interaction between the two physical phenomena is not accounted for. This is also true for linear superposition in the body forcing only. It is concluded that in the Southern North Sea nonlinear interaction between tide and wind are of essential importance for the residual flow.
- *feasibility of the direct model.* When applying a direct model in the case of tide and wind, a corresponding computation with a tidal model must be performed first to provide the proper forcing terms. For applications in the Southern North Sea the expected computational efficiency of the use of a direct model therewith disappears. Added to this the accuracy argument, it is concluded that for Southern North Sea circumstances the linear direct model of Eq. 16 does not present a practical alternative to the application of a tidal model followed by a filtering procedure to eliminate tidal contributions from the results.

Acknowledgement

The research leading to this contribution is part of the Dutch Government Public Works Department research program (TOW) on Flows and Transport Phenomena, in which the Public Works Department (Rijkswaterstaat) and the Delft Hydraulics Laboratory cooperate.

References

ALFRINK, B.J., 1980. Two-dimensional horizontal tide-averaged mathematical models. Literature survey (in Dutch), Delft Hydraulics Laboratory, TOW Report R 1469-I, 18 pp.

BACKHAUS, J.O., 1980. On currents in the German Bight, pp. 102-132. In: Lecture Notes on Coastal and Estuarine Studies, Springer Verlag, Berlin.

DIJKSTRA, T., 1980. Numerical computations of residual currents in the southern part of the North Sea. Dutch Government Public Works Department, DIV Report 1980627, 79 pp.

DRONKERS, J.J., 1964. Tidal computations in rivers and coastal waters. North-Holland Publishing Co., Amsterdam, 518 pp.

FLATHER, R.A., 1976. A tidal model of the North-West European continental shelf. Mém. Soc. R. Sc. Liège, 6° Série, tome X, pp. 141-164.

GERRITSEN, H., 1983. Residual currents. A comparison of two methods for the computation of residual currents with respect to their intrinsic properties and behaviour and their feasibility for the computation of residual currents in the Southern half of the North Sea. Delft Hydraulics Laboratory, TOW Report R 1469-III, 54 pp.

HEAPS, N.S., (1978). Linearized vertically-integrated equations for residual circulation in coastal seas. Dt.

hydrogr. Z., 31, pp. 147-169.

HUTHNANCE, J.M., 1981. Oh mass transport generated by tide and long waves. J. Fluid Mech., 102, pp. 367-387.

KREEKE, J. van de, & CHIU, A.A., 1981. Tide-induced residual flow in shallow bays. J. Hydr. Res., 19, no. 3, pp. 231-249.

LEENDERTSE, J.J., LANGERAK, A., & DE RAS, M.A.M., 1981. Two-dimensional tidal models for the Delta Works. pp. 408-450. In: Transport models for inland and coastal waters, H.B. Fisher ed., Academic Press, New York.

MAIER-REIMER, E., 1977. Residual circulation in the North Sea due to the M_2-tide and mean annual wind stress. Dt. hydrogr. Z., 30, pp. 69-80.

NIHOUL, J.C.J., 1980. Residual circulation, long waves and meso-scale eddies in the Norht Sea. Oceanologica Acta, 3, no. 3, pp. 309-316.

NIHOUL, J.C.J. & RONDAY, F.C., 1975. The influence of the tidal stress on the residual circulation. Tellus, 27, pp. 484-489.

NIHOUL, J.C.J. & RONDAY, F.C., 1976. Hydrodynamic models of the North Sea. Mém. Soc. R. Sc. Liège, 6° Série, tome X, pp. 61-96.

NIHOUL, J.C.J. & RUNFOLA, Y., 1981. The residual circulation in the North Sea. pp. 219-271. In: Ecohydrodynamics, Elsevier Publishing Co., Amsterdam.

PINGREE, R.D. & MADDOCK, L., 1977. Tidal residuals in the English Channel. J. Mar. Biol. Ass. U.K., 57, pp. 339-354.

PRANDLE, D., 1978. Residual flows and elevations in the Southern North Sea. Proc. R. Soc. Lond., A359, pp. 189-228.

RAMMING, H.G., 1976. A nested North Sea model with fine resolution in shallow coastal areas. Mém. Soc. R. Sc. Liège, 6° Série, tome X, pp. 9-26.

RIEPMA, H.W., 1980. Residual currents in the North Sea during the INOUT phase of JONSDAP'76. Meteor Forschungs Ergebnisse, Reihe A, 22, pp. 19-32.

TEE, K.T., 1976. Tide-induced residual current; a 2-D non-linear numerical tidal model. J. Mar. Res., 34, pp. 603-628.

On Lagrangian Residual Ellipse

R.T. Cheng

U.S. Geological Survey,
U.S.A.

Shizuo Feng & Pangen Xi

Shandong College of Oceanology,
Peoples Republic of China.

ABSTRACT

The tide-induced Lagrangian residual current has been analyzed using a small parameter expansion of κ which characterizes the relative importance of nonlinearity of the tidal system. The first order Lagrangian residual current has been shown to be the sum of the Eulerian residual current and the Stokes drift. In order to explain the dilemma that exists in the first order Lagrangian residual current, the second order solution has been obtained. The second order Lagrangian residual current has been shown to be an ellipse on a hodograph plane as expected. Additionally, the second order solution also gives an assessment of the accuracy in the usage when the mass transport velocity (LONGUET-HIGGINS, 1969) is used as an approximation to the Lagrangian residual current.

1. Introduction

The importance of residual current in studies of long-term transport of dissolved and suspended matter has long been recognized. Since progress in research of residual current has been relatively slow, confusion may still exist with regard to the definitions of various residual variables (ALFRINK and VREUGDENHIL, 1980). There is ample evidence that residual current has become the focus of research for many scientists who are concerned with the long-term transport processes in estuaries and in coastal seas. In the case of tide-induced residual current, the residual variables are an order of magnitude smaller than the tidal variables. Although the long-term transport of dissolved and suspended matter is of interest, the actual transport is characterized as a convection dominated process on a time scale on the order of the tidal period. Dominance of convection suggests that the actual transport is a Lagrangian process, the fate of dissolved solutes or suspended matter should be determined by a Lagrangian mean rather than by an Eulerian mean. In other words, the Eulerian residual current, the time averaged tidal current is not sufficient for the determination of the net mass transport. A mass transport velocity was introduced and shown to be the sum of the Eulerian residual current and the Stokes drift by LONGUET-HIGGINS (1969). He suggested that the mass transport velocity should be used to compute the net mass transport correctly, although the concept of the Lagrangian residual current was not given.

Following a Lagrangian point of view, a concise definition of the Lagrangian residual current was first introduced by ZIMMERMAN (1979). Using a numerical approach, the Lagrangian residual current has been calculated by CHENG and CASULLI (1982) for South San Francisco Bay, California. CHENG and CASULLI (1982) pointed out that the Lagrangian residual current is a function of the time (tidal current phase) when the water parcel is labelled and released. Further numerical experiments have been carried out by CHENG (1983) in an attempt to define the relation between the Lagrangian residual current and the tidal current phase. Unfortunately, because of the high degree of relative uncertainty in the computed residual currents, a definitive relation between the tidal and the residual currents was not attainable from the numerical solutions. Nevertheless, the computations have reinforced the fact that the Lagrangian residual current is a function of the tidal current phase.

In this paper, an analytical approach is used to study the properties of Lagrangian residual circulation. In a weakly nonlinear tidal system, the zeroth order solution of the governing equations gives the astronomical tides and tidal currents. The higher order solutions include the Eulerian residual current, the Stokes drift and the Lagrangian residual current; they are the results of nonlinear interactions among the lower order solutions. Using a Lagrangian approach, the Lagrangian residual current in an M_2 system has been obtained up to the second order. The first order Lagrangian residual current is shown to be the sum of the Eulerian residual current and the Stokes drift, and the second order solution of the Lagrangian residual current is shown to be an ellipse on a hodograph plane.

2. Governing equations

Circulation problems in well-mixed, tidal estuaries and shallow tidal embayments can be described by the familiar shallow water equations (STOKER, 1965). Let L_c and h_c denote the basin characteristic length and depth where L_c is on the order of $\sqrt{gh_c}/\omega_c$ with ω_c being the characteristic tidal frequency and g being the gravitational constant. Further, let u_c and ζ_c be the characteristic tidal current speed and tidal amplitude, then $l_c = u_c/\omega_c$ is essentialy the characteristic value of tidal excursion. In tidal estuaries, the ratios of l_c/L_c, $u_c/\sqrt{gh_c}$, ζ_c/h_c are of the same order of magnitude and a parameter κ can be introduced so that

$$\kappa = l_c/L_c = u_c/\sqrt{gh_c} = \zeta_c/h_c. \tag{1}$$

The value κ characterizes the relative magnitude of residual variables versus tidal variables. For many estuaries and shallow coastal seas around the world, κ has a value in the range of 0.0 - 0.2. Normalized by these characteristic values, the non-dimensional shallow water equations in two-dimensional Cartesian coordinates, (x,y), on the horizontal plane are the continuity equation

$$\frac{\partial \zeta}{\partial \theta} + \frac{\partial}{\partial x}[(h+\kappa\zeta)u] + \frac{\partial}{\partial y}[(h+\kappa\zeta)v] = 0, \tag{2}$$

the x-momentum equation,

$$\frac{\partial u}{\partial \theta} + \kappa[u\frac{\partial u}{\partial x} + v\frac{\partial u}{\partial y}] - fv = \frac{-\partial}{\partial x}(\zeta + \zeta_a) + \tau_x^w - \tau_x^b \tag{3}$$

the y-momentum equation,

$$\frac{\partial v}{\partial \theta} + \kappa[u\frac{\partial v}{\partial x} + v\frac{\partial v}{\partial y}] + fu = \frac{-\partial}{\partial y}(\zeta + \zeta_a) + \tau_y^w - \tau_y^b \tag{4}$$

where

(u,v) are the velocity components in the (x,y) directions;
θ is time;
ζ is the water surface elevation measured from mean sea level;

ζ_a is the water surface displacement due to atmospheric pressure;

(τ_x^w, τ_y^w) are the wind stress components in the (x,y) directions acting on the water surface;

(τ_x^b, τ_y^b) are the bottom stress components in the (x,y) directions;

f is the Coriolis parameter; and

h is the water depth measured from mean sea level.

In order to analyze the Lagrangian motions of labelled water parcels, the Lagrangian tidal velocity components (u_l, v_l) in the (x,y) directions are defined as the velocity of a labelled water parcel at time θ which is released from (x_0, y_0) at time θ_0. As time elapses, the position of the labelled water parcel can be given by

$$x = x_0 + \kappa \xi, \tag{5}$$
$$y = y_0 + \kappa \eta,$$

with the initial conditions that $\xi = \eta = 0$ at $\theta = \theta_0$. The Lagrangian water parcel displacement, (ξ, η), or simply the Lagrangian displacement is normalized by l_c. Thus the Lagrangian tidal velocity, (u_l, v_l), can be expressed in terms of the Eulerian tidal velocity as

$$u_l(x_0, y_0, \theta) = v(x_0 + \kappa x_i, y_0 + \kappa \eta, \theta), \tag{6}$$
$$v_l(x_0, y_0, \theta) = v(x_0 + \kappa x_i, y_0 + \kappa \eta, \theta),$$

where the initial position of the labelled water parcel is (x_0, y_0) at $\theta = \theta_0$. The Lagrangian displacement (ξ, ζ) can be obtained from integrating the differential equation of a streakline (parcel trajectory)

$$\frac{d\xi}{u_l} = \frac{d\eta}{v_l} = d\theta, \tag{7}$$

or,

$$(\xi, \eta) = \int_{\theta_0}^{\theta} [u_l(x_0, y_0, \theta'), v_l(x_0, y_0, \theta')] d\theta' \tag{8}$$

$$= \int_{\theta_0}^{\theta} [u(x_0 + \kappa \xi, y_0 + \kappa \eta, \theta'), v(x_0 + \kappa \xi, y_0 + \kappa \eta, \theta')] d\theta'.$$

Note that Eq. (8) is an integral equation of (ξ, η). With appropriate initial and boundary conditions, Eqs. (2) to (8) describe completely the dependent variables $(\zeta, u, v, u_l, v_l, \xi, \eta)$.

3. Method of Solution

3.1. The first order Lagrangian residual current - mass transport velocity

Because the ratio of residual current to tidal current, κ, is normally a small parameter for most estuaries and coastal seas, the weakly nonlinear solution of the tide-induced residual currents can be obtained by means of a small perturbation technique (FENG, 1977). In fact, the small parameter κ also represents the ratio of the net Lagrangian displacement to tidal excursion, or the ratio of tidal current speed to tidal wave celerity. The dependent variables can be expanded in ascending orders of κ such that

$$(\zeta, u, v, u_l, v_l, \xi, \eta) = \sum (\zeta_j, u_j, v_j, [u_l]_j, [v_l]_j, \xi_j, \eta_j) \kappa^j \tag{9}$$

where the subscript j indicates the j-th order solution of the dependent variables.

By substituting Eq. (9) into Eqs. (2) to (8), they become a set of independent linear systems of equations. Each subsystem of equations contains some elements which stem from Eqs. (2) to (7). The zeroth order equations determine the tides and tidal currents, and the first order subsystem of equations determines the Eulerian residual current and residual water level, and the Stokes drifts as a consequence of the nonlinear coupling of the zeroth order solutions.

For clarity and without losing generality, the flows in an M_2 tidal system are considered. Systems including several astronomical tides have essentially the same properties except that the algebra is much more complex. As before, κ is assumed to be a small parameter but not negligible. The solutions of tides and tidal currents for an M_2 system can be written as

$$u_0 = u'_0 \cos\theta + u''_0 \sin\theta,$$
$$v_0 = v'_0 \cos\theta + v''_0 \sin\theta, \qquad (10)$$
$$\zeta_0 = \zeta'_0 \cos\theta + \zeta''_0 \sin\theta,$$

where the amplitudes and the phases of M_2, (u_0, v_0, ζ_0), are indirectly given by (u'_0, u''_0), (v'_0, v''_0), and (ζ'_0, ζ''_0). The first order harmonics in an M_2 system are due to $(M_2 + M_2)$ and $(M_2 - M_2)$ interactions, and they are the M_4 and the first order Eulerian residual water level and residual current. Or,

$$u_1 = u'_1 \cos 2\theta + u''_1 \sin 2\theta + u_{er}^{(1)}$$
$$v_1 = v'_1 \cos 2\theta + v''_1 \sin 2\theta + v_{er}^{(1)} \qquad (11)$$
$$\zeta_1 = \zeta'_1 \cos 2\theta + \zeta''_1 \sin 2\theta + \zeta_{er}^{(1)}$$

$\qquad\qquad M_4 \qquad\qquad$ First order Eulerian Residual

Similarly, the second order solution include the shallow water harmonics with frequencies θ, and 3θ, and they are

$$u_2 = u'_2 \cos 3\theta + u''_2 \sin 3\theta + \sum_i (u'_{2,i} \cos\theta + u''_{2,i} \sin\theta)$$
$$v_2 = v'_2 \cos 3\theta + v''_2 \sin 3\theta + \sum_i (v'_{2,i} \cos\theta + v''_{2,i} \sin\theta) \qquad (12)$$
$$\zeta_2 = \zeta'_2 \cos 3\theta + \zeta''_2 \sin 3\theta + \sum_i (\zeta'_{2,i} \cos\theta + \zeta''_{2,i} \sin\theta)$$

$\qquad\qquad M_6 \qquad\qquad$ other harmonics

where the summation over i includes all other harmonics of frequency θ. As these do not contribute to the Lagrangian residual current up to the second order, the specific values of these harmonics are not essential. The complete solution of the M_2 system up to $0(\kappa^2)$ is, of course, the sum of the zeroth, first, and second order solutions.

A time averaging operator is defined as the time average of a dependent variable over one tidal period. Then, the time averaged tidal current, the sum of Eqs. (10)-(12), becomes the Eulerian residual current, (u_{er}, v_{er}), i.e.

$$u_{er} = u_{er}^{(1)} + 0(\kappa^2), \qquad (13)$$
$$v_{er} = v_{er}^{(1)} + 0(\kappa^2).$$

where superscript indicates the order of approximation of the variable with respect to the κ expansion. The Eulerian residual current, which is accurate up to $0(\kappa^2)$, is normalized by κu_c.

By applying the time averaging operator to the Lagrangian tidal velocities, the results lead naturally to the Lagrangian residual current, (u_{lr}, v_{lr}). By definition then,

$$u_{lr}(x_0,y_0,\theta_0) = \frac{1}{2\pi\kappa} \int_{\theta_0}^{\theta_0+2\pi} u_l(x_0,y_0,\theta')d\theta' \tag{14}$$

$$v_{lr}(x_0,y_0,\theta_0) = \frac{1}{2\pi\kappa} \int_{\theta_0}^{\theta_0+2\pi} v_l(x_0,y_0,\theta')d\theta'$$

with the aid of Eq. (8), the Lagrangian residual current in Eq. (14) can be written as

$$u_{lr}(x_0,y_0,\theta_0) = \frac{1}{2\pi\kappa}[x_0+\kappa\xi(\theta_0+2\pi)-x_0] = x_i(\theta_0+2\pi)/2\pi, \tag{15}$$

$$v_{lr}(x_0,y_0,\theta_0) = \frac{1}{2\pi\kappa}[y_0+\kappa\eta(\theta_0+2\pi)-y_0] = \eta(\theta_0+2\pi)/2\pi.$$

From a Lagrangian point of view, Eq. (8) describes the trajectories of labelled water parcels. By following the same water parcel to the end of a complete tidal cycle, one gets the net Lagrangian displacement of the labelled water parcel per tidal period. Thus, Eq. (15) becomes a concise definition of the Lagrangian residual current, and it also gives a lucid physical description of the properties of the Lagrangian residual current, (ZIMMERMAN, 1979; CHENG and CASULLI, 1982). It is worth noting that the Lagrangian residual current is a function of the time θ_0 when the labelled water parcel is released, Eq. (15).

Using the solutions of an M_2 tidal system given in Eqs. (10) to (12), and using the definition of the Lagrangian residual current, Eq. (15), the first order Lagrangian residual current can be shown to be

$$u_{lr} = u_{er}^{(1)} + \langle(\partial u_0/\partial x)_0\xi_0\rangle + \langle(\partial u_0/\partial y)_0\eta_0\rangle + 0(\kappa), \tag{16}$$

$$v_{lr} = v_{er}^{(1)} + \langle(\partial v_0/\partial x)_0\xi_0\rangle + \langle(\partial v_0/\partial y)_0\eta_0\rangle + 0(\kappa),$$

where $\langle\ \rangle$ is the notation for the time averaging operator, $()_0$ denotes that the argument in $()_0$ is evaluated at (x_0,y_0), and (ξ,η_0) is the zeroth order Lagrangian displacement. (x_{i0},η_0) can be calculated by

$$\xi = \int_{\theta_0}^{\theta} u_0(x_0,y_0,\theta')d\theta', \tag{17}$$

$$\eta_0 = \int_{\theta_0}^{\theta} v_0(x_0,y_0,\theta')d\theta',$$

The correlation terms in Eq. (16) are known as the Stokes drift, (u_{sd},v_{sd}), (LONGUET-HIGGINS, 1969; FENG et al, 1985), and

$$u_{sd} = \langle(\partial u_0/\partial x)_0\xi\rangle + \langle(\partial u_0/\partial y)_0\eta_0\rangle, \tag{18}$$

$$v_{sd} = \langle(\partial v_0/\partial x)_0\xi\rangle + \langle(\partial v_0/\partial y)_0\eta_0\rangle.$$

The first order Lagrangian residual current, Eq. (16), can then be rewritten as

$$u_{lr} = u_{er}^{(1)} + u_{sd} + 0(\kappa), \tag{19}$$

$$v_{lr} = v_{er}^{(1)} + v_{sd} + 0(\kappa).$$

At this order of the approximation, the Lagrangian residual current is shown to be the same as the mass transport velocity given by LONGUET-HIGGINS (1969). In the past, the mass transport velocity (the sum of the Eulerian residual current and the Stokes drift) has been interpreted as the *Lagrangian residual* current. The present analysis ascertains that this usage of the mass transport velocity is an acceptable approximation to the Lagrangian residual current. However, it is important to recognize that this usage is valid only as the first order approximation in a weakly nonlinear system. For this reason, the present authors deliberately identify the mass transport velocity as the first order Lagrangian residual current as in the sense introduced by LONGUET-HIGGINS (1969).

3.2. The dilemma in the Lagrangian residual current

The first order Lagrangian residual current presents a dilemma. From a physical point of view, the Lagrangian residual current is expected to be a function of the time when a water parcel is labelled and released, Eq. (15). When the labelled water parcels are released from the same position at different phases of the tidal current, each labelled water parcel inscribes a different trajectory in space over a tidal period. At the end of a tidal period the net Lagrangian displacements are not expected to be the same. Yet, the analytical results show that the first order Lagrangian residual current is not a function of time. Reported field experiments in the North Sea (MULDER, 1982) indicated that the measured net Lagrangian displacements cannot be adequately matched by the measured Eulerian residual current with the correction due to the Stokes drift.

To further illustrate this dilemma and the properties of the Lagrangian residual circulation, we resort to results from a numerical model of South San Francisco Bay, California (South Bay). The tidal current, the Eulerian residual and the Lagrangian residual currents in South Bay have been calculated by CHENG and CASULLI (1982) by means of a numerical model. South Bay is a shallow, semi-enclosed embayment which, for the most part of the year, is isohaline. The circulation in the Bay is strongly affected by the basin topography. South Bay bathymetry is characterized by a deep relict channel (>10 m) which connects to a broad shoal east of the channel. The isobaths shown in Figure 1 through 4 are 3 and 6 m at mean sea level. The shallow-water equations, Eq. (2) to (4), have been solved by a standard alternating-direction-implicit (ADI), finite difference method (LEENDERTSE, 1970; LEENDERTSE and GRITTON, 1971). Consult CHENG and CASULLI (1982) for a more detailed description of the numerical methods and other pertinent information relevant to the South Bay model. The numerical results presented here were obtained with the finite-difference grids of 500 m to give a more detailed spatial resolution of the circulation in South Bay than those reported in CHENG and CASULLI (1982).

Shown in Fig. 1 is a typical tidal circulation pattern in South Bay. An important and unique property of the tidal current distribution is that the tidal current, both in magnitude and in direction, is strongly affected by the basin bathymetry. The magnitude of the tidal current is roughly proportional to the local water depth, and the direction of the tidal current is generally tangent to the local isobath (CHENG and GARTNER, 1985). When we use this numerical model as the base-line flow field, the movements of labelled water parcels can be computed.

Shown in Fig. 2 are the parcel trajectories computed in the Lagrangian sense over a period of idealized mixed diurnal and semi-diurnal tides. Note that the parcel trajectories follow the isobaths, because the local tide currents are generally tangent to the isobaths. The numerical results given in Fig. 2 are in agreement with the basic assumption that the net Lagrangian displacement, the distance between symbols + and Δ, is an order of magnitude smaller than the tidal excursion. Thus, the small κ approximation should be valid for South Bay.

When the net Lagrangian displacement is computed for every grid point in the model, the final results are the Lagrangian residual circulation, Fig. 3. The scale for the residual velocity vectors in Fig. 3 is an order of magnitude smaller than the scale for the tidal velocity vectors in Fig. 2. Since the magnitudes for the velocity vectors in Figs. 2 and 3 are comparable, the actual ratio of residual current to tidal current, κ, is again

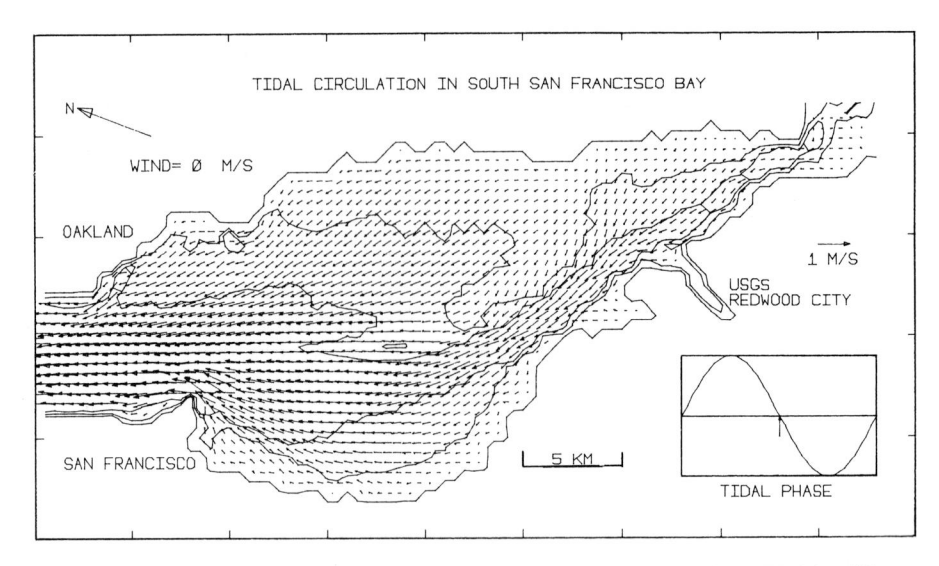

FIGURE 1. Simulated tidal circulation in South San Francisco Bay, California using ADI finite difference method. A semi-diurnal tide, 1 m. amplitude was specified at open boundary. Shown in the figure is the tidal current distribution at near maximum ebb.

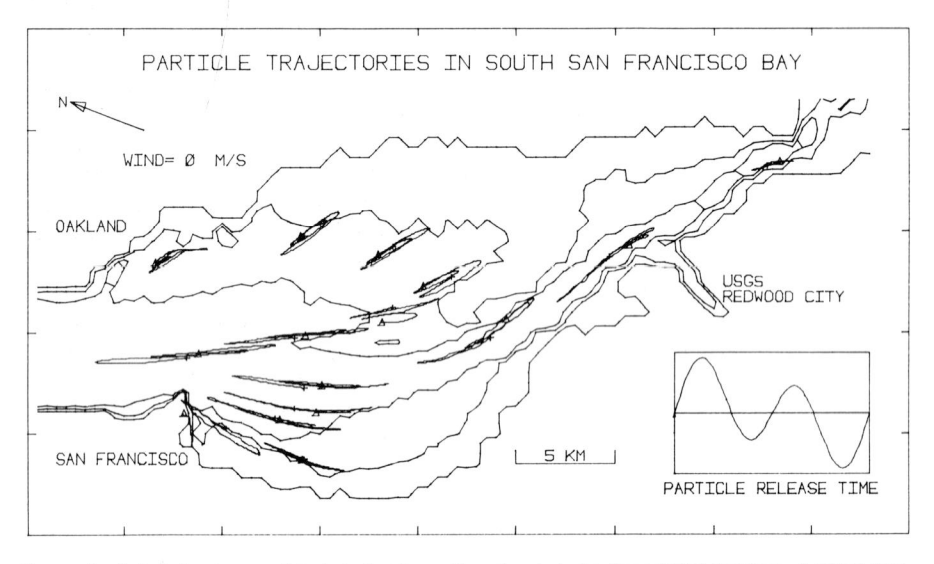

FIGURE 2. Labelled water parcel trajectories over a diurnal period at a few selected locations. A mixed semi-diurnal and diurnal tide was specified at open boundary. The water parcels were labelled and released from positions marked + and were found at the end of 24 hours at positions marked by Δ.

shown to be small. More importantly, however, in separate computations when the labelled water parcels are released at one hour intervals throughout a 12-hour period, the computed Lagrangian residual currents at a given position are not equal, but they are shown to be of the same order of magnitude (CHENG, 1983). Because the labelled water parcels are released at different phases of the tidal current, each water parcel inscribes a different trajectory. The net Lagrangian displacements are functions of the bathymetry

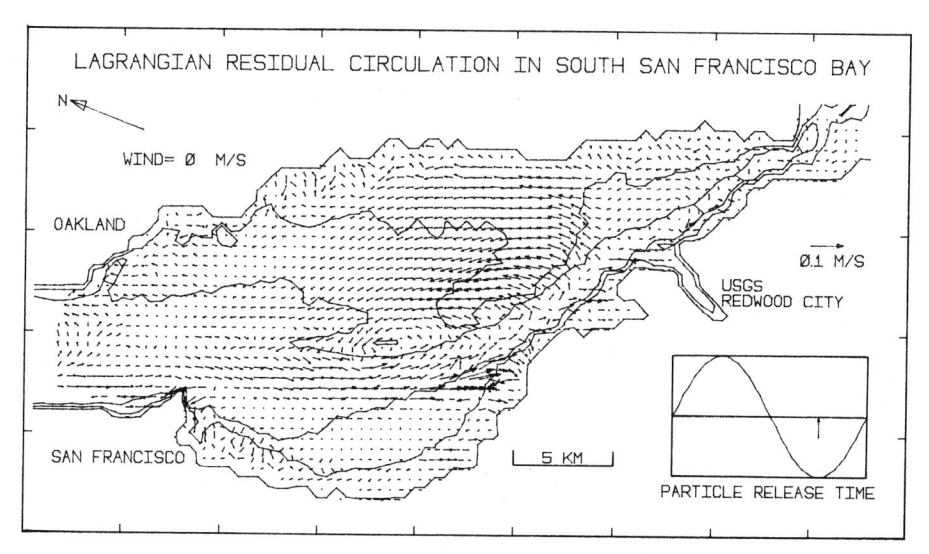

FIGURE 3. Lagrangian residual circulation in South San Francisco Bay due to a semi-diurnal tide. The computations for the Lagrangian residual circulation were initiated when low water occurred at the open boundary. When there is no velocity vector plotted, the tracers have moved outside of the computation. The Lagrangian residual current there is unknown, not zero.

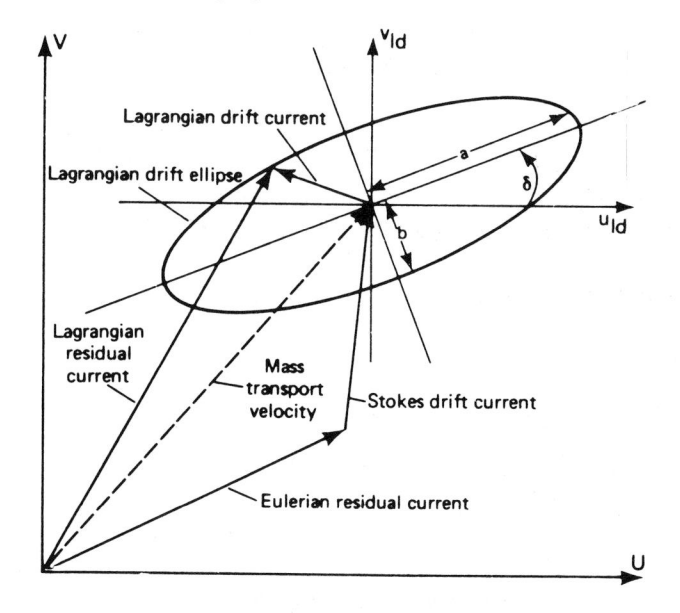

FIGURE 4. The schematic diagram of the interrelations between the Eulerian residual current, the Stokes drift, the mass transport velocity, the Lagrangian drift, and the Lagrangian residual current.

within the region enclosed by the water parcel trajectory. Thus, there is no reason to expect the final net Lagrangian displacements to be the same after a complete tidal cycle.

Clearly, the first order Lagrangian residual current or the mass transport velocity is not adequate to capture the essential properties of the Lagrangian residual current in a weakly nonlinear system even when κ is small. To explore further the properties of the Lagrangian residual current, it is necessary to examine the next order of the approximation.

3.3. The Lagrangian residual ellipse

By following the definition, Eq. (15), the second order Lagrangian residual current is the second order Lagrangian displacement divided by 2π. It can be shown that

$$u_{lr} = u_{er} + u_{sd} + \kappa u_{ld} + 0(\kappa^2), \tag{20}$$
$$v_{lr} = v_{er} + v_{sd} + \kappa v_{ld} + 0(\kappa^2),$$

where

$$u_{ld} = <(\partial u_1/\partial x)_0 \xi_0> + <(\partial u_1/\partial y)_0 \eta_0> + <(\partial u_0/\partial x)_0 \xi_1> + <(\partial u_0/\partial y)_0 \eta_1>$$
$$+ [<(\partial^2 u_0/\partial x^2)_0 \xi_0^2> + 2<(\partial^2 u_0/\partial x \partial y)_0 \xi_0 \eta_0> + <(\partial^2 u_0/\partial y^2)_0 \eta_0^2>]/2$$
$$v_{ld} = <(\partial v_1/\partial x)_0 \xi_0> + <(\partial v_1/\partial y)_0 \eta_0> + <(\partial v_0/\partial x)_0 \xi_1> + <(\partial v_0/\partial y)_0 \eta_1>$$
$$+ [<(\partial^2 v_0/\partial x^2)_0 \xi_0^2> + 2<(\partial^2 v_0/\partial x \partial y)_0 \xi_0 \eta_0> + <(\partial^2 v_0/\partial y^2)_0 \eta_0^2>]/2$$

In Eq. (20), the second order Lagrangian residual current depends on the first order Lagrangian displacement (x_{i1}, η_1) which is given as

$$\xi_1 = \int_{\theta_0}^{\theta} [u_1(x_0, y_0, \theta') + (\partial u_0/\partial x)_0 \xi_0 + (\partial u_0/\partial y)_0 \eta_0] d\theta', \tag{21}$$

$$\eta_1 = \int_{\theta_0}^{\theta} [v_1(x_0, y_0, \theta') + (\partial v_0/\partial x)_0 \xi_0 + (\partial v_0/\partial y)_0 \eta_0] d\theta',$$

with the initial condition $\xi_1 = \eta_1 = 0$ at $\theta = \theta_0$. By substituting the solutions of M_2 into Eqs. (19), (20), and (21), and after carrying out the integration with time, the expression for (u_{ld}, v_{ld}) becomes

$$u_{ld} = u'_{ld} \cos\theta_0 + u''_{ld} \sin\theta_0, \tag{22}$$
$$v_{ld} = v'_{ld} \cos\theta_0 + v''_{ld} \sin\theta_0,$$

where

$$u'_{ld} = u''_0 \partial(u_{er} + u_{sd})/\partial x + v''_0 \partial(u_{er} + u_{sd})/\partial y - (u_{er} + u_{sd}) \partial u''_0/\partial x - (v_{er} + v_{sd}) \partial u''_0/\partial y,$$
$$u''_{ld} = -u'_0 \partial(u_{er} + u_{sd})/\partial x - v'_0 \partial(u_{er} + u_{sd})/\partial y + (u_{er} + u_{sd}) \partial u'_0/\partial x + (v_{er} + v_{sd}) \partial u'_0/\partial y,$$
$$v'_{ld} = u''_0 \partial(v_{er} + v_{sd})/\partial x + v''_0 \partial(v_{er} + v_{sd})/\partial y - (u_{er} + u_{sd}) \partial v''_0/\partial x - (v_{er} + v_{sd}) \partial v''_0/\partial y,$$
$$v''_{ld} = -u'_0 \partial(v_{er} + v_{sd})/\partial x - v'_0 \partial(v_{er} + v_{sd})/\partial y + (u_{er} + u_{sd}) \partial v'_0/\partial x + (v_{er} + v_{sd}) \partial v'_0/\partial y,$$

In Eq. (22), (u_{er}, v_{er}) is the Eulerian residual current, and (u_{sd}, v_{sd}) is the Stokes drift which has been given in Eq. (18). It is important to observe that the second order correction to the Lagrangian residual, (u_{ld}, v_{ld}), is the result of the nonlinear interactions between the astonomical constituents, i.e., the zeroth order solution, and the Eulerian residual current and the nonlinear interactions between the astronomical constituents and the Stokes drift. At this order, (u_{ld}, v_{ld}) is shown to be a function of θ_0, Eq. (22), which is the tidal current phase when the labelled water parcel is released from a fixed point (x_0, y_0). In the analysis of the second order dynamics, the distinct Lagrangian property is finally brought forth. It thus seems proper to name (u_{ld}, v_{ld}) as the Lagrangian residual drift velocity, or simply the Lagrangian drift. Expressing the above results in words, then

| Lagrangian Residual Current | = | Eulerian Residual Current | + | Stokes' Drift Velocity | + | κ | Lagrangian Drift Velocity | (23) |

The unique property of the Lagrangian drift is that as the initial phase angle θ_0 varies from 0 to 2π, the Lagrangian drift velocities trace out an ellipse on a hodograph plane, Fig. 4. The properties of the Lagrangian residual ellipse can be given explicitly. The semi-major (+ sign) and semi-minor (- sign) axes are

$$1/\sqrt{2}\{u_{ld}'^2+u_{ld}''^2+v_{ld}'^2+v_{ld}''^2\pm[(u_{ld}'^2+u_{ld}''^2+v_{ld}'^2+v_{ld}''^2)^2 \qquad (24)$$
$$-4(u_{ld}'v_{ld}''-u_{ld}''v_{ld}')^2]^{1/2}\}^{1/2}.$$

The angle between the major axis of the residual ellipse and the x-axis is denoted by δ, (Fig. 4), and

$$\delta = 1/2\tan^{-1}[2\frac{u_{ld}'v_{ld}'+u_{ld}''v_{ld}''}{(u_{ld}'^2+u_{ld}''^2)-(v_{ld}'^2+v_{ld}''^2)}] \qquad (25)$$

Further, the phase angle θ_{max} which gives the maximum magnitude of the Lagrangian drift velocity is

$$\theta_{max} = 1/2\tan^{-1}[2\frac{u_{ld}'u_{ld}''+v_{ld}'v_{ld}''}{(u_{ld}'^2-u_{ld}''^2)+(v_{ld}'^2-v_{ld}''^2)}] \qquad (26)$$

The properties of the Lagrangian residual ellipse are very similar to the parent tidal current ellipse. The second order solution reveals the generation mechanism of the Lagrangian residual current, and it explains the dilemma in the first order Lagrangian residual current. With the second order solution at hand, it is now natural to expect that the water parcels released at different phases of the tides give slightly different values of the Lagrangian residual current. The inter-relations between the Lagrangian residual current, the Stokes drift, the mass transport velocity (the sum of the Eulerian residual current and the Stokes drift), and the Lagrangian drift are depicted in Fig. 4. Furthermore, the second order Lagrangian residual current gives an error assessment for using the mass transport velocity as an approximation to the Lagrangian residual current.

4. The Sverdrup wave

The properties of the Lagrangian displacement and the Lagrangian drift are further illustrated by means of an analytical example. Consider a two-dimensional Sverdrup wave, which is a harmonic wave in an infinite frictionless ocean of constant depth h_c (DEFANT, 1961). A typical form of the Sverdrup wave in dimensionless variables is

$$\zeta_0 = \sin(\theta-x) = \zeta_0'\cos\theta + \zeta_0''\sin\theta$$
$$u_0 = \sin(\theta-x) = u_0'\cos\theta + u_0''\sin\theta \qquad (27)$$
$$v_0 = f\cos(\theta-x) = v_0'\cos\theta + v_0''\sin\theta$$

where $0 < f < 1$, and the wave speed is $C = (1-f^2)^{-1/2}$, and where

$$\zeta_0' = u_0' = -\sin x, \quad \zeta_0'' = u_0'' = \cos x,$$
$$v_0' = f\cos x, \text{ and } v_0'' = f\sin x.$$

The Sverdrup wave given in Eq. (27) is a progressive wave in the direction of the wave propagation, whereas the phase of the velocity component normal to the wave direction lags the water level by 90°. The tidal current ellipse rotates clockwise and has an aspect ratio of f. The Eulerian residual current for a Sverdrup wave is zero, and the Stokes and the Lagrangian drifts can be computed straightforwardly to give

$$u_{sd} = 1/2, \quad v_{sd} = 0, \tag{28}$$

and from Eq. (22),

$$u_{ld} = 1/2 \sin x_0 \cos\theta_0 - 1/2 \cos x_0 \sin\theta_0,$$

$$v_{ld} = -f/2 \cos x_0 \cos\theta_0 - f/2 \sin x_0 \sin\theta_0.$$

Sverdrup Wave

$U_0 = \sin(\theta - x)$
$V_0 = f \cos(\theta - x)$
$f = 0.4$
$k = 0.2$

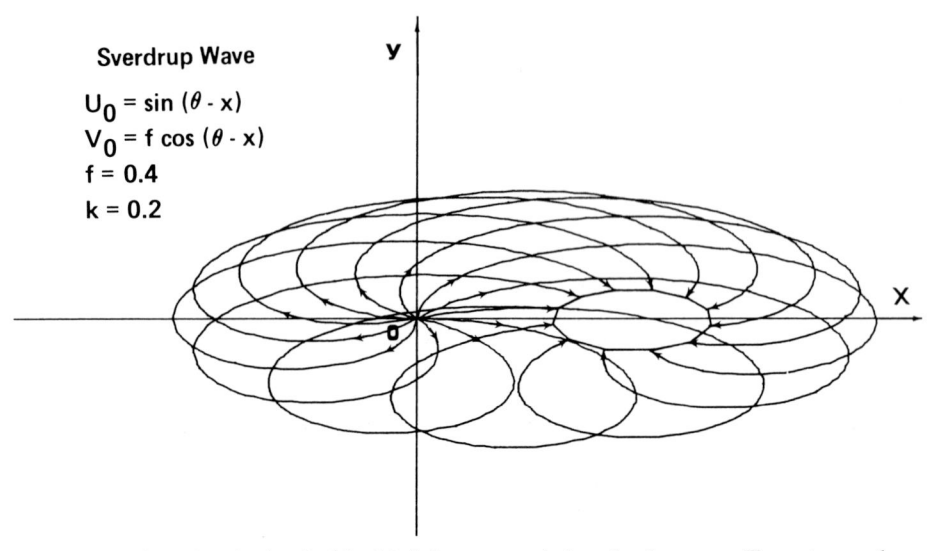

FIGURE 5. The trajectories inscribed by labelled water parcels in a Sverdrup wave. The water parcels are released from the origin, and their termini at then end of a complete tidal period form an ellipse in space.

The Lagrangian drift can be rewritten as

$$u_{ld} = 1/2 \sin(\theta_0 - x_0 - \pi), \tag{29}$$

$$v_{ld} = f/2 \cos(\theta_0 - x_0 - \pi),$$

which is a typical form of an ellipse on a hodograph plane. The Lagrangian residual current becomes

$$\begin{aligned} u_{lr} &= & 0 &+ & 1/2 & +\kappa 1/2 \sin(\theta_0 - x_0 - \pi), \\ v_{lr} &= & 0 &+ & 0 & +\kappa f/2 \cos(\theta_0 - x_0 - \pi). \end{aligned} \tag{30}$$

     Eulerian Stokes Lagrangian
     Residual Drift   Drift
     Current Velocity Velocity

Shown in Fig. 5 are the trajectories inscribed by the water parcels released at $x_0 = 0$, and θ_0 at intervals of one twelfth of a period. The termini of labelled water parcels at the end of a tidal period form an ellipse in space. The eccentricities for the tidal current and for the Lagrangian residual ellipses are the same, and they are equal to f. Both ellipses are clockwise rotating, but they are 180° out of phase.

5. Summary and conclusions

The dynamics of the tide-induced Lagrangian residual current has been discussed for a weakly nonlinear tidal system. The present analysis reveals the generation mechanism of the tide-induced Lagrangian residual current by means of a perturbation technique. Using a Lagrangian approach, the first order Lagrangian residual current has been shown to be the sum of the Eulerian residual current and the Stokes drift. Or, the mass transport velocity introduced by LONGUET-HIGGINS (1969) has been proven to be the first order Lagrangian residual current in a weakly nonlinear tidal system. The dilemma that exists in the first order Lagrangian residual current can now be explained by the solution of the second order Lagrangian residual current. The second order correction to the Lagrangian residual current is named as the Lagrangian residual drift which is an ellipse on a hodograph plane. The second order solution of the Lagrangian residual current confirms the expectation that the Lagrangian residual current is a function of the time when the labelled water parcels are released. Moreover, with the second order solution at hand, an assessment can be made of the adequacy in the usage that the mass transport velocity as an approximation to the Lagrangian residual current.

References

ALFRINK, B.J., & VREUGDENHIL, C.B., 1981, Residual currents, Rept. R-1469-11, Delft Hydraul. Lab., Delft, The Netherlands.

CHENG, R.T., 1983, Euler-Lagrangian computations in estuarine hydrodynamics, Proc. of the Third Inter. Conf. on Num. Meth. in Laminar and Turbulent Flow, (Eds.) C. Taylor, J.A. Johnson, and R. Smith, p. 341-352, Pinderidge Press.

CHENG, R.T. & CASULLI, V., 1982, On Lagrangian residual currents with applications in South San Francisco Bay, California, Water Resour. Res., v. 18, no. 6, p. 1652-1662.

CHENG, R.T., & GARTNER, J.W., 1985, Harmonic analysis of tides and tidal currents in South San Francisco Bay, California, to appear in Estuarine, Coastal, and Shelf Science.

DEFANT, A., 1961, Physical Oceanography, v. 1, Progamon Press, New York, 598 p.

FENG, S., 1977, A three-dimensional non-linear model of tides, Scientia Sinica, v. 20, no. 4, p. 436-446.

FENG, S., CHENG, R.T. & XI, P., 1985, On tide-induced Lagrangian residual current, submitted for publication.

LEENDERTSE, J.J., 1970, A water-quality simulation model for well-mixed estuaries and coastal seas, I. Principles of Computation, Resp. RM-6230-RC, Rand Corp., Santa Monica, California.

LEENDERTSE, J.J. & GRITTON, E.C., 1971, A water-quality simulation model for well-mixed estuaries and coastal seas, II. Computation Procedures, Rep. R-708-NYC, Rand Corp. Santa Monica, California.

LONGUET-HIGGINS, M.S., 1969, On the transport of mass by time-varying ocean currents, Deep Sea Res., v. 16, 431-447.

MUDLER, R., 1982, Eulerian and Lagrangian analysis of velocity of field in the southern North Sea, North Sea Dynamics, Ed. by Sundermann and Lenz, Springer-Verlag, Berlin, Heidelberg, p. 134-147.

STOKER, J.J., 1965, Water Waves, Interscience Publ., New York.

ZIMMERMAN, J.T.F., 1979, On the Euler-Lagrangian transformation and the Stokes drift in the presence of oscillatory and residual currents, Deep Sea Res., 26A, p. 505-520.

A Simulation of the Residual Flow in the Bohai Sea

Zhang Shuzhen
Wang Huatong
Feng Shizuo
Xi Pangen

Shandong College of Oceanology,
Peoples Republic of China

ABSTRACT

Removing the periodic tidal movement from the current meter records collected in coastal seas determines the so-called residual flow, which is the composition of tidal residual current, wind-driven current and thermohaline circulation. Experimentally, it is difficult to separate the residual flow into these three components. Therefore, a residual circulation model involving tide, wind and thermohaline effect was developed and applied to the Bohai Sea. Generally, simulation results are in agreement with the observed current fields.

1. Introduction

The Bohai Sea is an inland shallow water area situated in the northern part of China (Fig. 1). It has an average depth of approximately 20 m. The Bohai Strait in the east is the only passage connecting the sea to the oceanic Huanghai Sea (Yellow Sea). The East Asia monsoon dominates the meteorology of this region. Winds are northeasterly in winter, and southwesterly in summer; the spring and fall being transition periods with unstable winds.

There are several rivers emptying into the Bohai Sea. The biggest one, the Huanghe River (Yellow River), is taken into account in this paper. Because of the strong surface cooling and wind mixing the water body in the whole of the Bohai Sea basin is vertically well mixed and almost horizontally homogeneous in winter. While in summer both horizontal and vertical density gradients are observed. Therefore thermohaline effects are only present in summer circulation.

2. Governing equations

There are two different approaches to the derivation of residual currents, see CHENG (1981), FENG (1977), HEAPS (1978), NIHOUL (1975), RAMMING and KOWALIK (1980), TEE (1976), WANG HUATONG et al (1980). The first one is to filter or average the results obtained from a mathematical model. This is equivalent to processing current meter records with the same technique. In the second approach, the governing equations for the residual circulation are generated by applying filters, or time averages, over several tidal cycles to the hydrodynamic equations. The model of baroclinic residual currents presented in this paper is based on the second approach; XI PANGEN et al (1980), ZHANG SHUZHEN et al (1984).

On the basis of hydrodynamic equations and density diffusion equation, carrying out dimensional analysis and vertical integration with depth, then introducing stream function ψ, we obtain a 2-D (horizontal), baroclinic, nondimensional residual circulation

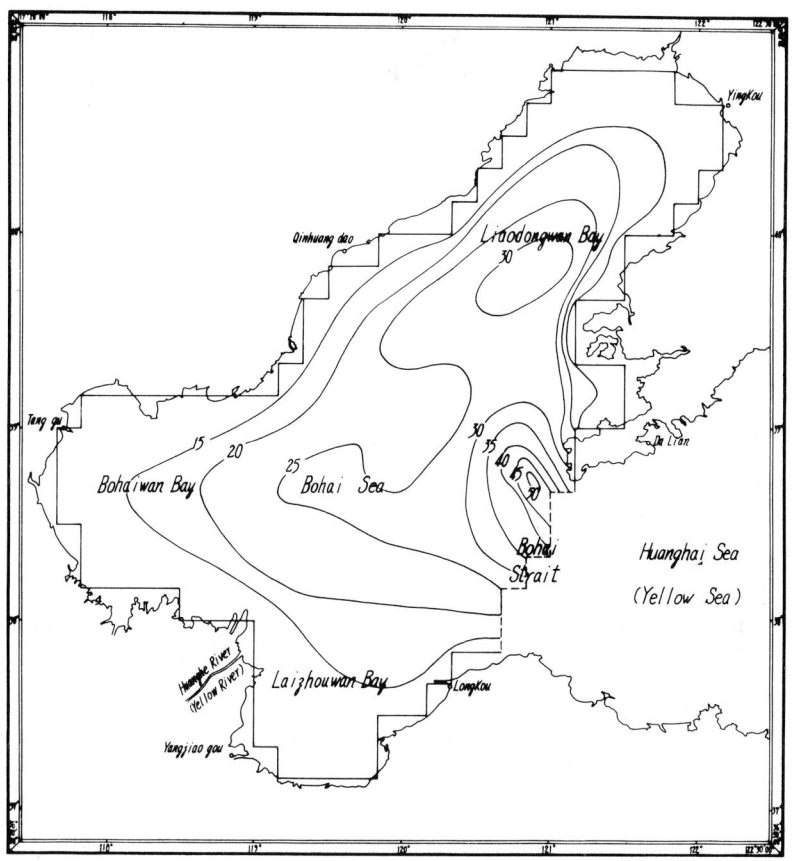

FIGURE 1. Depth of the Bohai Sea

transport model as follows:

$$\nabla \cdot (\frac{\beta}{h}\nabla\psi) - \frac{\beta}{h^2}\nabla h \cdot \nabla\psi - \frac{k_c}{h}\nabla h \cdot \vec{e}_3 \times \nabla\psi = \vec{e}_3 \cdot \nabla \times (\vec{\tau}' + k_\tau \frac{\delta_1}{k}\overline{\vec{\tau}}_a) \quad (1)$$
$$+ \frac{1}{h}\nabla h \cdot \vec{e}_3 \times x(\vec{\tau}' + k_\tau \frac{\delta_1}{k}\overline{\vec{\tau}}_a) - J\Upsilon_a + D;$$

the lateral boundary conditions:

along the shore boundary $C_1 - -\vec{n}_0 \cdot (\vec{e}_1 \overline{U} + \vec{e}_2 \overline{V}) = 0;$ \quad (2)

along the open boundary $C_2 - -\vec{n}_0 \cdot (\vec{e}_1 \overline{U} + \vec{e}_2 \overline{V}) = Q$ or $\zeta = S.$ \quad (3)

The right-handed rectangular coordinate system is adopted; the x-y plane coincides with the undisturbed water surface and z axis is positive upward. The f-plane approximation is used in this paper. Where

$$D = -(\frac{\epsilon}{k^2})h\left\{\frac{\partial}{\partial x}(\frac{1}{h}\int_{-hz}^{0}\int^{0}\frac{\partial\rho'}{\partial y}dz'dz) - \frac{\partial}{\partial y}(\frac{1}{h}\int_{-hz}^{0}\int^{0}\frac{\partial\rho'}{\partial x}dz'dz)\right\};$$

$$\tau'_x = -\left\{\tilde{\zeta}\frac{\partial\tilde{\zeta}}{\partial x} - \tilde{\zeta}\frac{\partial}{\partial x}(k_a\tilde{\zeta}_a) - \tilde{\zeta}\frac{\partial}{\partial x}(k_\tau\tilde{\zeta}_\tau) + \frac{\partial}{\partial x}(\frac{\tilde{U}^2}{h}) + \frac{\partial}{\partial y}(\frac{\tilde{U}\tilde{V}}{h})\right\};$$

$$\tau'_y = -\left\{\tilde{\zeta}\frac{\partial \tilde{\zeta}}{\partial y} - \tilde{\zeta}\frac{\partial}{\partial y}(k_a\tilde{\zeta}_a) - \tilde{\zeta}\frac{\partial}{\partial y}(k_T\tilde{\zeta}_T) + \frac{\partial}{\partial x}(\frac{\tilde{U}\tilde{V}}{h}) + \frac{\partial}{\partial y}(\frac{\tilde{V}^2}{h})\right\};$$

$$\beta = k_b K \frac{(\tilde{U}^2+\tilde{V}^2)^{\frac{1}{2}}}{h}; \quad \vec{\tau}' = \vec{e}_1\tau'_x + \vec{e}_2\tau'_y; \quad J = (\frac{k_c k_r}{P_r})\frac{\epsilon}{k^2}; \quad \vec{\nabla} = \vec{e}_1\frac{\partial}{\partial x} + \vec{e}_2\frac{\partial}{\partial y};$$

$$\vec{\tau}_a = \vec{e}_1\tau_{ax} + \vec{\tau}_2\tau_{ay}; \quad k_b = \frac{\tau_{b0}D}{\rho_0 v_0 V_0}; \quad k_\tau = \frac{\tau_{a0}D}{\rho_0 v_0 V_0}; \quad k_r = \frac{\Upsilon_{a0}D}{\rho'_0\gamma_0};$$

$$K = \frac{\rho_0 V_0^2}{\tau_{b0}}k; \quad \delta_1 = \frac{\tau_{a0}}{\tau_{a0}}; \quad P_r = \frac{v_0}{\gamma_0}; \quad k_c = R_0^{-1}E_u^{-1};$$

$$k = \frac{\zeta_0}{D}; \quad k_a = \frac{\zeta_{a0}}{\zeta_0}; \quad k_T = \frac{\zeta_{T0}}{\zeta_0}; \quad \epsilon = \frac{\rho'_0}{\rho_0};$$

$$R_0 = \frac{V_0}{fL};$$

h is water depth; $\vec{\tau}_{ax}, \vec{\tau}_{ay}$ denote the components of wind stress at the sea surface in x, y directions; $\vec{e}_1, \vec{e}_2$ and $\vec{e}_3$ denote unit vector along x, y and z axis; U and V are vertical integrations of velocity components along x and y directions; ζ is the elevation from the undisturbed sea surface; ζ_a and ζ_T represent the effects of atmospheric pressure and tide-generating force; $V_0, \tau_{a0}, \tau_{b0}, v_0, \zeta_0, D$ and L denote characteristic quantities of horizontal current velocity, wind stress at sea surface, bottom friction, eddy viscosity coefficient, surface elevation, vertical and horizontal scales of the Bohai Sea; ζ_{a0}, ζ_{t0} and Υ_{a0} denote characteristic quantities of ζ_a, ζ_t and Υ_a; k, γ, f and g are frictional coefficient, eddy diffusion coefficient, Coriolis parameter and gravitational acceleration; Q and $\vec{n}_0$ denote normal volume transport through the open boundary and unit vector of boundary normal; S denotes water elevation along the open boundary $\sim$ and $-$ represent tidal and residual parts, respectively; operator $\wedge$ denotes time average; ρ' and ρ denote the variation and reference constant densities of sea water; ρ'_0 is the characteristic quantity of ρ'; E_u is the Euler number; Υ_a denotes the thermohaline gradient.

3. Data and parameters for computation

Wind fields are cited from the synoptic and current map volumes; Royal Netherlands Meteorological Institute (1935) published in the Netherlands and other related data. The wind stress fields at the sea surface are calculated according to the following formula:

$$\vec{\tau}_a = \rho_a \gamma_a^2 |w|\vec{w} \tag{4}$$

Where w is the wind speed over the sea surface; the air density $\rho_a = 1.226 \times 10^{-3}$ g.cm^{-3}; $\gamma_a = 2.6 \times 10^{-3}$ (when wind force is above the Beaufort Wind Scale 4) or $\gamma_a = 0.85 \times 10^{-3}$ (when wind speed is below the Beaufort Wind Scale 4).

Eq. (1) is solved numerically using a space-staggered grid with a 10×10 km grid step. A central difference scheme is used for time-differencing.

4. Simulation results

The circulation patterns are quite different between winter and summer. There are two big gyres in winter circulation (Fig. 2), one clockwise and the other counterclockwise. This shows that the Warm Yellow Sea Current which enters the Bohai Strait branches into two asymmetric systems. One of these turns south and forms a counterclockwise gyre flowing out of the region through the southern part of the Bohai Strait. The other branch forms a clockwise gyre flowing out of the basin through the deep northern waterway of the Bohai Strait. Some small gyres can be found in this system: two small gyres in Laizhouwan Bay, one in Bohaiwan Bay and a long, narrow one in Liaodongwan Bay. The maximum current speed is 10 cm/sec in deep water and less than 1 cm/sec in shallow areas.

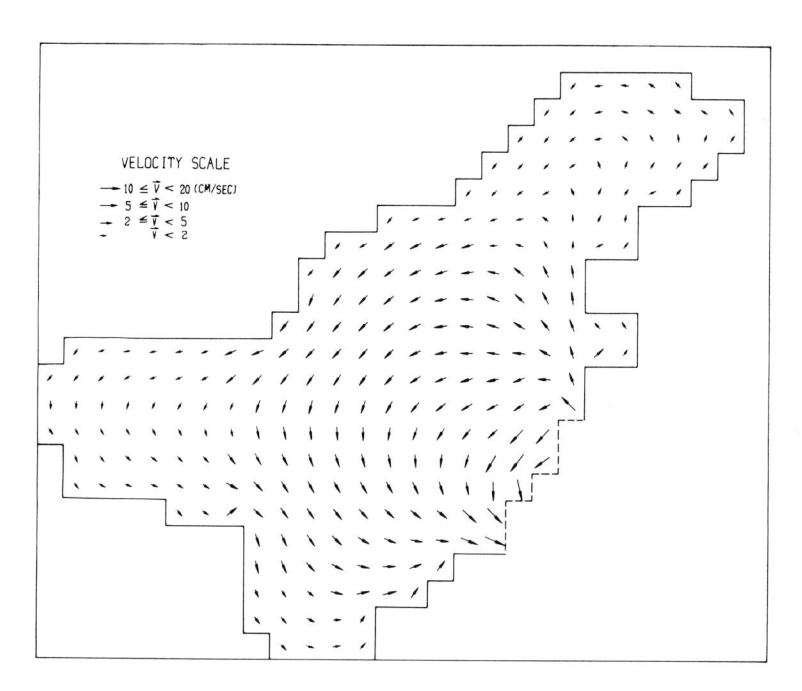

FIGURE 2. Winter Circulation

There is only one big counterclockwise gyre in the summer circulation pattern (Fig. 3). Current speeds in the summer circulation are lower than in winter.

The comparison between the computation and field current maps suggests that the corresponding circulation patterns are in agreement (see Figs. 4(a) and 4(b)).

5. Discussions and conclusions

In order to understand the roles of different factors affecting the circulation pattern, several sensitivity calculations have been carried out for the Bohai Sea. The preliminary results are covered in the subsequent discussion.

Wind stress plays an important role in forming the winter circulation, while for the summer circulation it is of less importance. If there is no wind stress in winter, the general circulation is still dominated by two big gyres, the differences are that the two gyres are almost symmetric and all the small gyres in shallow water vanish. This results points out that the influences of wind stress are stronger in shallow water than in deep water.

Variable bottom topography is one of the reasons why small gyres form in the winter circulation. A flat bottom circulation carried out for the Bohai Sea shows that these

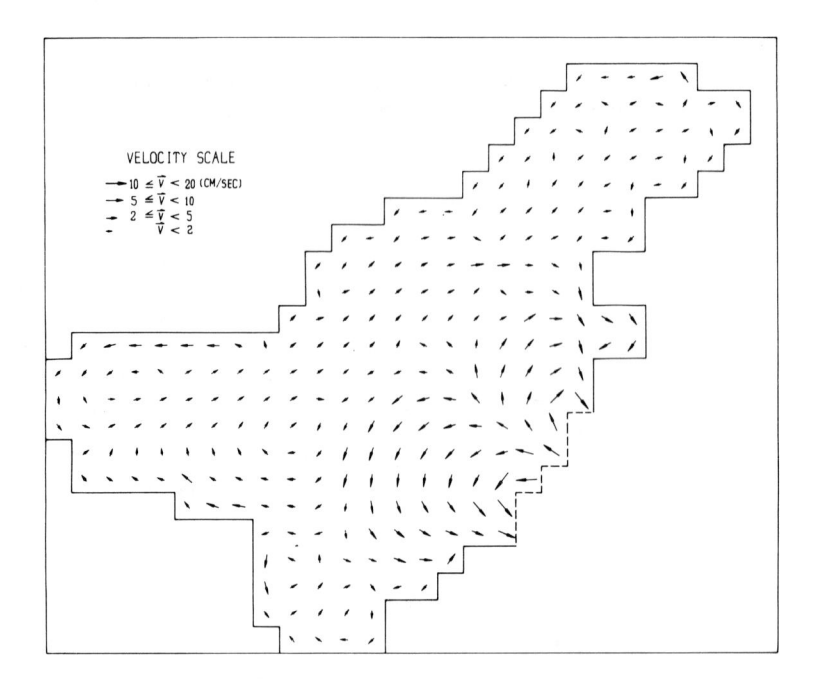

FIGURE 3. Summer Circulation

FIGURE 4(a). Observed Current Map in Bohai Sea in Winter

FIGURE 4(b). Observed Current Map in Bohai Sea in Summer

small gyres vanish. For summer circulation flat bottom computation suggests that the velocities are stronger than the corresponding actual current field, especially in deep water. Generally speaking, topographic effect is more important than thermohaline and wind stress effects in forming the summer circulation pattern.

Thermohaline force has no marked influence over the circulation pattern in summer, causing a 10% increment in the current speed. However, the thermohaline effect is more important than wind stress which causes 1% increment in the current speed for summer circulation.

Forcing through the open boundary plays a significant role in the winter and summer circulation patterns. In this experiment the open boundary in the east is closed. The summer circulation, for instance, has very low speeds.

The influence of tidal variation on the current pattern is slight. This phenomenon implies that the tidal-induced residual current is very weak in the Bohai Sea.

River runoff affects the current velocity only in the vicinity of the Huanghe River Estuary, it does not change the circulation patterns in the whole Bohai Sea both for summer and winter.

References

CHENG, R.T., 1981. Modeling of Tidal and Residual Circulation in San Francisco Bay, California, Seminar on 2D-Flows, HEC Rept., U.S. Army Corps of Eng., Davis, Calif., 1-14.

FENG SHIH-ZAO, 1977. A Three-Dimensional Nonlinear Model of Tides, Scientia Sinica, XX, No. 4., 436-446.

HEAPS, N.S., 1978. Linearized Vertically-Integrated Equations for Residual Circulation in Coastal Seas, Deutsche Hydrographische Z., 31, 147-169.

NIHOUL, J.C.J., 1975. Modeling or Marine Systems, Part I, CH.2, 41-66, ESPC N.Y.

RAMMMING, H.G. & Z. KOWALIK, 1980. Numerical Modelling of Marine Hydrodynamics, 57, ESPC N.Y.

TEE, K.T., 1976. Tide-induced Residual Current, a 2D Nonlinear Tidal Model, J. Mar. Res., 34, 603-628.

WANG HUATONG FANG XINHUA YANG DIANRONG KUANG GUORUI & CHEN SHIJUN, 1980. Numerical Modeling of the Circulation and Pollutant Dispersion in the Jiaozhou Bay, (I), J. of Shandong College of Oceanology, 10, No. 1, 26-63.

XI PANGEN, ZHANG SHUZHEN & FENG SHIZUO, 1980. Mathematical Modelling of Circulation in the East China Sea (I), J. of Shandong College of Oceanology, 10, No. 3, 13-25.

ZHANG SHUZHEN, XI PANGEN & FENG SHIZUO, 1984. Numerical Modeling of the Steady Circulation in the Bohai Sea, J. of Shandong College of Oceanology, 14, No. 2, 12-19.

ROYAL NETHERLANDS METEOROLOGICAL INSTITUTE, 1935. Oceanographic and Meteorological Observations in the China Seas and in the Western Part of the North Pacific Ocean, Government Printing Office, the Hague.

Principal Differences Between 2D- and Vertically Averaged 3D-Models of Topographic Tidal Rectification

J.T.F. Zimmerman

Neth. Inst. Sea Research
and
Inst. Meteorology and Oceanography,
Univ. of Utrecht, The Netherlands

ABSTRACT

It is shown that tilting of horizontal vortex lines, associated with the vertical shear of tidal currents, by differential vertical velocities induced by a depth-varying topography, gives rise to a rectification mechanism for the along-isobath vertically averaged flow. This mechanism is absent in the usual quasi two-dimensional formulations of topographic rectification in which a residual flow only arises from nonlinear corticity advection. By proper scaling, the former mechanism is shown to be of the same order of magnitude as the latter. Formulating the vorticity dynamics for a small amplitude topography, it appears that the effective Reynolds stresses associated with vertically averaged cross-products of deviations from the vertical mean, in the case of coherent topographically induced flow structures, form both a driving mechanism for residual circulation in the vertically averaged sense as well as an effective horizontal viscous damping term.

1. Introduction

In reviewing the physics of continental shelves, GARRETT (1983) classified the theoretical results concerning topographic tidal rectification as *conclusive*. These results, obtained during the second half of the 70's, all apply to quasi-two dimensional (horizontal) flow and have been reviewed a number of times in recent years (HUTHNANCE, 1981; ZIMMERMAN, 1981; ROBINSON, 1983). Indeed all the studies on topographic tidal rectification, though differing from each other in using different approximations in order to solve an essentially nonlinear problem, agree in their final conclusions. The most important one being that a spatially varying bottom slope gives rise to a rectified current along the isobaths, the strength of which depends on the ratio of tidal excursion amplitude to topographic length scale. For a one-dimensional topography, $h(x)$, such that $h \ll H$ (small amplitude topography), where H is the mean waterdepth, assuming a linear bottom friction coefficient r, the vorticity of the vertically averaged velocity field, $\bar{\zeta}$, induced by the tidal current $(\bar{u}_0, \bar{v}_0)$ and the topography obeys:

$$\frac{\partial \bar{\zeta}}{\partial t} + \bar{u}_0 \frac{\partial}{\partial x} \bar{\zeta} + \frac{r}{H} \bar{\zeta} = \left(\frac{\bar{u}_0 f}{H} - \frac{r \bar{v}_0}{H^2}\right) \frac{\partial h}{\partial x} \tag{1}$$

where f is the Coriolis parameter. A bar denotes the vertical average of quantities that are in general depth dependent:

$$\overline{(\ )} \equiv \frac{1}{D} \int_{-D(x)}^{0} (\ ) dz \tag{2}$$

except for $\bar{\zeta}$ which is defined as:

$$\bar{\zeta} = \frac{\partial}{\partial x} \bar{v}_1 \neq \frac{1}{D} \int_{-D(x)}^{0} \frac{\partial}{\partial x} v_1 dz \tag{3}$$

Here the subscripts 0 and 1 refer to the velocity in the flat-bottom reference state and the velocity induced by the topography in the small amplitude limit. Note the essential inequality in Eq. (3). Defining $\bar{\zeta}$ in this way obviously is more convenient if the along-isobath velocity profile, $\bar{v}_1(x)$, is the ultimate goal. Eq. (1) is discussed extensively in TEMME and ZIMMERMAN (1985) as the first order result in a double expansion in the small parameters h/H and L/L_0, L being the topographic length scale and L_0 the tidal wave length. The smallness of L/L_0 allows a rigid lid approximation for the topographically induced velocity field which explains the zero in the upper boundary of the integrals in Eqs. (2) and (3). As Eq. (1) shows, rectification arises solely from non-linear vorticity advection, the vorticity being first produced by topographic planetary vortex stretching and differential bottom friction over a sloping bed, a physical structure that is shared by all quasi two-dimensional theories of topographic tidal rectification.

An obvious extension of the theory now seems to be the incorporation of the vertical structure of the topographically induced velocity field. The vertical structure of tide-induced residual currents has been the subject of various investigations; for instance in a semi-enclosed tidal basin with a flat bottom (IANNIELLO, 1977; YASUDA, 1981) and over a uniformly sloping bottom along the coast (TEE, 1980; YASUDA, 1983). These studies all show the importance of vertical advection of horizontal vorticity as a rectification mechanism for the horizontal flow. However, these results cannot be extended directly to the situation described by (1) where in a vertically integrated sense the rectified current only arises if the *slope* of the bottom varies, which is not the case in the studies cited. Recently WRIGTH and LODER (1985) extended an earlier two-dimensional theory of LODER (1980) on the generation of residual currents at the sides of a submarine bank to three dimensions. As LODER's (1980) original model is compatible with Eq. (1) in the limit of small amplitude topography, their results are of importance in extending two-dimensional topographic rectification theories to three dimensions. Evidently such a theory gives the vertical shear of the along-isobath residual flow as well as a residual cross-isobath circulation that cannot be resolved by a 2D model, the depth averaged cross-isobath residual flow being zero. As to the depth averaged along-isobath residual current there is an intriguing conclusion of WRIGHT and LODER (1985) in that the depth-averaged residual current obtained from depth-averaging the solution of their 3D model, differs from the result obtained by using abinitio a genuine 2D model as LODER's (1980), the former giving a stronger current than the latter. Although WRIGHT and LODER (1985) ascribe this discrepancy primarily to the way bottom friction is parameterized, our purpose here is to draw attention to the fact that even if that problem is absent, a discrepancy between 2D and vertically averaged 3D models may exist.

Analysing and comparing 2D and 3D models from the vorticity point of view, we show here that the discrepancy can be due to the usual formulation of 2D models, in which an additional rectification mechanism is neglected. The mechanism at hand is vertical tilting of horizontal vortex lines by differential vertical velocities induced by the topography. The mechanism is related to the so-called REYNOLD's stesses associated with vertical averaged cross-products of velocity components deviating from their average, as they arise in derivation of the vertically averaged momentum equations (HEAPS, 1978). By using the 3D vorticity equations in the limit of small amplitude topography we show that if these REYNOLD's stresses, for coherent structures as are topographically induced residual circulations, are parametrized solely as an effective horizontal eddy viscosity something essentially different is left out. In fact our formulation shows that effective horizontal eddy viscosity acting on the vertically averaged flow is only part of the story, even an insignificant part, and that these REYNOLD's stresses are a generating rather than a dissipative agency for the depth averaged residual current. This conclusion comes near to a theorem derived by FLOKSTRA (1976, 1977) for stationary flow stating that in a fluid experiencing damping by bottom friction, the circulation of the depth averaged flow can only be maintained if the circulation is *driven* by the effective REYNOLD's stresses (in the absence of an applied curl of the surface stress, of course). Looking at the 3D vorticity equations we first identify the tilting mechanism and assess its strength by proper scaling of the relevant terms. After that we rederive the equation for the depth-averaged vertical vorticity component, expand this equation in h/H and finally propose a parameterization that leads to a modified form of Eq. (1). Thus our result points to a recitification mechanism different from and additional to vorticity advection. We show that results derived from Eq. (1) are not as conclusive as they seemed to be, particularly so for shallow tidal areas. Fortunately, we can show that, as a function of topographic wave number, the spectral response of the *new* rectification mechanism has the same behaviour as the *old* one — i.e. residual circulation is strongest for topographic length scales of the order of the tidal excursion. The additional rectification mechanism therefore only changes the strength of the rectified current but not its spatial structure.

2. Three dimensional vorticity dynamics

Our starting point are the equations of motion and continuity of a uniformly rotating homogeneous fluid in hydrostatic balance. Neglecting for the time being horizontal eddy viscosity — this will be shown to be a self-induced effect later on — and assuming a depth-independent vertical eddy viscosity, ν, these equations read:

$$\frac{\partial u}{\partial t}+u\frac{\partial u}{\partial x}+v\frac{\partial u}{\partial y}+w\frac{\partial u}{\partial z}-fv = -g\frac{\partial Z}{\partial x}+\nu\frac{\partial^2 u}{\partial z^2} \quad (4)$$

$$\frac{\partial v}{\partial t}+u\frac{\partial v}{\partial x}+v\frac{\partial v}{\partial y}+w\frac{\partial v}{\partial z}+fu = -g\frac{\partial Z}{\partial y}+\nu\frac{\partial^2 v}{\partial z^2} \quad (5)$$

$$\frac{\partial u}{\partial x}+\frac{\partial v}{\partial y}+\frac{\partial w}{\partial z} = 0 \quad (6)$$

where g is the acceleration of gravity, f the Coriolis parameter and Z the deviatioan of the sea surface from its equilibrium value. Defining the 3D vorticity components as:

$$\xi = \frac{\partial w}{\partial y}-\frac{\partial v}{\partial z}, \quad \eta = \frac{\partial u}{\partial z}-\frac{\partial w}{\partial x}, \quad \zeta = \frac{\partial v}{\partial x}-\frac{\partial u}{\partial y} \quad (7)$$

the vorticity equations derived from Eqs. (4) - (6) in the standard way are:

$$\frac{\partial \xi}{\partial t}+\vec{u}\cdot\nabla\xi = (f+\zeta)\frac{\partial u}{\partial z}+\xi\frac{\partial u}{\partial x}+\eta\frac{\partial u}{\partial y}+\nu\frac{\partial^2 \xi}{\partial z^2} \quad (8)$$

$$\frac{\partial \eta}{\partial t}+\vec{u}\cdot\nabla\eta = (f+\zeta)\frac{\partial v}{\partial z}+\xi\frac{\partial v}{\partial x}+\eta\frac{\partial v}{\partial y}+\nu\frac{\partial^2 y}{\partial z^2} \quad (9)$$

$$\frac{\partial \zeta}{\partial t}+\vec{u}\cdot\nabla\zeta = (f+\zeta)\frac{\partial w}{\partial z}+\xi\frac{\partial w}{\partial x}+\eta\frac{\partial w}{\partial y}+\nu\frac{\partial^2 z}{\partial z^2} \quad (10)$$

Obviously, with a one-dimensional topography $h(x)$ in mind, Eq. (8) is related to the shear of the along-isobath flow, Eq. (9) to the cross-isobath circulation and Eq. (10) must be the 3D analogue of Eq. (1). Already here it is worth noting that the second and third term in the r.h.s. of Eq. (10) denote the tilting of horizontal vortex lines by differential vertical velocities, a process that has no analogue in equation (1).

Using a double expansion, both in h/H and in two length scales (TEMME and ZIMMERMAN, 1984) — i.e. distinguishing the *slow* scale over which the tidal wave proper varies and the *fast* scale over which the topography varies — we have to zeroth order in

h/H (the flat bottom reference state), neglecting all horizontal derivatives that apply to the *slow* scale:

$$\frac{\partial \xi_0}{\partial t} = f\eta_0 + \nu \frac{\partial^2 \xi_0}{\partial z^2} \tag{11}$$

$$\frac{\partial \eta_0}{\partial t} = -f\xi_0 + \nu \frac{\partial^2 \eta_0}{\partial z^2} \tag{12}$$

$$\frac{\partial \zeta_0}{\partial t} = f\frac{\partial w_0}{\partial z} + \nu \frac{\partial^2 \zeta_0}{\partial z^2} \tag{13}$$

For a prescribed free surface variation $w_0 = \partial Z/\partial t \sim \exp(i\sigma t)$, where σ is the tidal frequency, Eqs. (11)-(13) give nothing more than oscillatory EKMAN dynamics, albeit in the not so usual formulation with vorticity as the dependent variable. In terms of velocity components, the solution of Eqs. (11)-(13) has already been given by SVERDRUP (1927). This solution shall not be repeated here as it is only of interest as the *unperturbed* reference state that shall be denoted by $u_0(z,\sigma), v_0(z,\sigma), w_0(z,\sigma)$. Note that in the present approximation $\xi_0(z,\sigma) \cong \partial v_0/\partial z$ and $\eta_0(z,\sigma) \cong \partial u_0/\partial z$. Our interest lies in the first order equations for the topographically generated vorticity components (ξ_1, η_1, ζ_1). As all y-derivatives must vanish for a one-dimensional topography depending only on x, we have:

$$\frac{\partial \xi_1}{\partial t} + u_0(z,\sigma)\frac{\partial}{\partial x}\xi_1 = f\eta_1 + \zeta_1\frac{\partial u_0(z,\sigma)}{\partial z} - \frac{\partial v_0(z,\sigma)}{\partial z}\frac{\partial u_1}{\partial u} + \nu\frac{\partial^2 \xi_1}{\partial z^2} \tag{14}$$

$$\frac{\partial \eta_1}{\partial t} + u_0(z,\sigma)\frac{\partial}{\partial x}\eta_1 = -f\xi_1 + \nu\frac{\partial^2 \eta_1}{\partial z^2} \tag{15}$$

$$\frac{\partial \zeta_1}{\partial t} + u_0(z,\sigma)\frac{\partial}{\partial x \zeta_1} = f\frac{\partial w_1}{\partial z} - \frac{\partial v_0(z,\sigma)}{\partial z}\frac{\partial w_1}{\partial x} + \nu\frac{\partial^2 \zeta_1}{\partial z^2} \tag{16}$$

This coupled set of vorticity equations has a much more rich physical contents than the 2D Eq. (1). Particularly Eqs. (14) and (15) bear information on respectively the vertical shear of the along isobath flow and the cross-isobath circulation. Interesting as these properties may be, in the rest of this paper we shall concentrate only on the possibility of reducing Eq. (16) to Eq. (1) and present a discussion of Eq. (14) and (15) elsewhere. Again it is interesting to compare the physical contents of Eq. (16) with Eq. (1). Terms in Eq. (16) having no analogue in Eq. (1) are underlined with a drawn line, whereas terms being the product of two (sinusoidally) time-varying components, giving rise to rectification, are underlined with a broken line. Evidently the tilting term $-\partial v_0/\partial z\, \partial w_1/\partial x$ does not have an anologue in Eq. (1) and is able to provide a rectification mechanism. Whether or not the term is of importance quantitatively can be assessed by scaling the two rectification terms in Eq. (16). Let L be the topographic length scale, U a scale of the tidal velocity amplitude, h the topographic height scale and H the undisturbed depth. If now the perturbation in the velocity is of order hU/H, then $\zeta_1 = O(hU/HL)$ so that $u_0\, \partial \zeta_2/\partial x = O(hU^2/HL^2)$. From the continuity equation we derive an order of magnitude for the topographically induced vertical velocity $w_1 = O(hU/L)$. Hence the tilting term $-\partial v_1/\partial z\, \partial w_1/\partial x = O(hU^2/HL^2)$, which is of the same order of magnitude as the advection term. Thus we now not only have evidence that an additional rectification mechanism could be present for the depth averaged flow, but also we have an indication that the mechanism can be of the same strength as vorticity advection. It now remains to be seen how this works out in the depth averaged sense.

3. Quasi two-dimensional vorticity equation

Defining, as in Eq. (3), the vertical vorticity component in quasi two-dimensional flow as the curl of the vertically averaged velocity, using a rigid lid approximation and boundary conditions:

$$\left\{ \begin{array}{ll} u = v = w = 0, & z = -D(x) = -H - h(x) \\ \dfrac{\partial u}{\partial z} = \dfrac{\partial v}{\partial z} = 0 & z = 0 \\ w = 0 & z = 0 \end{array} \right. \quad (17)$$

vertical averaging of the y-momentum Eq. (5) with the averaging operator being defined by equation (2), neglecting y-derivatives for an x-dependent topography, gives:

$$\frac{\partial \bar{v}}{\partial t} + \overline{u \frac{\partial u}{\partial x}} + \overline{w \frac{\partial v}{\partial z}} + f\bar{u} = -\frac{v}{D} \left(\frac{\partial v}{\partial z} \right)_{z=-D(x)} = \frac{\tau}{D} \quad (18)$$

Vertically averaging the continuity Eq. (6) in the same way gives:

$$\frac{\partial \bar{u}}{\partial x} = -\frac{\bar{u}}{D} \frac{\partial D}{\partial x} \quad (19)$$

Finally taking the x-derivative of Eq. (18), substituting Eq. (19) and using definition Eq. (3) we get:

$$\frac{\partial \bar{\zeta}}{\partial t} + \frac{\partial}{\partial x} \left[\overline{u \frac{\partial v}{\partial x}} + \overline{w \frac{\partial v}{\partial z}} \right] = \frac{f\bar{u}}{D} \frac{\partial D}{\partial x} + \frac{\partial}{\partial x} \left(\frac{\tau}{D} \right) \quad (20)$$

If we now write $u = \bar{u} + u'(z)$ and the same for v and w, whereas $\zeta = \bar{\zeta} + \zeta'(z)$, $\zeta'(z) \equiv \partial/\partial x \, v'(z)$, Eq. (20) reads:

$$\frac{\partial \bar{\zeta}}{\partial t} + \bar{u} \frac{\partial \bar{\zeta}}{\partial x} + \frac{\partial}{\partial x} \left[\overline{u'\zeta'} + \overline{(\bar{w}+w') \frac{\partial v'}{\partial z}} \right] = \left[\frac{f + \bar{\zeta}}{D} \right] \frac{\partial D}{\partial x} \bar{u} + \frac{\partial}{\partial x} \left(\frac{\tau}{D} \right) \quad (21)$$

where again use has been made of Eq. (19). As it stands Eq. (21) is the exact quasi-two-dimensional vorticity Eq. under the assumption of a rigid upper lid. In order to simplify a physical discussion of Eq. (21) and a comparison with Eq. (1) we now turn to an expansion of Eq. (21) in h/H. Denoting the zeroth order flat bottom reference state by $\bar{u}_0(\sigma)$, $u'_0(z,\sigma)$, and in the same way for v and w, where (σ) means fluctuating with the basic tidal frequency, a lengthy but straightforward expansion in h/H gives to first order, neglecting again all horizontal derivatives of variables that vary only on the *slow* scale of the tidal wave length:

$$\frac{\partial \bar{\zeta}_1}{\partial t} + + \bar{u}_0(\sigma) \frac{\partial}{\partial x} \bar{\zeta}_1 + \frac{1}{H} \int_{-H}^{0} u'_0(z,\sigma) \frac{\partial}{\partial x} \zeta'_1 dz$$

$$+ \frac{1}{H} \int_{-H}^{0} \frac{\partial}{\partial x} (\bar{w} + w') \frac{\partial}{\partial z} v'_0(z,\sigma) dz = f \frac{\bar{u}_0(\sigma)}{H} \frac{\partial h}{\partial x} \quad (22)$$

$$- \frac{v}{H} \left[\left(\frac{\partial}{\partial z} \zeta'_1 \right)_{z=-H} \frac{\partial h}{\partial x} \left\{ \left(\frac{\partial^2 v'_0(z,\sigma)}{\partial z^2} \right)_{z=-H} + \frac{1}{H} \left(\frac{\partial v_0(z,\sigma)}{\partial z} \right)_{z=-H} \right\} \right]$$

where the sum of three terms in the r.h.s., denoted by A, arises from an expansion of $\partial/\partial x \, \tau_D$ in h/H:

$$\frac{\partial}{\partial x} \frac{\tau}{D} = -\frac{v \left(\frac{\partial v}{\partial z} \right)_{z=-H-h}}{H+h} \quad (23)$$

$$= -\frac{\nu}{H}\frac{\partial}{\partial x}(1-\frac{h}{H}+\cdots)(\frac{\partial v}{\partial z}|_{z=-H}-h\frac{\partial^2 v}{\partial z^2}|_{z=-H}+\cdots),$$

subsequently substituting $v=v_0+v_1$, neglecting x-derivatives of v_0 and collecting all terms of first order in h/H.

Equation (22) is the quasi two-dimensional vorticity equation in the limit of small amplitude topography. It suggests a closure problem, in fact similar fo the one discussed by FLOKSTRA (1977), in that the terms denoted by A, B and C contain the deviations of variables from their vertical average. However closure in the small amplitude limit is facilitated by the fact that variables with a subscript 0 are known quantities from the zeroth order reference state. Even then, finding an exact solution of Eq. (22) is in general a formidable task that we shall not pursue here. Instead we shall discuss the parameterization of terms A, B and C in Eq. (22) from a more physical, but also more sketchy point of view.

4. Parametrization

Denoting the inverse frictional damping time scale ν/H^2 by r, an obvious parameterization of the term A in Eq. (22) is:

$$\frac{c_1 r}{H}\bar{\zeta}_1 + \frac{c_2 r}{H^2}\bar{v}_0(\sigma)\frac{\partial h}{\partial x} \tag{24}$$

where c_1 and c_2 depend on the particular vertical profiles of respectively ζ_1 and v_0. Note that in contrast to Eq. (1) we must expect that in general $c_1 \neq c_2$, a fact that has also been observed in relating three dimensional to vertically integrated two-dimensional flow in a different context (KOMEN and RIEPMA, 1981). Yet we see that the bottom friction term in Eq. (22), parameterized as in Eq. (24), describes the same physics as in Eq. (1); i.e. part of the bottom friction term acts as a damping term for the vorticity of the vertically integrated flow, whereas in the presence of a tidal velocity component parallel to the depth contours, vorticity is produced by differential bottom friction.

More interesting is the interpretation of B and C in Eq. (22). To start with B we observe that

$$\frac{1}{H}\int_{-H}^{0} u'_0(z,\sigma)\frac{\partial \zeta'_1}{\partial x}dx$$

is the vertically integrated effect of differential horizontal advection of deviations of ζ_1 from its vertical average by the prescribed zeroth order velocity component $u'_0(z,\sigma)$. As $u'_0(z,\sigma)$ is prescribed and ζ'_1 is subject to vertical eddy diffusion parameterized by ν, the combined effect of differential horizontal advection and vertical diffusion acts as an effective *horizontal* dispersion on the vertical average of ζ_1. Evidently B represents an oscillatory shear-dispersion process, that has been well-studied in a tidal context (BOWDEN, 1965; OKUBO, 1967). We therefore parametrize this term as:

$$\nu_e \frac{\partial^2 \bar{\zeta}_1}{\partial x^2} \tag{25}$$

where ν_e depends in general on the ratio of the vertical diffusion time scale and the tidal period (OKUBO, 1967). If we assume the former to be small compared to the latter ("*shallow*" water), ν_e is simply proportional to $\bar{u}_0 H$, the constant of proportionality depending on the particular vertical profile of $\bar{u}_0$ (BOWDEN, 1965).

Finally we are left with the term C in Eq. (22):

$$\frac{1}{H}\int_{-H}^{0}\frac{\partial}{\partial x}(\overline{w}_1+w'_1)\frac{\partial v'_0(z,\sigma)}{\partial z}dz \tag{26}$$

the interpretation of which is evident in view of the discussion before. It is the vertical average effect of the tilting of horizontal vortex lines associated with the vertical shear of the along-isobath component of the tidal current by differential vertical velocities induced by the topography. As to its parameterization in terms of vertically averaged quantities we first observe that a straightforward integration of the first part of Eq. (26) gives

$$\frac{1}{H}\int_{H}^{0}\frac{\partial \overline{w}_1}{\partial x}\frac{\partial v'_0}{\partial z}dz = \frac{1}{H}\frac{\partial \overline{w}_1}{\partial x}\{v'_0(0,\sigma)-v'_0(-H,\sigma)\} \tag{27}$$

$$= \frac{1}{H}\frac{\partial \overline{w}_1}{\partial x}\{v'_0(0,\sigma)+\overline{v}_0(\sigma)\}$$

Evidently $v'_0(0,\sigma) \sim \overline{v}_0(\sigma)$ and $\partial \overline{w}_1/\partial x \sim \overline{u}_0 \partial^2 h/\partial x^2$, so that Eq. (27) is proportional to $-\overline{u}_0(\sigma)\overline{v}_0(\sigma)/H \, \partial^2 h/\partial x^2$. Assuming now that the remaining part of Eq. (26) has the same parameterization we are finally left with:

$$\frac{1}{H}\int_{0}^{H}\frac{\partial w_1}{\partial x}\frac{\partial v'_0}{\partial z}dz = -c_3\frac{\overline{u}_0(\sigma)\overline{v}_0(\sigma)}{H}\frac{\partial^2 h}{\partial x^2} \tag{28}$$

Hence, unless $\overline{u}_0(\sigma)$ and $\overline{v}_0(\sigma)$ are exactly in quadrature as in a circularly polarized current or in an elliptically polarized one with one of the axes perpendicular to the gradient of the bottom topography. Eq. (28) provides an additional rectification mechanism for the vorticity of the depth averaged flow. Thus the quasi two dimensional vorticity equation in the limit of small amplitude topography reads:

$$\frac{\partial}{\partial t}\overline{\zeta}_1 + \overline{u}_0(\sigma)\frac{\partial}{\partial x}\overline{\zeta}_1 - \frac{c_1 r}{H}\overline{\zeta}_1 - \nu_e \frac{\partial^2 \overline{\zeta}_1}{\partial x^2} \tag{29}$$

$$= \left[\frac{f\overline{u}_0(\sigma)}{H} + \frac{c_2 r}{H^2}\overline{v}_0(\sigma)\right]\frac{\partial h}{\partial x} - c_3\frac{\overline{u}_0(c)\overline{v}_0(\sigma)}{H}\frac{\partial^2 h}{\partial x^2}$$

This is the final result to be compared with Eq. (1). It shows that the effective REYNOLD's stressses neglected in Eq. (1) not only give rise to a self-induced horizontal dispersion term for tidal flow over a depth varying sea bed, but that these stresses also act as a production term for the vorticity of the vertically averaged velocity field due to coherent effects of vertical tilting of horizontal vortexlines by the topographically induced vertical flow.

5. Spectral response

The advantage of analyzing tidal rectification in the small amplitude limit of the topography is that it allows a Fourier decomposition of the topography, giving a specific response to each wave number independent of the other ones. After that, back transformation gives the vorticity- or velocity field for any shape of the topography. For rectification by horizontal vorticity advection as in Eq. (1) the response functions have been discussed extensively before (ZIMMERMAN, 1978 and 1980; TEMME and ZIMMERMAN, 1984). They have the peculiar and crucial characteristic of giving a residual velocity proportional to the *slope* of the topography for a topographic wave length large compared to the tidal excursion, whereas for a relatively small wave length the response drops proportional to the decreasing wavelength. The optimum response then occurs for a wavelength having the order of the tidal excursion amplitude. For a step-like topography these response characteristics result in producing a residual jet centered at the break in the topography which extends, in absence horizontal diffusion, exactly to a distance of twice the tidal excursion amplitude on both sides of the break (TEMME and ZIMMERMAN, 1985). As we now have a modified equation, Eq. (29), with an additional rectification mechanism it remains to be seen whether that *new* mechanism has the same spectral response as the *old* one. Instead of solving the Fourier transformed version of Eq. (29) exactly, as has been done for Eq. (1) in ZIMMERMAN (1978, 1980) and TEMME and ZIMMERMAN (1985), we shall use the approximate method of harmonic truncation in the frequency domain. The method has been introduced in this context by LODER (1980). It can be shown to give the same asymptotic behaviour for the spectral response functions as an exact solution does, without having to deal with complicated products of Bessel functions. However, it should be stressed here that although the method gives the approximately right response for a single wave number in the topography, it produces spurious horizontal vorticity diffusion if it is used in a back transformation for a steep step-like topography (TEMME and ZIMMERMAN, 1985). The result therefore should be used only for a simple sinusoidal topography, representing say, a sequence of tidal sand ridges.

Consider then Eq. (29), making it dimensionless with the following scaling:

$$t \to \sigma t, \quad \zeta \to \zeta/\sigma, \quad x \to x\sigma//U, \quad h \to h/H,$$

where U is the amplitude of a rectilinear tidal current making an angle θ with the topographic wave number k, scaled as $k \to kU/\sigma$. The topography itself is represented by $h \exp(ikx)$. Introduce the time scale ratios:

$$\tau_1 = \frac{c_1 r}{H\sigma}, \quad \tau_2 = f/\sigma, \quad \tau_3 = \frac{c_2}{c_1} \frac{\tau_1}{\tau_2},$$

Then the Fourier transformed dimensionless version of Eq. (2) reads:

$$\frac{\partial}{\partial t}\zeta_1 + (ik \cos\theta \sin t + \tau_1)\zeta_1 \qquad (30)$$

$$= \tau_2 ikh \sin t \{\cos\theta + \tau_3 \sin\theta\} + c_3 hk^2 \sin\theta \cos\theta \sin^2 t$$

where we have neglected the self-induced horizontal dispersion term so as to stress that the response peaks hat are present in the final spectral results arise even in the absence of horizontal diffusion. Harmonic truncation now assumes that we substitute a harmonic series in the frequency domain for ζ, truncating the series after a number of terms. The simplest nontrivial truncation is only to retain the zeroth (residual) and first (basic tidal) harmonics, such that:

$$\zeta = \zeta_1^{(0)} + \zeta_1^{(1)} e^{it} + \zeta_1^{*(1)} e^{-it} \qquad (31)$$

neglecting all higher harmonics that are in fact generated in Eq. (30) as well. The *old* solution is given by considering only the first driving term in the r.h.s. of Eq. (30). Substituting Eq. (31) in Eq. (30), the result for the residual vorticity reads:

$$\zeta_1^{(0)} = \frac{\tau_2 f(\theta)\delta(\tau_1)hk^2\cos^2\theta}{1+\delta(\tau_1)k^2\cos^2\theta} \qquad (32)$$

where $f(\theta) = 1 + \tau_3 + g\theta$ and $\delta(\tau_1) = \frac{1}{2}(1+\tau_1^2)^{-1}$. As the residual velocity is given by $\zeta_1^{(1)} k^{-1}$, it is easy to see from Eq. (32) that it has the following asympotic behaviour:

$$\left.\begin{array}{l} v_1^{(0)} \to kh, \quad k \to 0 \\ v_1^{(0)} \to k^{-1}h, \quad k \to \infty \end{array}\right\} \qquad (33)$$

which is the result mentioned before, suggesting the response peak around $k = \mathcal{O}(1)$. The same method can now be applied to the second driving term in the r.h.s. of Eq. (3) describing the effect of vortex tilting. The result then reads:

$$\zeta_1^{(0)} = \frac{c_3}{\tau_1} h \frac{k^2 \sin\theta \cos\theta}{2(1 + \delta(\tau_1)k^2\cos^2\theta)} \tag{34}$$

This evidently has the same asymptotics for $v_1^{(0)}$ as in Eq. (33). We thus see that the additional rectificatioan mechanism may alter the *strength* of the residual circulation as calculated from Eq. (1), but *not its spatial structure*.

6. Conclusion

Vertically averaged 3D models of topographic tidal rectification may show residual currents along bottom contours that are different in strength from those obtained by a genuine 2D model. The difference can be explained by vertical tilt of horizontal vortex lines a mechanism that is absent in the usual 2D formulations. Moreover vertical diffusion of vertical vorticity, combined with a vertical shear of the tidal current, induces an effective horizontal dispersion process that tends to smear out any horizontal residual current velocity profile. This mechanism, however, is quantitatively unimportant. Because of the fact that the spectral response function of the residual velocity profile for a fluctuating seabed has the same shape, both for production by vorticity advection as for production by tilting of vortex lines, the overall shape of the residual current velocity field over a specific topography is not affected by this additional mechanism.

Acknowledgements

This work was started during a stay at the Dept. of Oceanography, Dalhousie University, Halifax. The stay has been made possible by a Strategic Grant from the Natural Sciences and Engineering Research Council of Canada to Prof. C.J.R. Garrett. I am very grateful to him for inviting me and to Dan Wright and John Loder of Bedford Institute of Oceanography for showing me a first draft of their manuscript and for subsequent discussions. I thank Prof. C.B. Vreugdenhil of the Delft Hydraulics Laboratory for a discussion on the derivation of the vertically averaged vorticity equation removing a mistake in an earlier draft of the manuscript.

References

BOWDEN, K.F. (1965) Horizontal mixing in the sea due to a shearing current. Journal of Fluid Mechanics, 21, 83-95.

FLOKSTRA, C. (1976). Generation of two-dimensional horizontal secondary currents. Research Report S163 II, Delft Hydraulics Laboratory, 1-33.

FLOKSTRA, C. (1977). The closure problem for depth-averaged two-dimensional flow. Proceedings 17th International Association Hydraulic Research, Baden-Baden, 2, 247-256.

GARRETT, C.J.R. (1983). Coastal dynamics, mixing and fronts. In: P.G. BREWER (ed.), *Oceanography: the present and the future,* Springer Verlag, Heidelberg, 69-86.

HEAPS, N.S. (1978). Linearized vertically integrated equations for residual circulation in the coastal seas. Deutsche Hydrographische Zeitschrift, 31, 147-169.

HUTHNANCE, J.M. (1981). On mass transports generated by tides and long waves. Journal of Fluid Mechanics, 102, 367-387.

IANIELLO, J.P. Tidally induced residual currents in estuaries of constant breadth and depth. Journal of Marine Research 35, 755-786.

KOMEN, G.J. and H.W. RIEPMA (1981),. The generation of residual vorticity by the combined action of wind and bottom topography in a shallow sea. Oceanologica Acta, 4, 267-277.

LODER, J.W. (1980). Topogarphic rectification of tidal currents on the sides of Georges Bank. Journal of Physical Oceanography, 10, 1399-1416.

OKUBO, A. (1967). The effect of shear in an oscillatory current on horizontal diffusion from an instanteneous source. International Journal of Oceanography and Limnology, 1, 194-204.

ROBINSON, I.S. (1983). Tidally induced residual flow. In: B. JOHNS (ed.), *Physical Oceanography of coastal and shelf seas.* Elsevier, Amsterdam, 321-356.

SVERDRUP, H.U. (127). Dynamics of tides on the North Siberian shelf. Geofysika Publikationer, 4, (5), 1-75.

TEE, K.T. (1980). The structure of three-dimensional tide induced currents, II: residual currents. Journal of Physical Oceanography, 10, 2035-2057.

TEMME, N.M. and J.T.F. ZIMMERMAN (1985). On the mathematical theory of tidal rectification. SIAM Journal of Applied Mathematics (submitted).

WRIGHT, D.G. and J.W. LODER (1985). A depth dependent study of the topographic rectification of tidal currents. Geophysical and Astrophysical Fluid Dynamics 31, 169-220.

YASUDA, H. (1981). Tide-induced vertical circulation in a bay with homogeneous water: theory and experiment. Journal of the Oceanographical Society of Japan, 37, 74-86.

Yasuda, H. (1983). HORIZONTAL CIRCULATIONS CAUSED BY THE BOTTOM OSCILLATORY BOUNDARY LAYER IN A BAY WITH A SLOPING BED. JOURNAL OF THE OCEANOGRAPHICAL SOCIETY OF JAPAN, 40, 123-134.

ZIMMERMAN, J.T.F. (1978). TOPOGRAPHIC GENERATION OF RESIDUAL CIRCULATION BY OSCILLATORY (TIDAL) CURRENTS. GEOPHYSICAL ASTROPHYSICAL FLUID DYNAMICS, 11, 35-47.

ZIMMERMAN, J.T.F. (1980). VORTICITY TRANSFER BY TIDAL CURRENTS OVER AN IRREGULAR TOPOGRAPHY. JOURNAL OF MARINE RESEARCH, 38, 601-630.

ZIMMERMAN, J.T.F. (1981). DYNAMICS, DIFFUSION AND GEOMORPHOLOGICAL SIGNIFICANCE OF TIDAL RESIDUAL EDDIES. NATURE, 290, 549-555.

IV

LOW-FREQUENCY MOTIONS

Subtidal Current Variability in the Lower Hudson Estuary

Robert E. Wilson

Ronald J. Filadelfo

Marine Sciences Research Center
State University of New York, USA

ABSTRACT

Current records approximately 40 days in length from a single mooring in the lower Hudson River Estuary are analyzed through frequency domain regression to determine the response to coastal sea level fluctuations, local wind stress, and fresh water inflow. Data from both an early spring period characterized by high river inflow and moderately intense meteorological activity and a late summer period characterized by low river inflow and weak meteorological activity are considered. For the spring period, variability at periods longer than 5 days was associated with a baroclinic response to river forcing, and at periods shorter than 5 days primarily with a depth independent response to coastal sea level fluctuations although a depth dependent response to local wind stress did account for an appreciable fraction of current variance. For the late summer period current spectra showed a marked spectral peak at 2.5 days associated primarily with a depth independent response to coastal sea level fluctuations.

1. Introduction

This study arose initially out of our interest in subtidal sea level variability in the lower Hudson Estuary (Fig. 1). We were especially interested in determining the relative importance of the strong coastal sea level fluctuations in the New York Bight and local wind stress and river inflow in producing subtidal velocity fluctuations in the lower estuary. Subtidal velocity fluctuations in the lower estuary are of considerable importance for exchange between compartments of the estuary and for exchange between the estuary and the Apex of the New York Bight because their standard deviation often exceeds the mean.

Analyses of subtidal sea level fluctuations within the New York Bight by WANG (1979) have shown that sea level at Sandy Hook (Fig. 1) exhibits the highest modal amplitude of all coastal stations. For the one year records (1975) analyzed in that study the dominant sea level fluctuations occurred at time scales of 4 days and most of the sea level change was driven by alongshore winds. Wang showed also that there were large seasonal variations in sea level fluctuations at Sandy Hook. Maximum variance occurred in March and the minimum in August; the total variance differed by more than a factor of five. These seasonal variations in sea level were for the most part associated with variations in alongshore winds.

WILSON et al. (1985) showed that sea level within the lower estuary was highly coherent with coastal sea level. Using a simple two input frequency domain regression model for Battery sea level (representative of sea level in the lower estuary) forced by longitudinal wind stress and coastal sea level, they showed that coastal sea level made the dominant contribution to the sea level variance in the estuary with minor contributions by the local wind stress and its cross spectrum with coastal sea level. Wilson et al. also found that sea level variance decreased by a factor of approximately 5 from winter

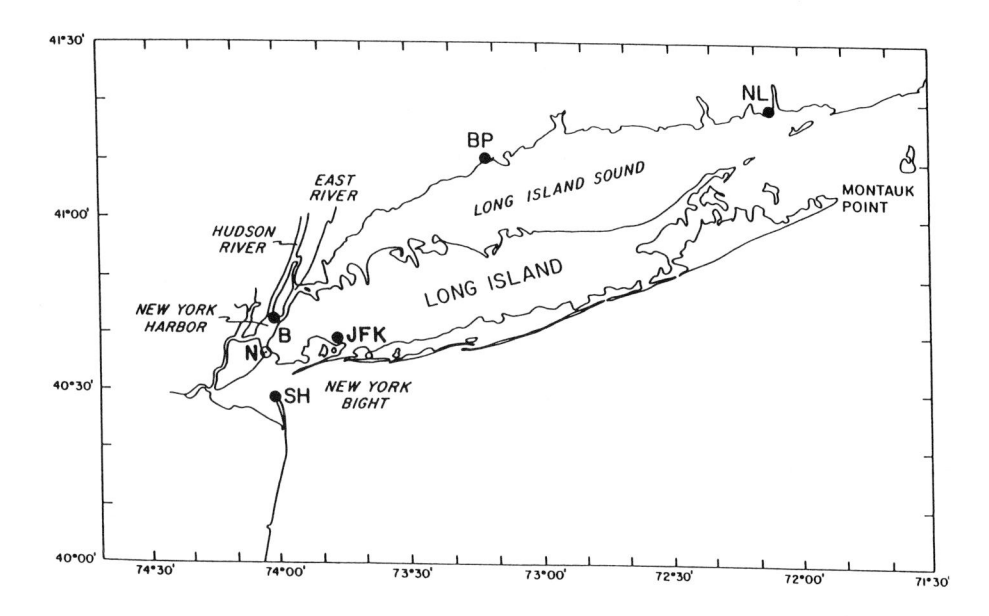

FIGURE 1. Hudson Estuary, Long Island Sound and New York Bight area. Sea level stations Sandy Hook, NJ (SH) and Battery, NY (B). Meteorological stations Bridgeport, CT (BP) and John F. Kennedy Airport (JFK). Current mooring is at the Narrows (N).

to summer, and that local wind forcing became relatively more important in summer. The characteristics of the cross spectrum between local wind stress and coastal sea level led to a decrease in the variance at Battery relative to that at Sandy Hook, especially in summer.

The present study addresses the following specific questions:

1) Do we find the expected depth independent current response to coastal sea level forcing, and depth dependent response to longitudinal wind forcing and river inflow?
2) What fraction of current variance is represented by these two modes of response?
3) Can we separate the influence of coastal sea level fluctuations, local winds and river forcing on currents at a given depth?
4) How do our results depend on frequency and season?

Our approach has been to examine current velocity time series for the Narrows and concurrent sea level, surface wind stress and river inflow time series. Relationships between these series are first established through simple time domain Empirical Orthogonal Function (EOF) analysis and then more definitively through frequency domain regression.

2. Data sets

Current velocity time series were obtained from a mooring in the Narrows deployed during the 1980-1981 National Ocean Service Circulatory Survey of New York Harbor. This survey provided the most complete collection to date of long term current moorings

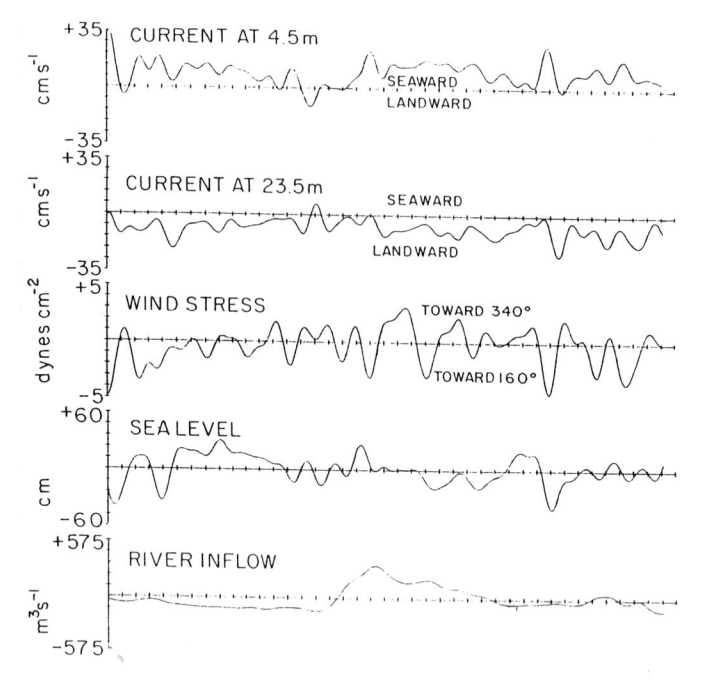

FIGURE 2. Low pass filtered series for (from top to bottom) principal axis component of current at 4.5m (ordinate ranges from -35 to +35 cm s^{-1}); principal axis component at current at 23.5m (ordinate ranges from -35 to +35 cms^{-1}); along-channel wind stress component (ordinate ranges from -5 to +5 dynes cm^{-2}); SH sea level (ordinate ranges from -60 to +60 cm); fluctuations in Hudson River inflow gauged at Green Island, NY (ordinate ranges from -575 to +575 m^{3}s^{-1}). Origin for abscissa is 0000 hr 03/14/81; tick marks represent 1 day.

within the estuary; the Narrows mooring was chosen for the present analyses because of its strategic position. The complete current velocity data set from the 1980-1981 survey is described by FILADELFO (1984). We confine our attention to a late spring - early summer period of 10 March to 30 April 1981 and a late summer - early fall period of 22 September to 15 November 1980 for which unbroken records were available for at least two depths. These records were augmented with concurrent series for sea level at Sandy Hook and BatteryB, surface wind stress estimated from wind records from the National Weather Service Stations at either John F. Kennedy Airport or Bridgeport, CT and river inflow records from the USGS flow gauge on the upper Hudson at Green Island, NY.

Fig. 2 shows filtered series for the 1981 deployment. Current velocity is presented as scalar series for the principal axis (alongchannel) component (positive seaward) for depths of approximately 4.5 m and 23.5 m below MLW; the MLW channel depth is approximately 26 m in the vicinity of the mooring. The surface wind stress component presented is directed along 340°T - 160°T and unlike the current is taken positive towards 340°T which is up the estuary. Fig. 3 shows similar series for the 1980 deployment for which current velocity measurements were available for 4.5m and 17m below MLW. A river inflow series is not presented because its variance is extremely low. With the exception of the river inflow the filtered series in Fig. 2 were obtained using a Lanczos filter with a half power point at 34 hours. The series for river inflow in Fig. 2 represents mean daily discharge values.

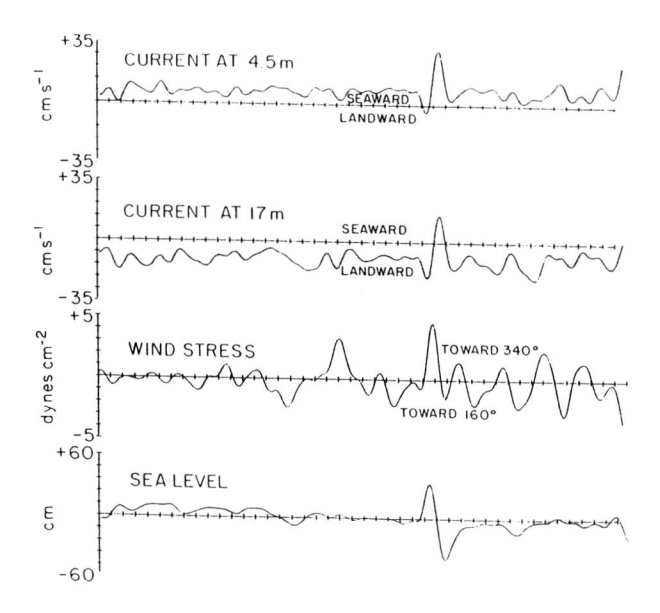

FIGURE 3. Low pass filtered series for (from top to bottom) principal axis component of current at 4.5m (ordinate ranges from -35 to +35 cm s^{-1}); principal axis component at current at 17m (ordinate ranges from -35 to +35 cm s^{-1}); alongchannel windstress component (ordinate ranges from -5 to +5 dynes cm^{-2}); SH sea level (ordinate ranges from -60 to +60 cm). Origin for abscissa is 0000 hr 09/22/80; tick marks represent 1 day.

3. Statistical analyses

To address questions 1 and 2 posed in the introduction, we first applied time domain EOF analysis to the series in Fig. 2 and 3 to determine the fraction of variance accounted for by depth independent and depth dependent modes of current response, and the relationships between each mode to coastal sea level, local wind stress and river inflow fluctuations. Time domain EOF methods are described in more detail bu KUNDU and ALLEN (1976). These methods cannot accomodate phase shifts between series nor can they easily describe frequency dependent relationships.

In order to separate the influence of coastal sea level, local wind stress and river inflow on currents at a given depth (question 3) and to establish the frequency dependence of these relationships, we elected to use frequency domain regression. This is the frequency domain analog to time domain regression for a multiple input - single output model. This technique can accomodate coherent inputs which is important in our case because local longitudinal wind and coastal sea level (forced primarily by alongshore winds) are coherent.

A possible alternative statistical technique would be frequency domain EOF. This method can accommodate phase shifts between series and it provides for response functions between individual series and each modal series, but it does not provide the definitive amplitude and phase relationships between forcing functions and current provided by the frequency domain regression.

4. Results of EOF analyses

In this section we provide a summary of results of the EOF analyses which provide insight into the partitioning of variance between depth dependent and depth independent modes of response for both the early spring and late summer deployments. Before proceeding, however, we note that current spectra for the March deployment (Fig. 4) show considerable variance at long periods (periods longer than 5 days) which is associated primarily with forcing by long period fluctuations shorter than 5 days, which is associated primarily with forcing by long period fluctuations in river inflow. This is in addition to spectral peaks at periods shorter than 5 days. Current spectra for the September deployment (Fig. 5) show little variance at periods longer than 5 days and a marked spectral peak at approximately 2.5 days. To accomodate different relationships between series in different frequency bands for the March series, we applied time domain EOF analysis to low low pass filtered series (periods longer than 5 days) and band pass filtered series (periods between 5 days and 1.6 days) separately.

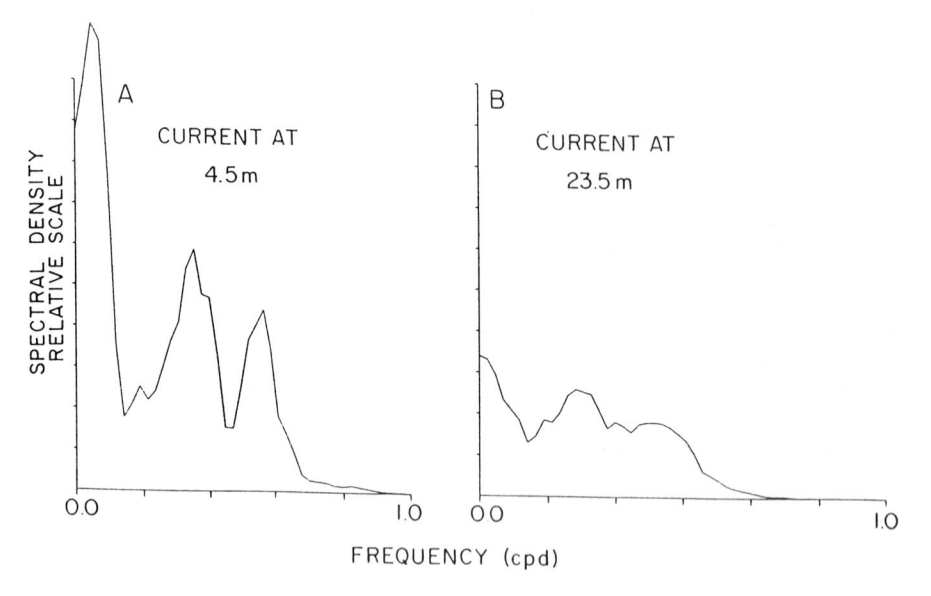

FIGURE 4. Auto spectra for the principal axis component for current velocity in Figure 2 at 4.5m (A) and 23.5m (B). Ordinate is relative spectral density on a linear scale; abscissa is frequency on a linear scale from 0 to 1 cycle per day. Degrees of freedom 18 and band width .18 cpd.

For the March 1981 deployment for frequencies lower than 0.2 cycles per day, the first EOF mode accounted for 79% of the variance in the river inflow, 64% of the variance in current at 4.5m depth and only 4% of the variance in current at 23.5m. This suggests that river inflow affects the upper water column only; an increase in river inflow corresponds to enhanced outflow which is confined to the upper water column. The second EOF mode accounts for 55% of the variance in longitudinal wind stress, 24% of the variance in current at 4.5m and 85% of the current at 23.5m. A wind stress towards 340°T produces an up estuary flow in the upper water column and a seaward flow at depth in response to the pressure gradient set up by the wind. At periods longer than 5 days both river inflow and local wind forcing is important. The upper water column responds most strongly to river forcing; most of the current variance at depth is wind forced.

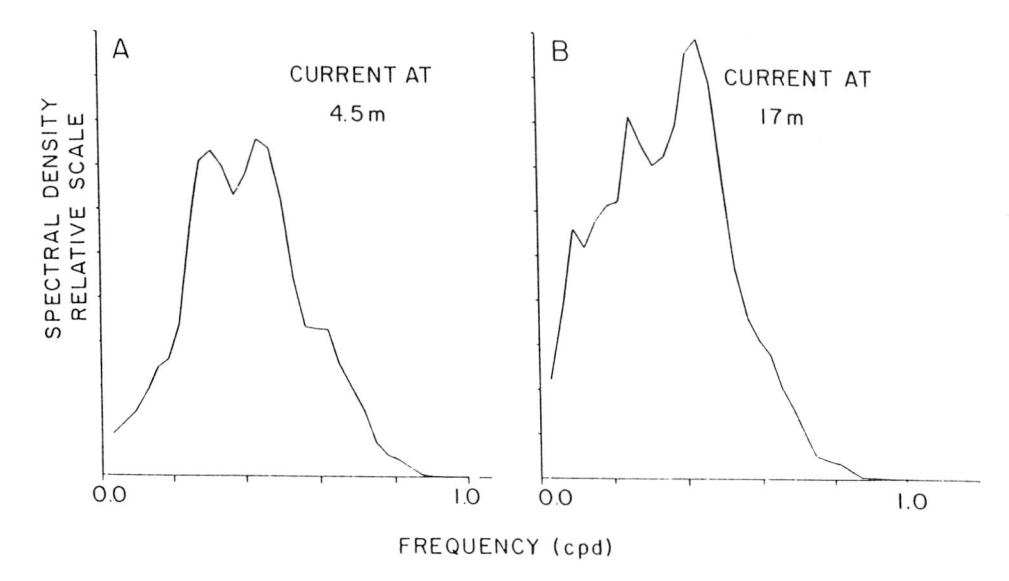

FIGURE 5. Auto spectra for the principal axis component for current velocity in Figure 3 at 4.5m (A) and 17m (B). Ordinate is relative spectral density on a linear scale; abscissa is frequency on a linear scale from 0 to 1 cycles per day. Degress of freedom 18 and band width .18 cpd.

For frequencies between 0.2 and 0.6 cycles per day the first EOF mode represented depth independent velocity fluctuations forced by coastal sea level fluctuations. It accounted for 74% of the current variance at 4.5m, 64% of the current variance at 23.5m and 79% of the variance in the time rate of change in sea level. A drop in coastal sea level produces a seaward flow at all depths. This mode also accounted for 67% of the variance in local longitudinal wind stress emphasizing the coherence between longitudinal wind stress and coastal sea level mentioned earlier. The second mode accounted for 10% of the current variance at 4.5m, 28% of the variance at 23.5m and 23% of the variance in longitudinal wind stress. It describes a depth dependent response to local wind forcing.

For the September 1980 deployment EOF analysis resolved one very strong mode of response: a depth independent response to coastal sea level fluctuations. This mode accounted for 84% of the variance in current at 4.5m, 81% of the variance at 17m and 89% of the variance in the time rate of change in sea level. Recall that the variance in current for this deployment was narrow banded and centered on a period of 2.5 days.

5. Results of frequency domain regression analyses

Frequency domain regression provides more definitive amplitude and phase relationships between current fluctuations and coastal sea level, local wind stress and river inflow. The technique is described in detail in Filadelfo (1984). Very basically, frequency domain regression examines the simple linear model

$$\hat{u} = h_1 h\hat{\eta} + h_2 \hat{\tau} + h_3 \hat{Q} + \hat{\epsilon} \qquad (1)$$

where $\hat{u}, \hat{\eta}, \hat{\tau}$, and Q are the Fourier transforms of the current, coastal sea level, longitudinal wind stress, and river inflow, respectively; $\hat{\epsilon}$ represents an incoherent residual and h_1, h_2 and h_3 are complex frequency response functions.

Response functions describing the amplitude and phase relationships between the currents at both 4.5m and 23.5m depth and each of the 3 input series for the March 1981 deployment are shown in Fig. 6. 75% confidence intervals are shown for the amplitude only although they can be determined for the phase as well. It is important to note that when the confidence intervals on the amplitude begin to diverge, the phase becomes unreliable and the relationship between the series becomes poorly defined.

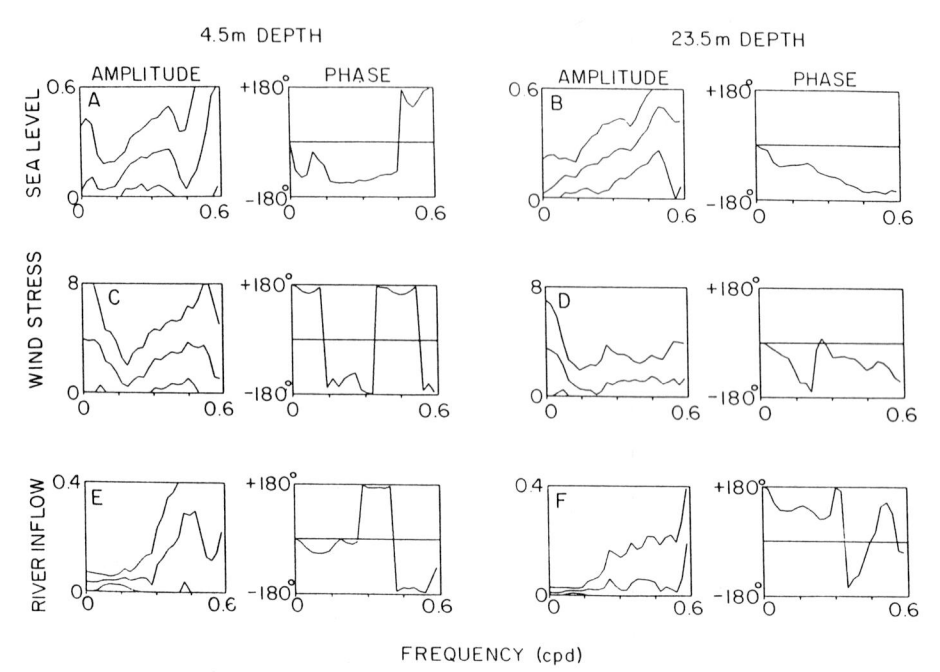

FIGURE 6. Partial response functions (amplitude and phase) for the currents at 4.5m and 23.5m in Figure 2 to SH sea level (A) and (B) (amplitude ordinate ranges from 0 to 0.6 (cm s^{-1}) m^{-1}); to alongchannel wind stress (C) and (D) (amplitude ordinate ranges from 0 to 8 (cm s^{-1}) dyne cm^{-2}); to fluctuations in Green Island discharge (E) and (F) (amplitude ordinate ranges from 0 to .4 (cm s^{-1}) m^{3}s^{-1}). Common phase ordinates range from -180 to +180 degrees. Abscissae are frequency from 0 to .6 cpd. 75% confidence intervals are provided for amplitude estimates.

Response functions between currents at 4.5m and 23.5m and coastal sea level (Fig. 6a and 6b) show a well defined relationship between periods of approximately 6 days and 2 days. For these periods the amplitude of the response function is approximately the same for currents at both depths and the phase is approximately 90°, suggesting that forcing by coastal sea level produces a nearly depth independent response with the characteristics of a standing wave.

Response functions describing the relationship between currents at both depths and along-channel wind stress (Fig. 6c and 6d) show clearly that the confidence intervals on the amplitude for currents at 23.5m depth converge at longer periods and tend to be narrower than confidence intervals for currents at 4.5m depth. The phase between wind stress and surface currents tends to be 180° while that at depth tends to be closer to 0°; the aberation in the phase in Fig. 6d for periods near 2 days occurs at periods where the amplitude (and thereby the phase) is poorly defined. The response to along-channel wind stress is strongly depth dependent and better defined for currents at 23.5m depth than at 4.5m; it is characteristic of a simple setup response.

Response functions for currents at 4.5m and 23.5m to river inflow (Fig. 6e and 6f) show clearly that the amplitude for currents at 4.5m is substantially higher than that for currents at 23.5m and that currents at 4.5m tend to be in phase with the river inflow. Time lags can be determined from these diagrams as a function of frequency; the time lag is simply the phase divided by the frequency. Confidence intervals on the amplitude at both depths diverge for periods shorter than 6 days suggesting that the relationship becomes poorly defined.

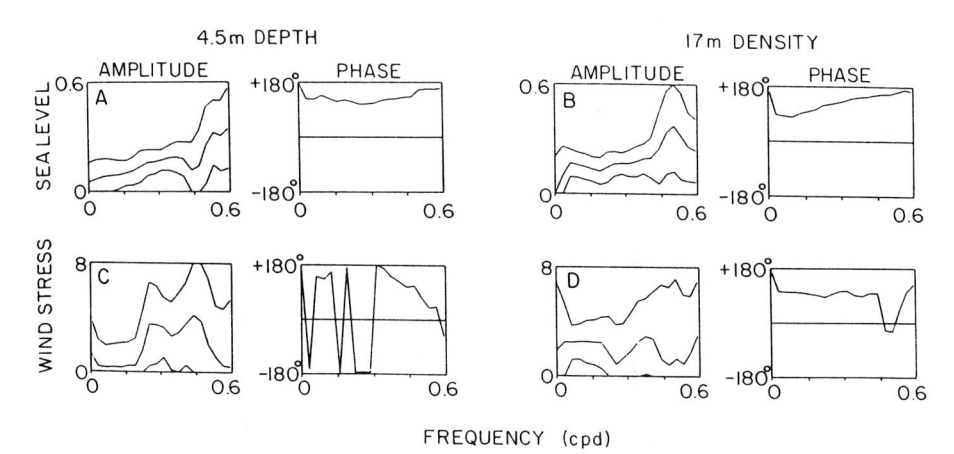

FIGURE 7. Partial response functions (amplitude and phase) for the currents at 4.5m and 17m in Figure 3 to SH sea level (A) and (B) (amplitude ordinate ranges from 0 to 0.6 (cm s^{-1}) m^{-1}); to alongchannel wind stress (C) and (D) (amplitude ordinate ranges from 0 to 8 (cm s^{-1}) dyne cm^{-2}). Common phase ordinates range from -180 +180 degrees. Abscissae are frequency from 0 to .6 cpd. 75% confidence intervals are provided for amplitude estimates.

Response functions between currents at 4.5m and 17m and coastal sea level for the September 1980 deployment are shown in Fig. 7a and 7b. The confidence intervals on the amplitude are realtively narrow for periods as short as 2 days. The phase is approximately 90 degrees at both depths. These amplitude and phase relationships are consistent with those of a nearly depth independent standing wave response. Response functions between currents and alongchannel wind (Fig. 7c and 7d) show that the confidence intervals on the amplitude remain relatively broad suggesting that direct wind forcing is not effective during this period.

These multiple input models can be used to examine the details of the relative contributions to the variance in current made by each input as a function of frequency. More specifically, multiplication of Eq. (1) by its complex conjugate yields an expression for current variance in terms of the spectra and cross spectra of the inputs. For the three input model used for the March 1981 time series we have

$$<\hat{u}\hat{u}\ast> = <h_1 h_1\ast><\hat{\eta}\hat{\eta}\ast> + <h_2 h_2\ast><\hat{\tau}\hat{\tau}\ast> + <h_3 h_3\ast><\hat{Q}\hat{Q}\ast>$$
$$+ 2R_e<h_1 h_2\ast><\hat{\eta}\hat{\tau}\ast> \qquad (2)$$
$$+ 2R_e<h_1 h_3\ast><\hat{\eta}\hat{Q}\ast>$$
$$2R_e<h_2 h_3\ast><\hat{\tau}\hat{Q}\ast> + <\hat{\epsilon}\hat{\epsilon}\ast>$$

Fig. 8 shows the contributions made by the auto spectra for wind stress, coastal sea level, and fresh water inflow to the total variance in current at 4.5m depth for the March 1981 deployment. The major contribution to the variance at periods longer than 5 days

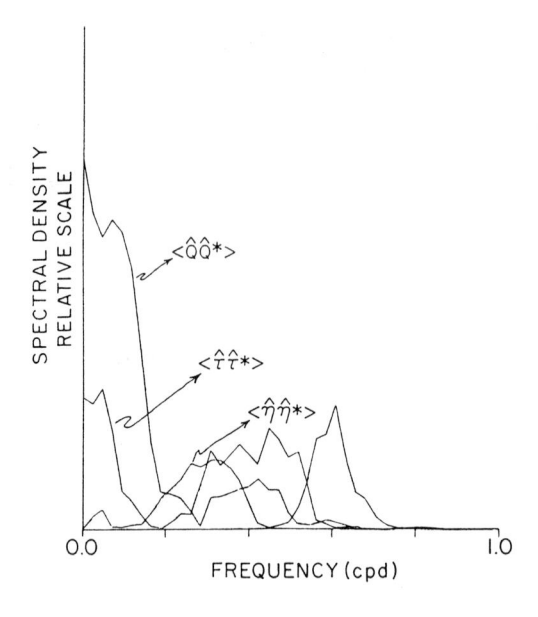

FIGURE 8. Spectral model estimates for the relative contributions to the variance in current at 4.5m for the March 1981 deployment made by the auto spectra for coastal sea level $<\hat{\eta}\hat{\eta}*>$, alongchannel wind stress $<\hat{\tau}\hat{\tau}*>$ and river inflow $<\hat{Q}\hat{Q}*>$. Cross spectral contributions are not shown.

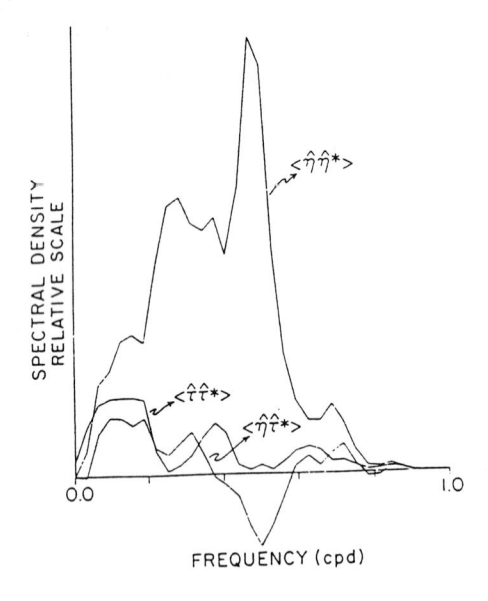

FIGURE 9. Spectral model estimates for the relative contributions to the variance in current at 17m for the September 1980 deployment made by the auto sepctra for coastal sea level $<\hat{\eta}\hat{\eta}*>$ and alongchannel wind stress $<\hat{\tau}\hat{\tau}*>$. Cross spectral contributions are not shown.

is by the auto spectrum for river inflow. At shorter periods, the wind stress spectrum and coastal sea level spectrum both make significant contributions. The contribution by the

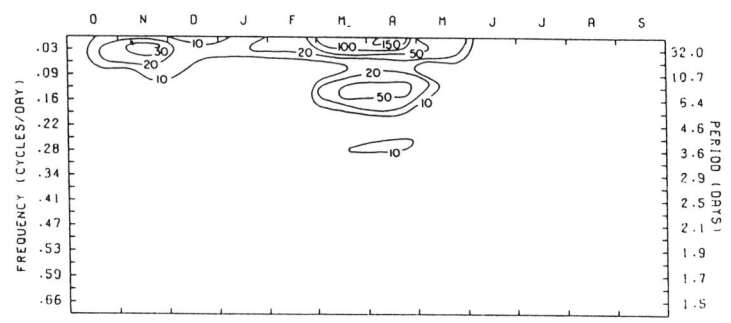

FIGURE 10. Seasonal spectra for fluctuations in Hudson River inflow gauged at Green Island, NY. Contours are in $(m^3 s^{-1})^2$ cpd^{-1} × 10^4. Degrees of freedom 9.6.

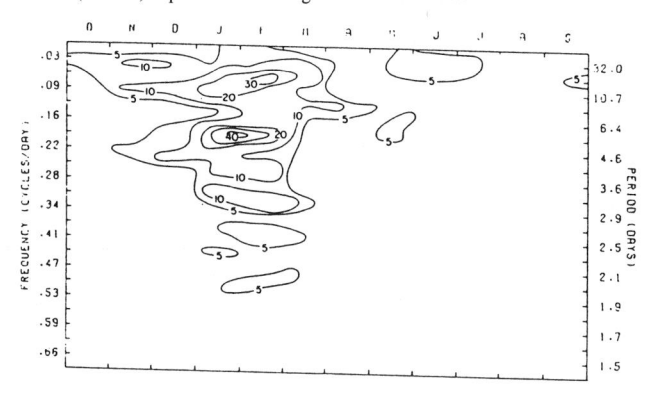

FIGURE 11. Seasonal spectra for fluctuations in SH sea level. Contours are in cm^2 cpd^{-1} × 10^2. Degrees of freedom 9.6.

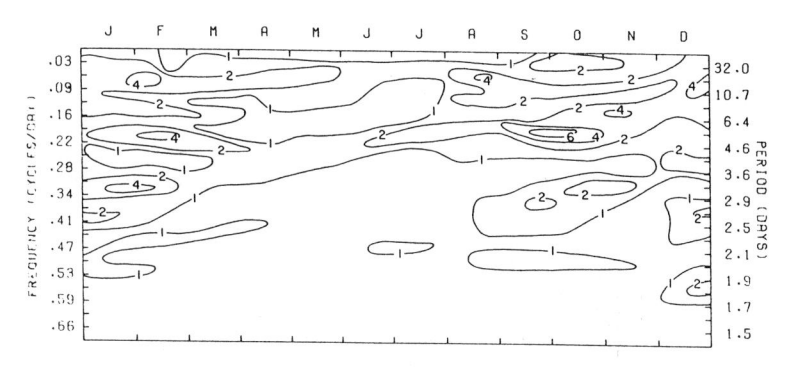

FIGURE 12. Seasonal spectra for fluctuations in longitudinal wind stress estimated from winds at BP. Contours are in (dyne cm^{-2})^2 cpd^{-1}. Degrees of freedom 9.6.

cross spectrum between wind stress and coastal sea level (not shown) is significant at periods near 2 days. At 23.5m depth, the major contribution to current variance at long periods is made by the wind stress spectrum; at shorter periods the coastal sea level spectrum makes the major contribution.

Similarly, spectral models obtained using the two input regressions for the September 1980 time series show the dominant contribution by the spectrum of coastal sea level to the current variance at both 4.5m and 17m depth (Fig. 9). For currents at 4.5m, there is also a secondary contribution by the cross spectrum between wind stress and coastal sea level at periods near 3 days.

In order to put the results for these two current deployments (22 September to 15 November1980 and 10 March to 30 April 1981) in some climatological perspective, it is useful to examine the seasonal spectra for the forcing functions: river inflow, coastal sea level and longitudinal wind stress (Fig. 10, 11 and 12). These were computed for the four year period 1968 through 1972 and then ensemble averaged. This period was chosen because it was one for which relatively long unbroken sea level records were available. The degrees of freedom indicated for Figs. 10, 11 and 12 is the result of ensemble averaging and some band averaging. The river spectrum shows a minor peak between November and December. This is followed by a period of low variance and then a strong spectral peak in April - May. The coastal sea level and longitudinal wind stress spectra exhibit some variance in late fall and a very strong spectral peak between late December and February. This is a period of relatively low variance in river inflow.

It is clear from these figures that neither of our deployments encompassed a period of minimum variance in either river inflow or meteorological forcing as represented by coastal sea level and longitudinal wind stress.

6. Summary and conclusions

The complex current response in the lower estuary can be effectively described through frequency domain regression. The response functions obtained from two separate deployments suggest that processes are somewhat stationary.

Seasonal spectra for the forcing functions can lead to a climatology for current response although cross spectra should be described as well.

Acknowledgements

We are pleased to acknowledge the support of NOAA under Contract #NA81 RAH 00002 and the Hudson River Foundation under Grant #1083A47.

References

FILADELFO, R.J. 1984. Subtidal sea level and current variability in the Hudson Raritan Estuary. Ph.D. thesis, 183 pp., State University of New York, Stony Brook, NY.

KUNDU, P.K. & J.S. ALLEN. 1976. Some three dimensional characteristics of low frequency current fluctuations near the Oregon coast. Journal of Physical Oceanography, 6, 181-199.

WANG, D-P. 1979. Low frequency sea level variability in the Middle Atlantic Bight. Journal of Marine Research, 37, 683-697.

WILSON, R.E., K-C. WONG & R.J. FILADELFO. 1985. Low frequency sea level variability in the vicinity of the East River Tidal Strait. Journal of Geophysical Research, 90, 954-960.

Subtidal Exchanges between Corpus Christi Bay and Texas Inner Shelf Waters

Ned. P. Smith

Harbour Branch Foundation, Inc.
U.S.A.

ABSTRACT

Air pressure and water level data from the Texas coast, along the northwestern rim of the Gulf of Mexico, are used to infer surface windstress and investigate subtidal, meteorological forcing in a cause-and-effect sense. A 49-day study period in 1972 is selected to investigate the response of shelf waters to windstress forcing. The cross-shelf component of windstress, as inferred from the surface pressure gradient, forces bay-shelf exchanges over time scales of 2-4 days; the longshore component of windstress is coherent with bay-shelf exchanges over longer time scales, suggesting a cross-shelf EKMAN transport. A 196-day study in 1975 documents inverse barometric forcing over time scales in excess of about two days. A one-dimensional numerical model simulates the tidal and nontidal influx of Gulf water into Corpus Christi Bay as a response to coastal water level variations. Results suggest that tidal exchanges dominate subtidal exchanges in this coastal setting. Tidal forcing pumps Gulf water into the Bay at an average rate of 17 million cubic meters per day; nontidal forcing imports water to the Bay at an average rate of approximately 14 million cubic meters per day.

1. Introduction

The exchange of water between an estuary and the adjacent inner continental shelf, over any time scale, is important not only to maintain water quality, but also for importing and exporting planktonic plant and animal life. Therefore, exchanges are of central importance in the ecology of the coastal zone. Mass transport between an estuary and the inner continental shelf occurs in two forms. Fresh water, arriving from the drainage basin, follows a one-way path through the estuary and onto the continental shelf. In most estuarine settings, this is superimposed onto a two-way exchange of water which occurs in response to the rise and fall of coastal sea level.

To the extent that the water level in the estuary depends upon the time history of coastal water levels, the exchange process lends itself readily to modelling. A one-dimensional, barotropic model needs only a coastal water level record to drive the calculations, and only enough estuarine water level data to verify the simulations. Insofar as one is able to decompose the coastal water level record into its constituent parts, one can then manipulate the input time series to investigate the magnitudes and relative importances of the individual physical processes driving estuarine-shelf exchanges.

Along most coastlines, the tide plays the primary role and provides a baseline value of sorts. In many situations, however, forcing at subtidal frequencies enhances tidal exchanges substantially and thus deserves attention. MONTGOMERY (1937) among others has listed eight causes of long-period fluctuations in sea level. Discounting the long-period tides, in most physical settings this list can be reduced to four entries:

(a) dynamic effects associated with currents along or across the shelf;
(b) wind effects on shelf waters;

(c) inverse barometric forcing; and

(d) thermal effects (both local and advective gains and losses).

The time scales associated with these processes characteristically range from a few hours to a year; the relative importance of the processes varies with location to such an extent that generalizations are inappropriate.

This paper describes an attempt to assess the importance of subtidal estuarine-shelf exchanges, relative to tidal processes, along the Gulf coast of South Texas. Results from field studies conducted in Corpus Christi Bay indicate that windstress forcing is important in this setting; the inverse barometer effect plays a decidedly secondary role. A one-dimensional numerical model suggests that subtidal forcing accounts for approximately 40% of the total influx of Gulf water into Corpus Christi Bay.

2. Data and methods of analysis

Water level records from a 49-day period of time (January 18 through March 8, 1972), and from a later 196-day period of time (January 1 through July 15, 1975) at the Aransas Pass, on the Texas Gulf coast (Fig. 1), form the primary data base for the investigation of sub-tidal coastal water level variations and bay-shelf exchanges. Analog data were obtained with a Stevens Model A recorder. Water levels were digitized hourly to the nearest 3 mm (0.01 foot); time checks were put on the chart at approximately monthly intervals.

Wind effects were investigated by using the surface pressure gradient, as determined from measurements at three locations forming a triangle which enclosed the study area (Fig. 1). SCHWING and BLANTON (1984) have discussed the need to adjust winds measured at inland stations to represent over-water conditions better. RICHARDS et al. (1966) have presented tabular data, and HSU (1981) has suggested a simplified expression to accomplish this when wind speeds are in excess of 2 m/sec. Neither approach, however, is satisfactory when the wind speeds recorded inland are damped significantly by a radiation inversion, for example, which does not exist in the lowest layers of the marine atmosphere. The spatial coherence of the surface wind field has not been examined in detail for the Texas Gulf coast region. However EIGSTI (1978) and YU and WAGNER (1970) have compared both instantaneous winds and monthly means measured at overwater, coastal and inland weather stations. Coastal and over-water wind speeds were found to be significantly higher, due in part to a midnight maximum in surface wind speed which occurred at a time when inland wind speeds were experiencing a minimum in the diurnal cycle. In this study, neither coastal nor over-water wind measurements were available from the study area. The approach therefore was use to surface pressure gradient to infer the magnitude and direction of the geostrophic wind, to assume that the surface wind was 70% of the geostrophic wind speed (BAKUN, 1975), and to assume that the cross-isobaric flow was a constant 30°.

Surface air pressure data, recorded during the 1975 study period at the International Airports in Corpus Christi, Texas, and Tampa, Florida (Fig. 2), were used to investigate the importance of the inverse barometer effect on coastal sea level records. Surface pressures were read to the nearest 0.1 millibar every three hours. Hourly air pressures were obtained for comparison with the water level record by fitting a natural cubic spline function through the available data and then reading hourly values from the curve. With a second water level record from Pass-a-Grille, the inverse barometer effect could be examined as a response to the cross-Gulf pressure difference:

$$\Delta \eta / \Delta x = -\rho g \Delta p / \Delta x.$$

This reduces to $\Delta \eta = -0.860 \Delta p$ when the density of the air, ρ, is assumed a constant

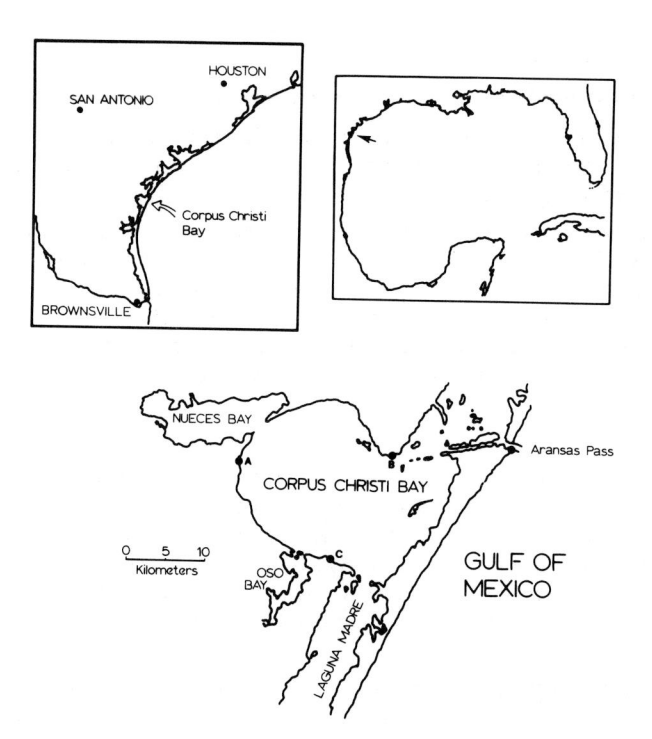

FIGURE 1. Tide gauge locations surrounding Corpus Christi Bay, Texas. Insert maps show the three weather stations at which surface pressure data were obtained (January 19 - March 8, 1972), and the study site in the northwestern Gulf of Mexico. [After SMITH (1977); used with permission].

1.188×10^{-3} gm / cm^3 (air temperature 20°C, air pressure 1000 mb).

Both tidal and subtidal exchanges through the Aransas Pass were quantified with a one-dimensional numerical model which simulated barotropic forcing. The model was conceptually similar to that described by MEHTA (1978). Exchanges were computed as a response to the pressure gradient associated with bay-shelf water level differences; frictional resistance to exchanges was modelled using a quadratic friction term:

$$\Delta U / \Delta t = g \, \Delta \eta / \Delta x - FU|U|.$$

The development and use of the model is described at length elsewhere (SMITH, in press). The friction term, F will have units of m^{-1} when the current speed, U, is in m/sec. A 49-day calibration period was used to compare bay volumes simulated using coastal water levels with bay volumes computed from water level records collected around the bay (see SMITH 1977). Although the model calculations were verified only with volumes of Corpus Christi Bay itself, the movement of water through Corpus Christi Bay and into adjacent sub-estuaries (Fig. 1) was incorporated into the model as well. Water level records from these adjacent bodies of water were used to verify inter-bay exchanges.

The relative importance of tidal and nontidal forcing in flushing Corpus Christi Bay

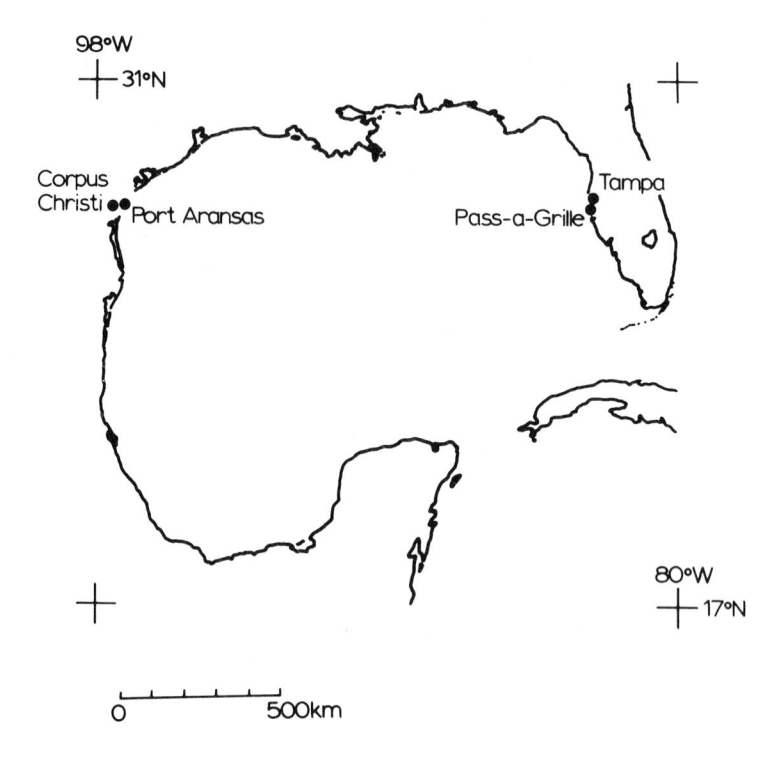

FIGURE 2. Sampling sites for surface pressure and water level data in the northwestern and eastern Gulf of Mexico, January 11 - July 15, 1975. [After SMITH (1979); used with permission].

was quantified by computing the influx of Gulf water into the Bay in response to variations in coastal sea level. Corrections were made for bay water returning to the bay at the start of the flood tide. This is an important refinement, because the channel connecting the bay with the Gulf contains an estimated 37.3 million cubic meters of water.

The 196-day water level record was decomposed into tidal and non-tidal components for analysis by the model. A least squares harmonic analysis of the data provided the harmonic constants of the principal tidal constituents; hourly water levels were then generated with a tidal prediction program (PORE & CUMMINGS 1967). The time series of purely tidal water level variations was used to quantify tidal exchanges. Subtidal variations in coastal sea level were obtained by passing the time series of total sea level variation at the coast through a 40-hour low-pass filter.

The time series of subtidal water level variations was modified further by removing the inverse barometer effect. This was done in two ways. First, the response to variations in the surface pressure field was assumed to occur as a local response, and air pressure data from the International Airport at Corpus Christi were used to adjust water levels recorded at Aransas Pass. Second, the cross-Gulf surface pressure difference were used to correct the Aransas Pass water level record. Pressure difference were first smoothed with a 40-hour low-pass filter. Earlier work (SMITH 1979) documented statistically significant coherences in cross-Gulf pressure difference and water level difference records over time scales in excess of 40 hours. It was assumed that 63% of the cross-Gulf pressure difference was due to local variations in surface pressure in the vicinity of Corpus Christi. During the 196-day study, the standard deviation of the surface pressures recorded at Corpus Christi was 5.7 mb, while the standard deviation of the Tampa pressures was 3.4 mb.

The time series from which both tidal effects and inverse barometric forcing had been removed could still reflect the effects of several unrelated physical processes. In view of the weather characteristics of the South Texas Gulf coast in winter months, however, it is probable that the time series is dominated by wind effects.

3. Results

Windstress forcing as a mechanism for producing subtidal bay-shelf exchanges was investigated by comparing components of the square of the surface pressure gradient vector with the volume of Corpus Christi Bay. The method for computing bay volumes from water level records from around the bay (see Fig. 1) has been described in detail in a previous paper (SMITH 1977). Components of the inferred surface windstress vector were compared with bay volumes to determine the time scales for which a coherent relationship existed. Results are shown in Fig. 3. Two regions of statistically significant coherence appear at periodicities normally associated with meteorological forcing. The first, computed with the 060-240° component of the surface pressure gradient, occurs over time scales of approximately 2-4 days. The second, more broadly distributed between the 120-300° component and the 180-000° component, is found over time scales in excess of about a week.

Any component of the pressure gradient can be translated into a corresponding component of the surface windstress vector, if one knows or can make an assumption regarding the cross-isobaric flow in the surface wind field. This will vary with surface roughness (wave height, and hence wind speed), and with the air-water temperature difference (stability or instability in the lowest layers of the atmosphere). Assuming cross-isobaric flow 30° to the left of the geostrophic wind, and noting the 033-213° orientation of the coastline at Aransas Pass, the 060-240° component of the pressure gradient represents the cross-shelf component of the surface windstress. By a similar argument, the 150-330° component of the surface pressure gradient can be taken to represent the longshore component of the surface windstress.

Results therefore suggest that bay volume changes are coherent with the set-up and set-down of coastal sea levels caused by variations in the cross-shelf component of the surface windstress over time scales of approximately 2-4 days. Similarly, bay volumes respond to variations in the longshore component of the surface windstress over time scales in excess of about seven days. The coupling mechanism in this latter case cannot be verified with the available data, but the relationship is consistent with the idea of a near-surface EKMAN transport across the shelf. A process involving the circulation of shelf waters would logically require longer time scales, and probably several inertial periods (the inertial period is 26.0 hours at latitude 27°37'N).

Inverse barometric forcing as a mechanism producing bay-shelf exchanges was investigated by comparing cross-Gulf water level differences with cross-Gulf surface air pressure differences. By considering this process in terms of spatial gradients, no corrections were made for temporal variations in pressure which might have been affecting the Gulf of Mexico as a whole, and thus producing only a negligible expansion or compression of the water column. The cross-Gulf approach restricts the analysis to the east-west component of the surface pressure field — the north-south component, and especially east-

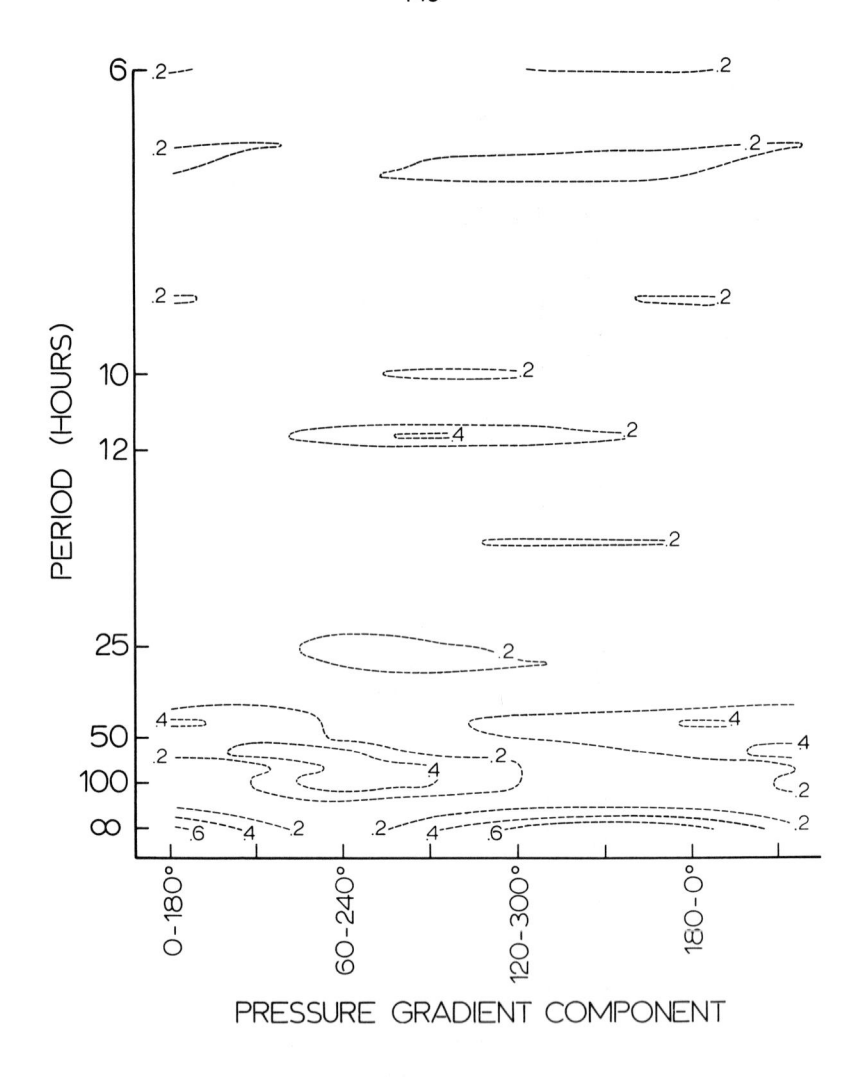

FIGURE 3. Composite of coherence spectra, computed using three-hourly surface pressure gradient components and computed columns for Corpus Christi Bay, January 19 - March 8, 1972. Frequency resolution is 0.00333 cph. [After SMITH (1977); used with permission].

west variations in the north-south component, will contaminate the relationship. The analysis is not restricted to synoptic scale features which are large enough to encompass both recording stations, however. Smaller-scale features, which may have influenced only the northwestern corner of the Gulf, for example, would have affected sea level locally as well, and the relationship between the cross-Gulf differences would be maintained. Figure 4 is a composite which summarizes results of the time series analysis. Energy density spectra at the top of the figure confirm the presence of significant variability in both pressure differences and water level differences over the longest time scales. The

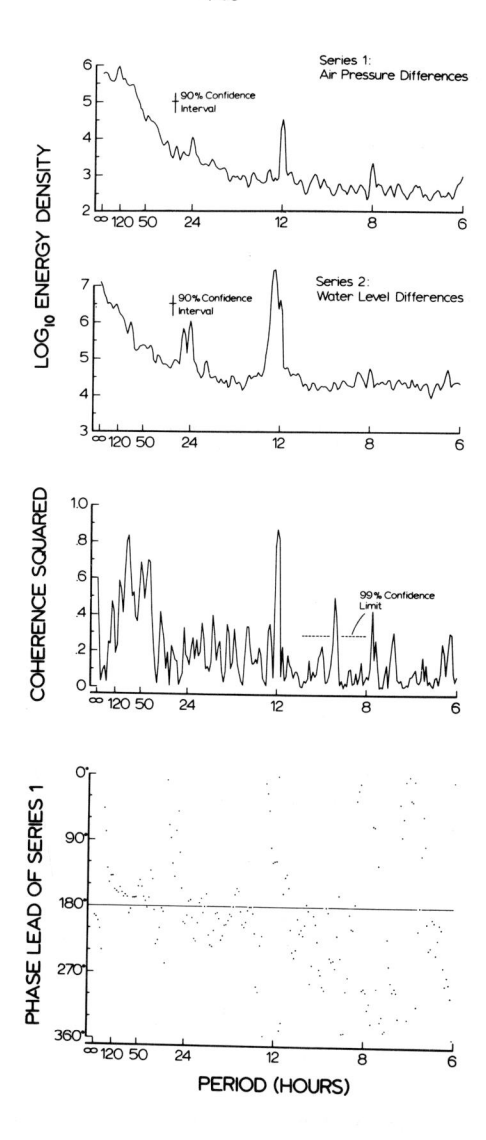

FIGURE 4. Composite of energy density spectra, in mb^2/cph and cm^2/cph, coherence and phase spectra, computed from three-hourly cross-Gulf surface pressure and water level differences, January 11 - July 15, 1975. [After SMITH (1979); used with permission].

coherence spectrum in the center of the plot is of particular interest. A band of statistically significant coherence values stands out at time scales of between approximately two and six days. The phase spectrum at the bottom of the figure indicates an approximately 180°, out-of-phase relationship between the two time series — a result consistent with the inverse nature of this forcing mechanism.

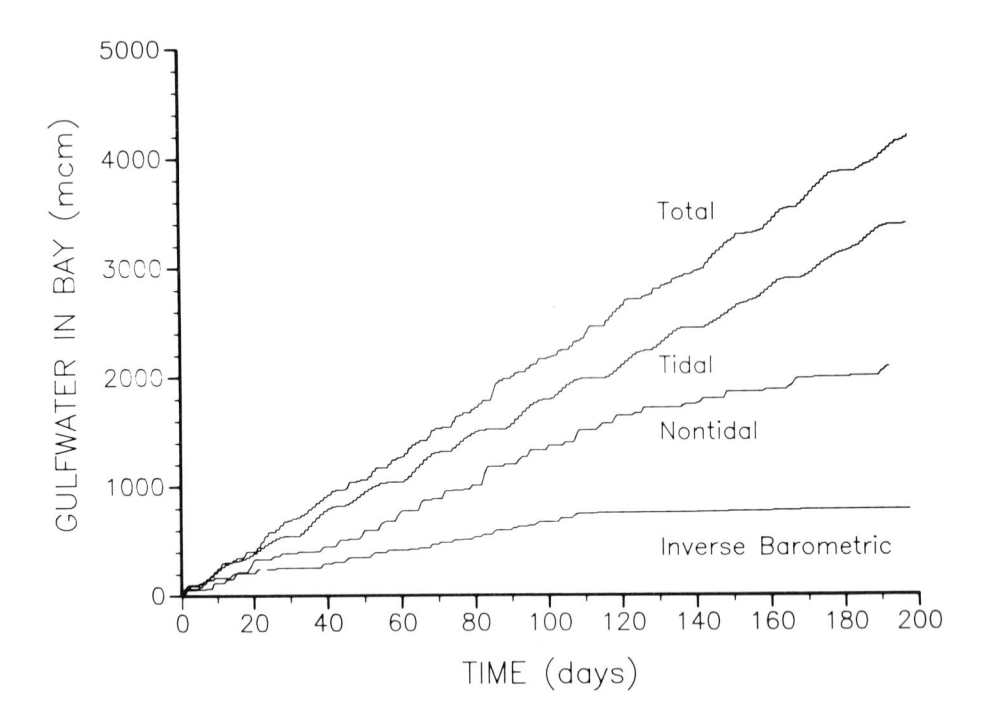

FIGURE 5. Composite showing the accumulation of Gulf of Mexico water in Corpus Christi Bay, in millions of cubic meters, January 1 through July 15, 1975. Curves represent total, tidal, nontidal and inverse barometric forcing individually.

The numerical model described in the previous section was fed three times series to determine the rate at which Gulf water enters the bay, and the accumulation over the 196-day study period. Results, shown in Fig. 5, quantify the relative importance of total forcing, tidal and subtidal forcing individually, and inverse barometric forcing alone. The top curve represents the accumulation of Gulf water in Corpus Christi Bay in response to the total water level variation at the coast, and once all bay water in the connecting channel has been forced back into the bay. In spite of temporal variations in the individual forcing mechanisms, the long term accumulation remains relatively close to a steady rate of 21 million cubic meters per day. Over time scales on the order of a few days, accumulation rates may increase sharply or decrease to near zero. When the time series of tidal water level variations alone is used to drive the model, the accumulation rate decreases slightly to 17 million cubic meters per day, and a distinct periodicity appears. At times of equatorial tides, Gulf and bay water merely move back and forth within the connecting channel, occupying complementary fractions of the total volume. With no Gulf water entering the bay, the accumulation rate decreases to zero. Conversely, at times of tropic tide, Gulf water floods into the bay at an average rate of 30-35 million cubic meters per day.

The nontidal influx of Gulf water indicates a degree of seasonality. From the start of the study period through mid April (approximately day 125), the nontidal accumulation of Gulf water averages about 14 million cubic meters per day. For the remainder of the study period, the nontidal accumulation of Gulf water decreases to an average of only 6 million cubic meters per day. Inverse barometric forcing, the only nontidal process singled out by the model, shows a very distinct seasonality. Again, through mid April Gulf water accumulates at an average rate of approximately 7 million cubic meters per day. For the rest of the record, however, the curve flattens out and indicates essentially no additional accumulation. Inverse barometric forcing becomes negligible in forcing bay-shelf exchanges during the relatively quiescent summer months.

4. Discussion

The results which are summarized in this paper, while site-specific in the strictest sense, demonstrate the importance of subtidal forcing. Although the relative importance of tidal and nontidal processes may vary substantially, subtidal exchanges probably play a significant supplementary role in other estuarine systems as well. A further decomposition of the nontidal component of the total into its several parts was not possible with the available data base, but it is clear that subtidal exchanges are worthy of attention in the case of Corpus Christi Bay.

In spite of the diversity in the time scales and the nature of the forcing mechanisms influencing bay-shelf exchanges, it is encouraging to note that the net effect can be quantified and characterized by a simple, one-dimensional model. A two-dimensional model would be needed to describe motion within the bay (PENUMALLI, et al. 1975) and to examine the mixing of bay and Gulf waters theoretically, but the one-dimensional model described here appears to be satisfactory for describing the two-way exchange of water along the channel connecting the bay with the Gulf.

Figure 5 demonstrates clearly that both tidal and nontidal exchanges tend to occur in discrete bursts which characteristically persist over time scales on the order of a week. These injections of Gulf water into the bay are separated by periods of variable length during which essentially no net exchanges occur. This is a consequence of the temporary storage of water in the channel connecting the bay with the Gulf (see Fig. 1). At times of equatorial tides, and at times of relatively weak meteorological forcing, Gulf and bay water simply move back and forth within the channel. Mixing across an interface within the channel is not included in the calculations and the modelling results suggest that the bay may be isolated from the Gulf for periods of several days. If true, this would have important implications in terms of maintaining and restoring water quality.

5. Conclusions

The main point to come out of this work is that subtidal forcing must be included in a comprehensive investigation of bay-shelf exchanges. In the Corpus Christi Bay work, tidal exchanges are of primary importance, and they constitute the baseline level of exchange. Tidal flushing may occur only at times of tropic tides, but it is predictable and dependable. Meteorological forcing in the form of discrete events makes substantial, if isolated, contributions. Whether or not subtidal forcing is or can be further decomposed into its constituent parts, it must be quantified and included in both biological and physical studies because of the role it plays in maintaining water quality, and in ferrying plant and animal populations between estuarine and shelf waters.

Acknowledgements

I would like to thank Mr. Mark Schmalz, applications programmer, for his effort in writing and running the computer program which quantified the total, tidal and nontidal exchanges. Figure 2 and 4 were used with permission from Estuarine, Coastal and Shelf Science, Vol.8, copyright 1979, Academic Press, Inc. (London) Limited; Figure 1 and 3 were used with permission from Estuarine, Coastal and Shelf Science, Vol.5, copyright 1977, Academic Press, Inc. (London) Limited.

Contribution No.446 of Harbour Branch Foundation, Inc.

References

BAKUN, A. 1975. Daily and weekly upwelling indices, west coast of North America, 1967-73. NOAA Tech. Rept. NMFS SSRF-693. Seattle Washington. 7 pages

EIGSTI, S.L. 1978. The coastal diurnal wind cycle at Port Aransas, Texas. Report No. 48. Atmos. Sci. Group, The Univ. of Texas, Austin, Texas.

HSU, S.A. 1981. Model for estimating offshore winds from onshore meteorological measurements. Boundary-Layer Meteor. 20:341-351.

MEHTA, A.J. 1978. Inlet hydraulics. pp. 83-161 In: Stability of Tidal Inlets. Elsevier Sci. Publ. Co., New York.

MONTGOMERY, R.B. 1937. Fluctuations in monthly sea level on eastern U.S. coast as related to dynamics of western North Atlantic Ocean. Journ. of Mar. Res. 1:165-185.

PENUMALLI, B.R., R.H. FLAKE and E.G. FLAKE and E.G. FRUH. 1975. Establishment of operational guidelines for Texas coastal zone management. Spec. Rept. I: Water quality modelling and management studies for Corpus Christi Bay: A large systems approach. Research in Water Resources, The Univ. of Texas, Austin, Texas, p.223.

PORE, N.A. and R.A. CUMMINGS 1967. A FORTRAN program for the calculation of hourly values of atronomical tide and height of high and low water. U.S. Weather Bureau (now National Wea. Serv.). Tech. Memorandum TDL-6. Silver Spring, Maryland, 17 pages.

RICHARDS, T.L., H. DRAGERT and D.R. MCINTYRE 1966. Influence at atmostpheric stability and over-water fetch on winds over the lower Great Lakes. Monthly Wea. Review 94:448-543.

SCHWING, F.B. and J.O. BLANTON 1984. The use of land and sea based wind data in a simple circulation model. Journ. Of phys. Oceanogra. 14:193-197.

SMITH, N.P. 1977. Meteorological and tidal exchanges between Corpus Christi Bay, Texas, and the northwestern Gulf of Mexico. Estuarine and Coastal Marine Science 5:511-520.

SMITH, N.P. 1979. Meteorological forcing of coastal waters by the inverse barometer effect. Estuarine and Coastal Marine Science 8:149-156.

SMITH, N.P. (In press). Numerical simulation of bay=shelf exchanges with a one-dimensional model. Contributions in Marine Science, Vol.28.

YU, T-S. and N.K. Wagner 1970. Diurnal variation of kenetic and internal energy in onshore winds along the upper Texas coast. Report No.19, Atmos. Sci. Group, The Univ. of Texas, Austin, Texas.

V

COASTAL CIRCULATION

Water Circulation in a Topographically Complex Environment

Eric Wolanski

Australian Institute of Marine Science
Australia

ABSTRACT

Examples are presented of field studies of the circulation around headlands, islands and reef passages in the shallow waters of the Great Barrier Reef. The water circulation was studied using aerial and satellite images, moored current meters, radar-tracked drogues and shipborne instruments such as temperature-conductivity probes. The data reveal that flow separation occurs at the tips of headlands, islands and reefs, and generates tidal jets and topographically trapped tidal eddies in shallow water. These eddies and jets absorb a significant fraction of the kinetic energy of the upstream flow. They also generate fronts as well as mixing, vertically by upwelling and downwelling as an organized secondary circulation in the eddy, and horizontally by the effects of secondary circulation and trapping. They also selectively move sediment, and are believed to play an important role in the biology and ecology of these shallow shelf waters.

1. Introduction

Eddies and jets shed by side boundary irregularities, e.g. islands or headlands with steep slopes in shallow water, have often been reported but the current data was too sparse and insufficient for testing of analytical or numerical models. For instance (WOLANSKI et al., 1984a), computer-enhanced LANDSAT images of reefs or islands on the Great Barrier Reef, of length 500 to 2000 m, in depth 10 to 40 m and with currents 0.3 to 0.8 m s^{-1}, reveal oblong wakes similar in shape to small REYNOLDS number (of order 20) wakes in two-dimensional laboratory experiments.

The *standard* numerical models are generally unable to yield a similar picture (FISCHER, 1981), probably because the algorithms do not attempt to represent the large vorticity introduced in the water in a thin sub-grid scale boundary layer at the tip of the headland. Without this vorticity source, the computed fluid motion is roughly the balance between bottom friction and pressure gradients, and resembles potential flow.

In view of the lack of field data on eddies shed by headlands, detailed field studies of the recirculating flow around Rattray Island (the barotropic case) and Myrmidon Reef (the baroclinic case) were recently completed. In addition, remote-sensing data on unsteady tidal jets have also been collected. Some pertinent results, and their interpretation, are summarized below.

2. The barotropic island wake

An intensive study of the water circulation around Rattray Island (Fig. 1), northeast Australia, was recently completed (see WOLANSKI et al., 1984b), for further details). The waters are about 20 m deep and vertically well-mixed. A well defined steady stationary eddy was found, both from synoptic maps of currents measured with 26 moored current meters and with radar-tracked drogues (Fig. 1), and from the temperature distribution (not shown here), the incoming tidal waters being slightly warmer (by 0.5°C) than the

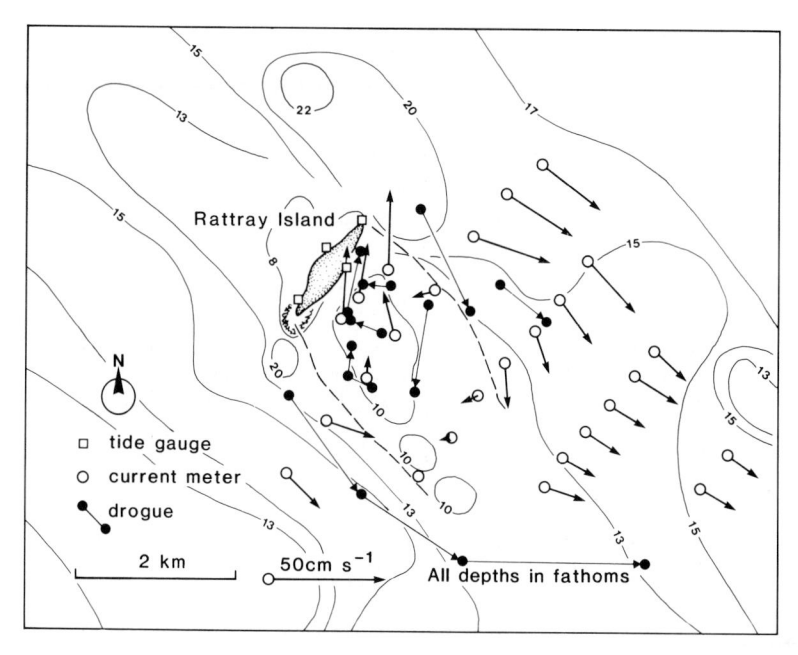

FIGURE 1. Map of Rattray Island at 20°, northeast Australia. Also the synoptic distribution of the water currents measured with moored current meters (open circles) and with radar-tracked drogues (full circles) at peak tidal currents. Reproduced with permission of Academic Press from Wolanski et al. (1984a)

displaced waters (including those trapped in the eddy). The eddy waters are very turbid and the eddy is readily visible from the air (Fig. 2).

The shape of the eddy suggests similarity with two-dimensional flow around obstacles in laboratory experiments at small values (of the order 10-30) of the REYNOLDS number. However, a number of discrepancies between field and laboratory data are found, namely that the Rattray Island Reynolds number based on the vertical eddy viscocity is of order 10^3, that a thin boundary layer exists at the tip of Rattray Island contrary to the laboratory results at low values of the Reynolds number, and that the ratio of the peak velocity inside and outside the wake is of order 1 at Rattray Island and of order 0.01 in the laboratory. Further, a strong upwelling is also observed in the island wake. Finally, the Rossby number is large, implying that earth rotation effects are not very important.

These various observation conflict with some of the assumptions behind the island wake models of GARRETT and LOUCKS (1976), PINGREE and MADDOCK (1979) and GUO and XIA (1984). To reconcile these differences, it has been proposed (WOLANSKI et al., 1984b) that the stable eddy is in solid body rotation, except near the sea floor where it is assumed that the no-slip condition applies (see a sketch in Fig. 3). As a result of the centrifugal acceleration in the bulk of the eddy, a radial seal level gradient is established. This balance between centrifugal acceleration and radial pressure gradient cannot hold near the bottom because the velocities are smaller there. The excess pressure gradient

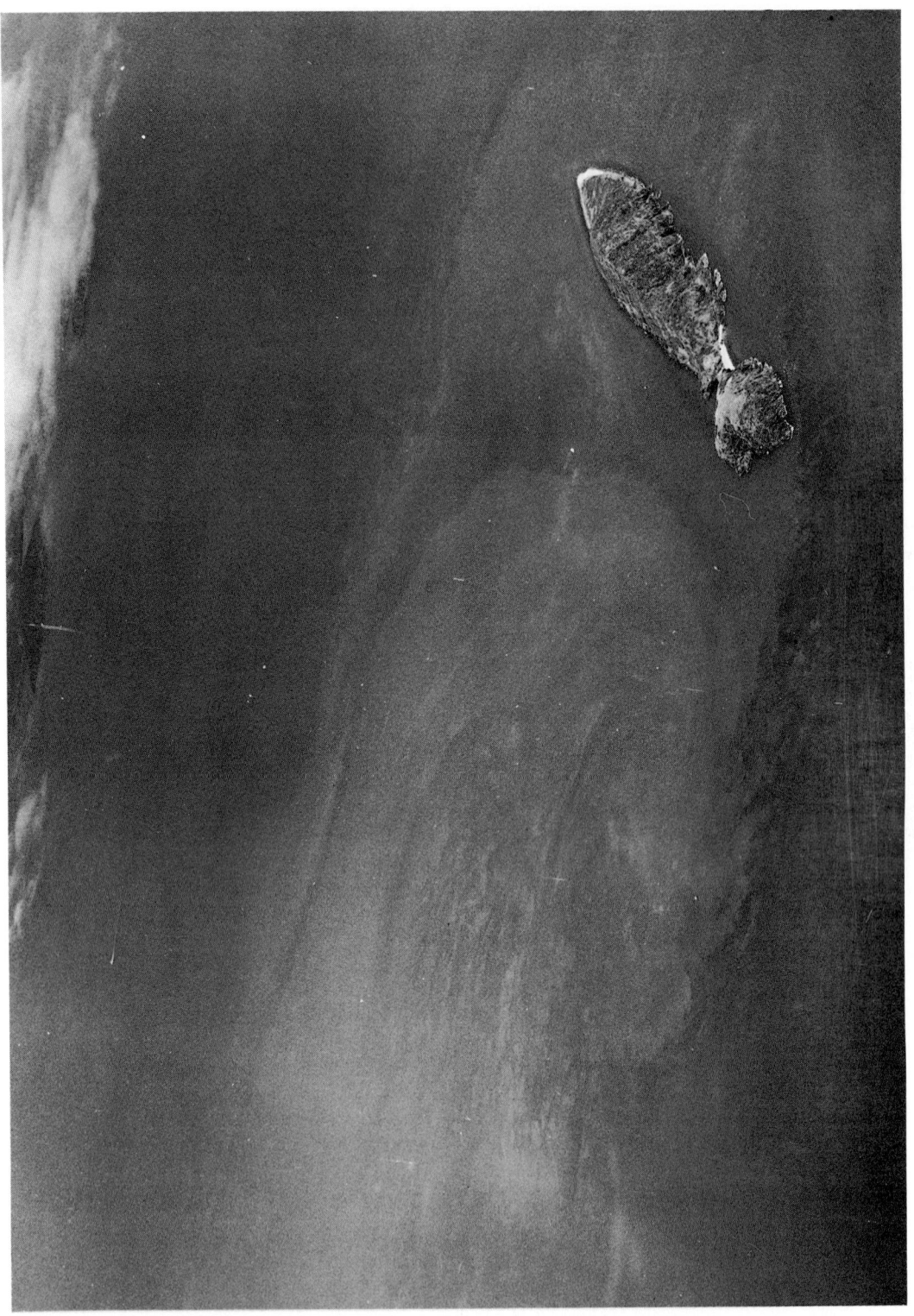

FIGURE 2: Aerial view of Rattray Island at peak flood tidal currents. The eddy is made visible by the high turbidity of the wake waters.

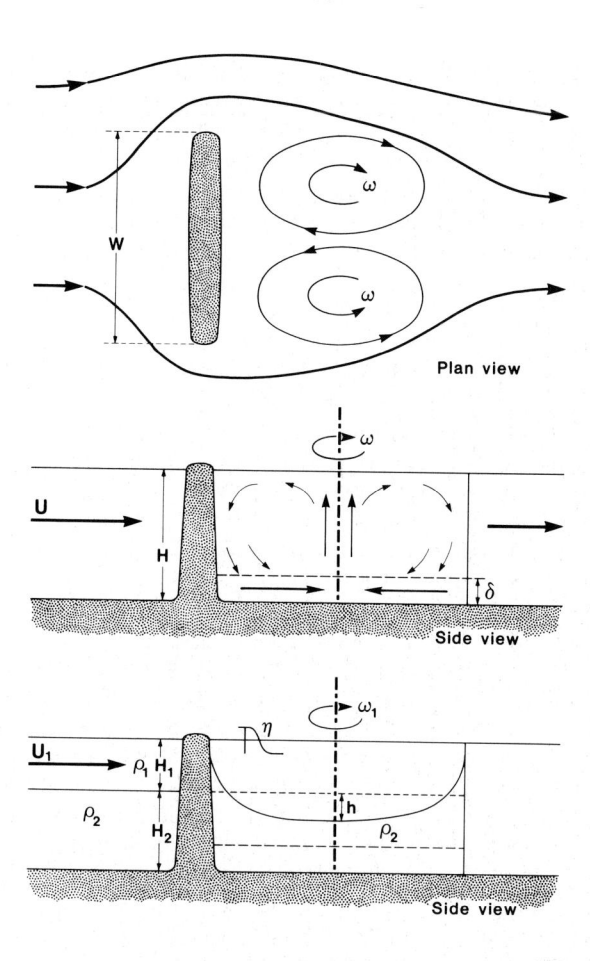

FIGURE 3: TOP: Plan view of eddies shed by an island of width W in a flowing stream of velocity U. There exists a thin shear layer at the island tips.
MIDDLE: In the barotropic situation, the bulk of the water is in solid body rotation. At the sea floor, there exists an Ekman benthic boundary layer characterized by strong radial currents towards the eddy centre. This water is sucked from the eddy core and upwelled near the centre.
BOTTOM: For the two-fluid system, idealising the thermal stratification and current structure near Myrmidon Reef, the prevailing current U_1 is confined to the top layer. The rotation in the lee of Myrmidon Reef in the top layer deflects the surface by a small amount η and the interface by a much larger value h.

generates a radial flow near the bottom and converging towards the eddy centre. The radial flow is of such magnitude that the radial viscous and pressure forces balance each other. The water drawn by this self-induced EKMAN suction is upwelled, together with the fine sediment it contains, near the eddy centre (Fig. 3).

The radius of the eddy can be computed from a balance at quasi-steady-state between the vorticity flux through the thin boundary layer at the separation point on one hand, and, on the other hand, the vorticity flux of opposite sign introduced by the EKMAN suction at the sea floor. The eddy radius is found to be

$$R = UH^2 / K \qquad (1)$$

where H is the depth, U the free stream velocity and K the vertical eddy diffusion coefficient.

Provided that K is replaced by the kinematic viscosity, equation (1) is identical to that discovered by J. IMBERGER (unpubl. data) for flows around obstacles in a HELE-SHAW cell, further validating the hypothesis that frictional effects are important in the dynamics of island wakes in shallow waters.

In equation (1), R is independent of the island length, so that equation (1) is also applicable to eddies shed by headlands. If an island is present, its length W (see Fig. 3) introduces a length scale that determines the eddy size. If $W/R >> 1$, no eddies are formed at the scale of the island. If $W/R \simeq 1$, then a stable eddy is formed such as at Rattray Island (Fig. 2). If $W/R << 1$, the vorticity flux at the island tip cannot be entirely negated by that of opposite sign in the wake eddy, and vorticity is transported further downstream. As a result, on can expect increasing instabilities downstream, for decreasing values of the ratio W/R. For a small vorticity imbalance, wavy streaklines are observed (Fig. 4a). For a moderate vorticity imbalance, these meanders become unstable at their crests and troughs, leading to a pattern reminiscent of a Karman vortex street (Fig. 4b). For a very large vorticity imbalance, the flow downstream is disturbed and turbulent for long distances from the obstacles and no organized flow structure is apparent in the wake (Fig. 4c).

Though the Ekman layer suction implies a three-dimensional motion, the aspect ratio $R/H > 100$, so that one would hope that it might still be possible to use two-dimensional depth-averaged models to simulate these eddies. Indeed, such a model was used to simulate the water circulation around Rattray Island (FALCONER et al., 1984). The grid size was of order 150 m, i.e. much larger than the thickness of the shear layer (of order 30 m) at the tip of the island at the separation point. The momentum equations used in the model are based on the depth integration of the Navier-Stokes equations. The equations include the sub-grid scale parameterization of the shear stresses associated with the free shear in the lateral mixing layer along the wake shed by the headland. Although in most flow situations the magnitude of the turbulant shear stress is small in comparison with the pressure gradient and bed friction, the neglect of the lateral shear stress does not lead to wake eddies in numerical models, and in fact this lateral REYNOLDS stress is essential to obtain vortex shedding in the numerical models. To avoid using the much too coarse grid to estimate the lateral shear and, from these, the REYNOLDS stress, use was made of an experimentally determined lateral velocity distribution in the shear layer to compute the REYNOLDS stress.

The results are quite encouraging (see FALCONER et al., 1984) and it appears that flow separation effects can indeed be included in two-dimensional depth-averaged numerical models of water circulation in a topographically complex environment.

When the current does not reverse sign with the tide, water can be trapped for long periods in the wake. When the current reverses sign with the tide, considerable

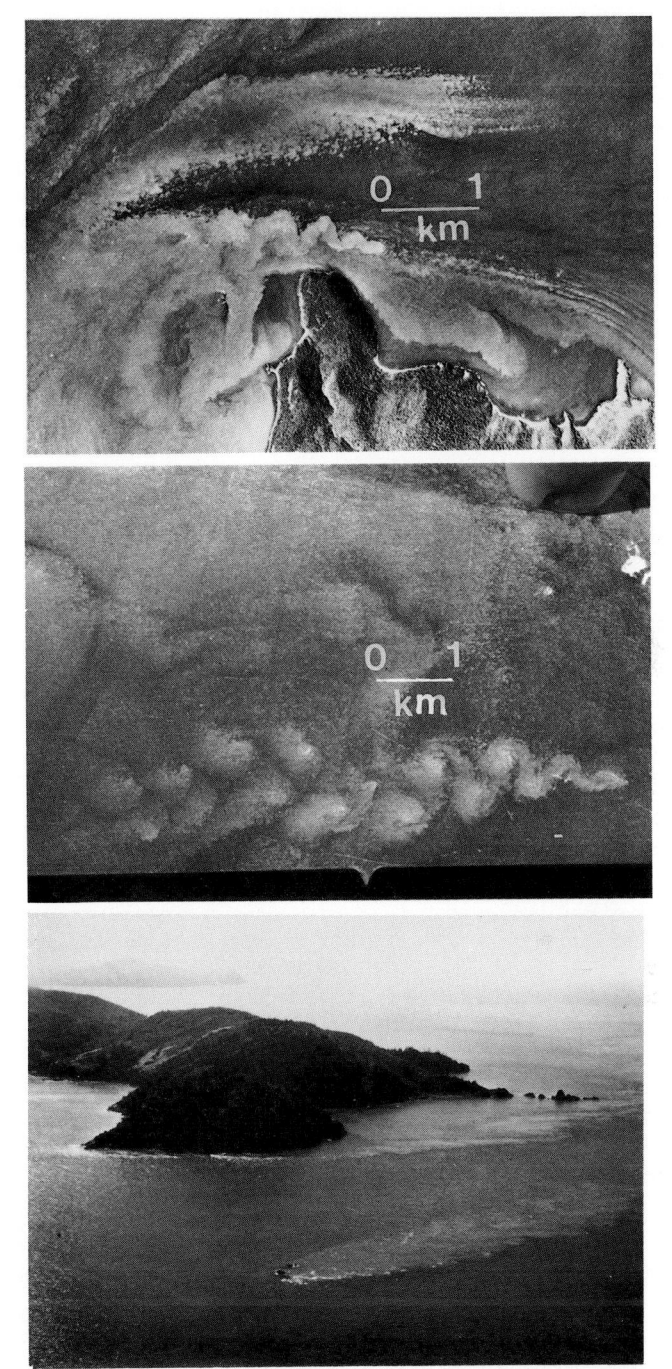

Figure 4. a to c are aerial views of island wakes in the Whitsunday area, at 20°S, northeast Australia, for increasing vorticity imbalance. Water depth ≃ 10 m.

horizontal mixing results from the joint action of the resulting large tide-averaged residual circulation (ZIMMERMANN, 1978; UNCLES, 1982), and the trapping of water patches in an oscillating water body (OKUBO, 1973).

The internal circulation in the eddy is an advected sediment sorter, as all movable particles near the bottom will be advected towards the eddy centre, but, while the coarse particles will stay there, the fine clay particles will be upwelled and moved away towards the outside of the eddy. The distribution of, respectively, the fine (diameter $< 2.8\,\mu m$) and the coarse (diameter $> 62\,\mu m$) sediment around Rattray Island (Fig. 5) seem to confirm this mechanism.

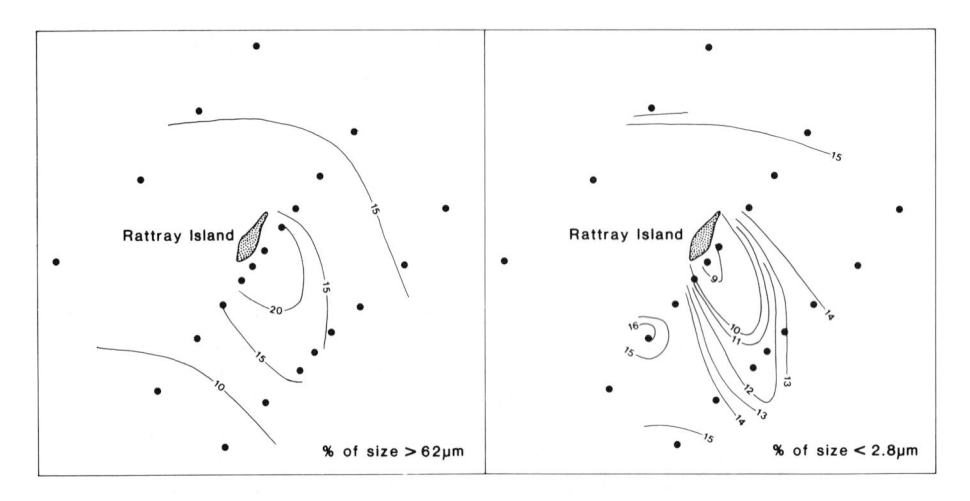

FIGURE 5. Distribution (in % per dry weight) of the fine and coarse sediment on the sea floor around Rattray Island.

FIGURE 6. Aerial view of Ross Islet, a small island in 10m of water at 20° S, northeast Australia. Note the tear-shaped coral reef in the lee of the rocky slab-shaped island.

PINGREE and MADDOCK (1979) have previously shown that eddies shed by headlands can lead to the formation of offshore tidal banks. Further, aerial observations reveal that, when a uni-directional current prevails, coral grows in the lee of a rocky island in shallow water, and forms a fringing coral reef that can take the exact shape of the tear-shape wake that one would expect in the absence of the coral (Fig. 6; see also HAMMER and HAURI, 1977).

3. Baroclinic island wakes

If the water is vertically stratified in density, the island wake mechanism described earlier is not necessarily valid. For instance, in a two-layer fluid (see a sketch in Fig. 3), if the upstream current is confined to the upper layer, the bottom layer being at rest upstream, flow separation at the tip of the island or headland will still generate an eddy downstream in the upper layer (BRIGHTON, 1978). Interfacial stresses between the two layers are small and the eddy in the upper layer does not have a strong tendency to be spun down by EKMAN suction. The eddy in the top layer could thus become rapidly very elongated and probably unstable.

Is there a theoretical limit to the eddy size so that a steady state situation still prevails? In hydrostatic response to the sea level change, η, due to rotation in the top layer, the interface will be displaced by an amount h, to yield zero radial pressure gradient in the bottom layer which ideally could remain at rest (see a sketch in Fig. 3). However, rotation is introduced, after sufficient time, in the bottom layer both by interfacial stresses and by the conservation of potential vorticity on a rotating earth, since the vortex lines under the background earth rotation and in the bottom layer are stretched by the interface deformation. That this mechanism causes a spin-up in a two-layer fluid was demonstrated recently in laboratory experiments by LINDEN and HEIJST (1984). As a result of friction at the sea floor, an EKMAN layer may also be generated. This frictional spin-down occurs in the bottom boundary layer. Presumably there may result, at steady state, a vorticity balance between, on the one hand, the vorticity flux entering the top layer at the point of separation through the boundary layer, and, on the other hand, the friction-driven flux of vorticity in the bottom layer. Such eddy dynamics can only exist in quasi-steady conditions, i.e. in practice when the currents do not reverse direction with the tides.

The few satellite and aerial images of wakes in the lee of islands and reefs in stratified waters of the Great Barrier Reef show wakes much longer than those in the barotropic case and also showing sign of instabilities, so that the shape of the wakes resembles those of dispersion clouds from a point source in a turbulent environment.

A recent intensive field investigation of the circulation around Myrmidon Reef, located at the shelf break of the Great Barrier Reef (Fig. 7), northeast Australia, revealed the existence of complex unsteady flow patterns and trapping effects. The interpretation of the data is more difficult than in the barotropic case in view of the influence of large amplitude internal tides (some of them clearly are controlled by the wake effect) and by unsteady events in the Coral Sea. Nevertheless, evidence was found for strong wake eddies generated by flow separation at the reef, both from drogue data and synoptic maps of the currents measured with moored current meters (Fig. 7), and also from the temperature distribution across the eddy (Fig. 8). In practice, these vortices are particularly apparent (from the current meter data) in the top layer, as the prevailing undisturbed flow is largely confined to the top layer. However, weaker vortices are also observed in the bottom layer and it is likely that these are generated by the mechanisms described above.

Assuming around Myrmidon Reef (see Fig. 3) a two fluids system of density ρ_1 and ρ_2 respectively (the subscripts 1 and 2 indicate respectively the top and bottom layers) and of depth H_1 and H_2, assuming an undisturbed velocity U_1 in the top layer only,

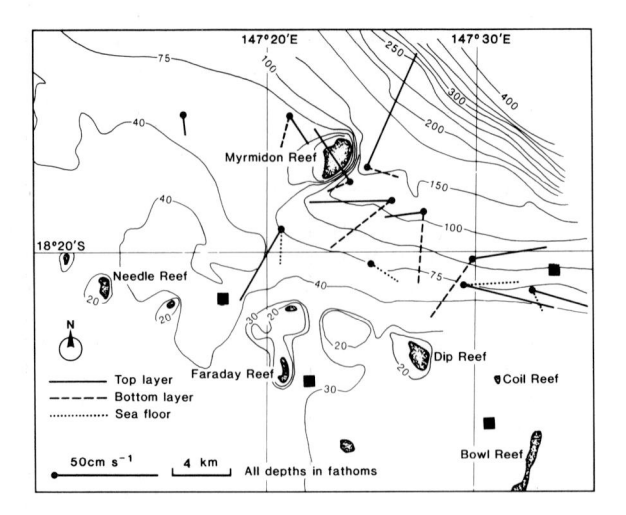

FIGURE 7: Map of the shelf break area around Myrmidon Reef, at 19°S, northeast Australia, with depths in fathoms (1 fathom = 1.8 m). Also, synoptic distribution of currents measured with moored current meters (full circles) at various depths. In addition, temperature loggers were also deployed at the sites marked with a square.

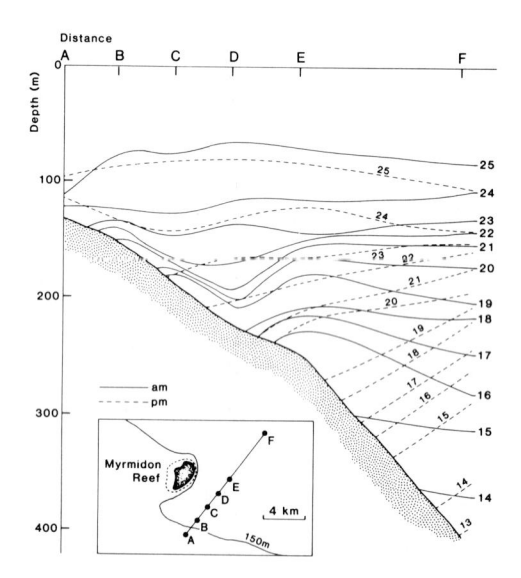

FIGURE 8: Cross-sectional distribution of temperature a few kilometers downstream of Myrmidon Reef, northeast Australia. Note the strong doming of the isotherms in the eddy in the morning, and the large differences between morning and afternoon results due to internal tides.

and for solid body rotation in the top layer eddy, the vorticity balance described above yields after some algebra (WOLANSKI, unpubl. data) an eddy radius

$$R = U_1^4 K^{\frac{1}{2}} / f^{1.5} H_1 H_2^{1.5} [g(\rho_2 - \rho_1)/\rho_2]^{1.5} \qquad (2)$$

where f is the CORIOLIS parameter and g the acceleration due to gravity. For the case of Myrmidon Reef, equation (2) yields $R \gg$ the reef width, suggesting that the eddy is unstable. In any case, the tidal time scale prevents a steady state situation to prevail, as can be seen in Fig. 8 from the differences between the morning and afternoon cross-sectional distribution of temperature. These differences are due to the large internal tides that are observed at the shelf break.

The unsteadiness and instability of the eddy in the baroclinic case lead to considerable patchiness and horizontal mixing. No numerical model appears to have yet been developed for this situation, though presumably baroclinic eddies occur also in salt-wedge estuaries in the presence of headlands.

4. Tidal jets

At the mouth of creeks, estuaries, bays and harbours, and in situations where the opening between reefs or islands and headlands is small, remote-sensing images reveal that flow separation occurs at both side boundaries (JOSHI and TAYLOR, 1983; ONISHI, 1984; WOLANSKI et al., 1984a). An unsteady tidal jet is thus formed with a vortex pair at its leading edge.

Further examples of such flows in the Great Barrier Reef are shown in Figs. 9 and 10 to illustrate the differences due to buoyancy effects. A well-defined vortex pair is apparent in the barotropic case for both fairly deep waters (40 m deep in Fig. 9) and in fairly shallow coastal waters (5 m deep in Fig. 10b). When the buoyancy effects are large enough that the jet lifts off from the sea floor (so that the bottom friction is reduced), but small enough that buoyancy effects do not greatly influence the rate of lateral spread, the vortex pair appears to be inhibited (Fig. 10a). For very large buoyancy differences, the jet lifts off from the bottom, the vortex pair appears to be completely inhibited and the front spreads by buoyancy nearly radially away from the source (e.g. the river mouth, IMBERGER, 1983).

WILKINSON (1978) has shown from laboratory experiments in the two-dimensional case, that the natural length scales for the barotropic jets are the source width and $(mT^2/\rho)^{1/3}$, where T is the tidal period and m the maximum momentum flux at the source. He also noticed that vortices are present at the leading edge of the jet and propagate with it away from the source.

As in the case of the island wake problem, basic differences appear to exist between barotropic jet-generated eddies in the laboratory and in the field. Indeed, in the laboratory, most of the circulation in the vortex pair involves ocean fluid uncontaminated with water discharged through the opening (WILKINSON, 1978). In the field, the remote sensing images suggest that most of the water in the vortex originates from the opening. No current meter or drogue data are yet available to assess the dynamics of barotropic jet-generated eddies in the field.

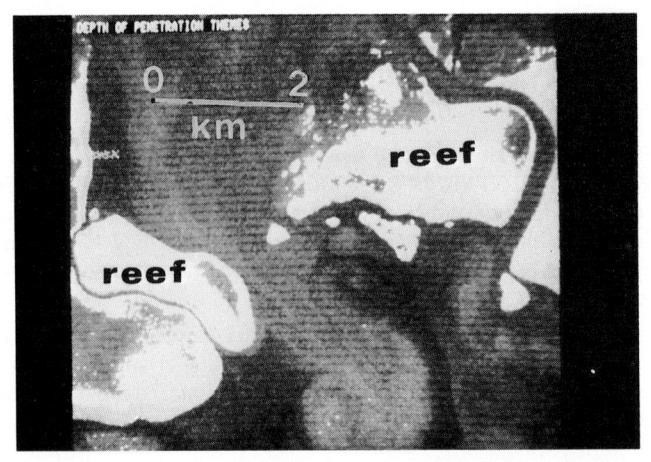

FIGURE 9. Computer-enhanced LANDSAT view of a tidal jet and the vortex pair at its leading edge off a Reff Passage near 20°S on the Great Barrier Reef, northeast Australia. Water depth $\simeq 40$m.

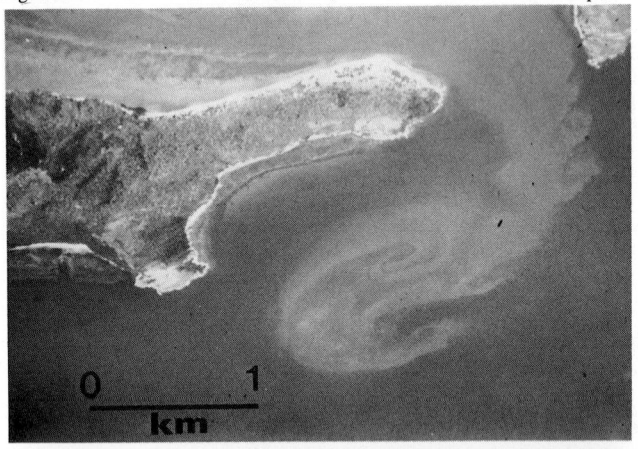

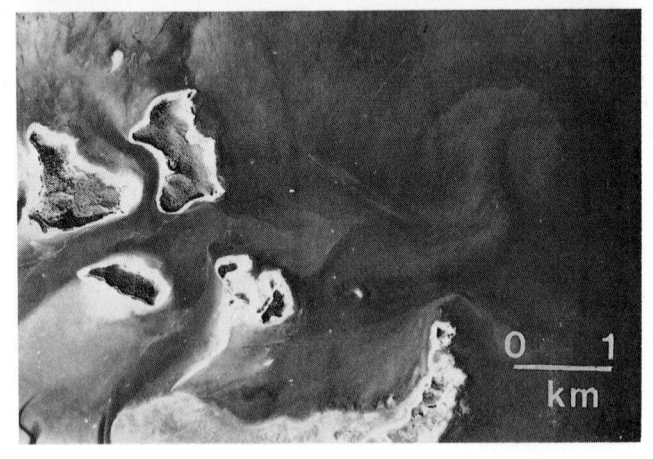

FIGURE 10: Aerial observation of a tidal jet and vortex pair off the Palm Islands (a, water depth $\cong 20$ m) and the Whitsunday Islands (b, water depth $\cong$ 5-10 m), northeast Australia.

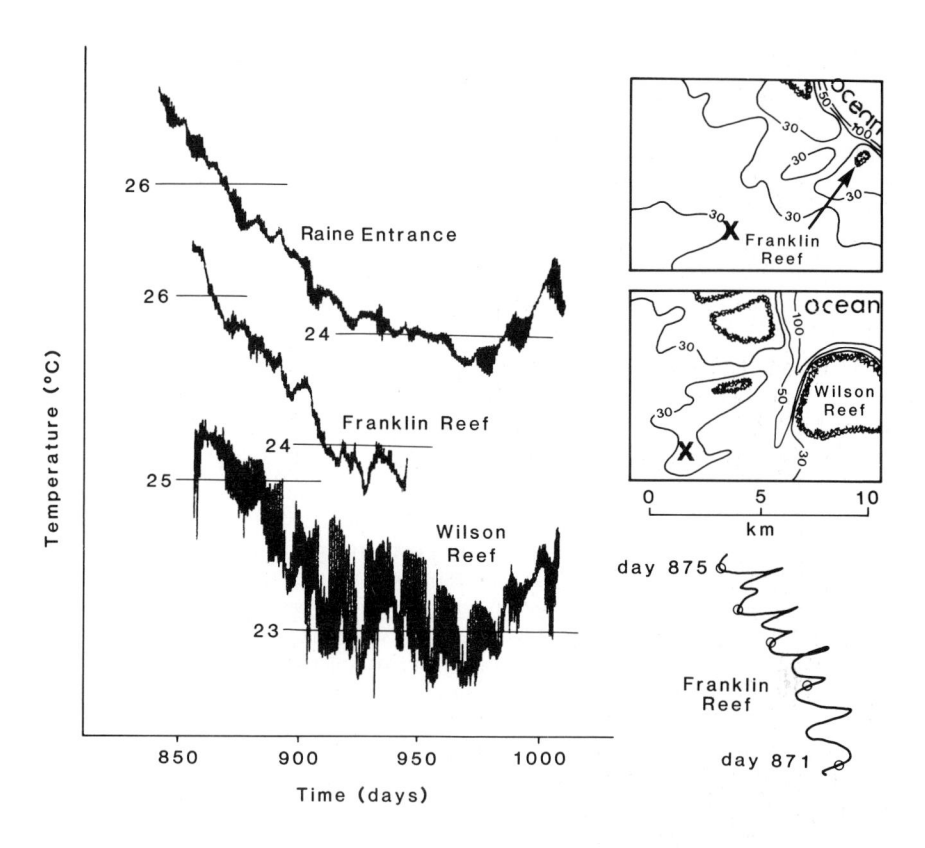

FIGURE 11: On the left are shown time series plot of the temperature 5 m above the bottom on the shelf near Raine Entrance, and the passages near Franklin Reef and Wilson Reef. The trend of decreasing temperature with time is due to seasonal cooling. The semi-diurnal temperature fluctuations indicate the tidal upwelling. On the right are shown maps, with depth in meters, of the topography near Franklin and Wilson Reefs. The topography (not shown) on the shelf near Raine Entrance is, like that at Franklin Reef, quite flat. The crosses indicate the mooring sites. The graph at the bottom right hand corner shows the progressive vector diagram at the mooring site near Franklin Reef. Note the strong cross-shelf tidal excursions. The time is marked in day no. from January 1, 1980. The temperature profiles show, as in Fig. 8, a well-mixed layer down to about 80m, i.e. well below the elevation of the shelf.

5. Conclusions

Because of their potential for trapping water and particulates, eddies shed by headlands in estuarine and coastal waters are a topic of increasing interest by engineers interested in siting of waste outfall and by numerical modellers and sedimentologists. These eddies are still very difficult to model numerically at reasonable costs primarily because of the importance of a thin boundary layer at the separation points, and because the flow appears to be three-dimensional though the width to depth ratio is very large (WOLANSKI et al. 1984b). They also appear to be important in the formation of tidal banks and shoals (PINGREE and MADDOCK, 1979) and also of coral reefs (see Fig. 6; also HAMNER and HAURI, 1977).

Partially because of the vertical mixing and the upwelling they generate, these eddies also appear to be important to the ecology of coastal waters in a topographically complex environment. They have been shown to control the distribution, and lead to the aggregation of, plankton, fish egg and benthic invertebrates, and the determination of coastal fisheries (UDA and ISHINO, 1958; HAMNER and HAURI, 1977, 1981; ALLDREDGE and HAMNER, 1980; LEIS, 1982; SIMPSON et al., 1982).

In areas where the flow is measurably obstructed by headlands and islands, and where the currents reverse sign with the tides, the eddies are periodically formed and flushed away. As a result, the values of the bulk friction coefficient and of the horizontal mixing coefficient are greatly increased (WOLANSKI et al., 1984a and 1984b).

Tidal jets and jet-generated vortices are important locally as they can control the formation of tidal bars (JOSHI and TAYLOR, 1983), and lead to patchiness, trapping, enhanced mixing and a reduced value of the tidal return coefficient of an estuary (ONISHI, 1984).

In the Great Barrier Reef at least, one biologically important effect of tidal jets is the forced upwelling. Nutrient-rich water from below the thermocline is sucked upwards by the BERNOULLI effect into the fast flowing waters converging into the reef passage at rising tide (THOMSON and GOLDING, 1981; THOMSON and WOLANSKI, 1984). This enriched water then propagates on the shelf with the tidal jet and its vortex pair. The intensity of this upwelling at tidal frequencies can be measured by the magnitude of the temperature fluctuations at tidal frequencies experienced by current meters moored on the shelf in front of reef passages. It turns out that the upwelling intensity is very dependent on the sea floor topography. For instance, as is shown in Fig. 11, the upwelling intensity at tidal frequencies is much smaller at Raine Entrance and at the passage near Franklin Reef, where in both cases the sea floor is flat up to the steep shelf slope, than at the passage near Wilson Reef, where the sea floor in the passage is inclined towards the ocean and forms a small canyon. The water currents are otherwise comparable at the three sites. It is interesting to note that the calcareous green alga, *Halimeda*, which grows in segmented branches up to 25 cm long, forms profuse meadows on the sea floor of the shelf in front of the reef passages in the areas where the tidal jets and jet-forced vortices penetrate on the shelf (see Fig. 9). The sedimentary deposits of *Halimeda* are known to form domes, several hundreds of metres wide, 15 m high in 50 m of water, and located on either side of the channel axis, i.e. precisely in the areas where, from satellite images, the jet-forced vortices are presumed to be present (E. DREW and P. DAVIES, unpubl. data). The *Halimeda* meadows are most prevalent near passages forming a small canyon, and appear not to exist in areas where the shelf elevation is too high for nutrient-rich deep water to be upwelled by tidal jets.

Acknowledgements

It is a pleasure to thank Drs. J.S. BUNT and R. STRICKLER who made this study possible and Dr. D.L.B. JUPP who provided the LANDSAT image.

References

ALLDREDGE, A.L. and HAMNER, W.M. 1980. Recurring aggregation of zooplankton by a tidal current. Estu. Coastal. Mar. Sci., 10, pp. 31-37.

BRIGHTON, P.W.M., 1978. Strongly stratified flow past three-dimensional obstacles. Q.J.R. Meteor. Soc., 104, pp. 289-307.

FALCONER, R.A. WOLANSKI, E. and MARDAPITTA-HADJIPANDELI, L., 1984. Numerical simulation of secondary circulation in the lee of headlands. Proc. 19th Int. Coastal Eng. Conf., ASCE, Houston, Texas, (in press).

FISCHER, H.B. (ed.), 1981. Transport models for inland and coastal waters. Academic Press, New York, 542 pp.

GARRETT, C.J.R. and LOUCKS, R.H., 1976. Upwelling along the Yarmouth shore of Nova Scotia. J. Fish. Res. Board Can., 33, pp. 116-117.

GUO, B. and XIA, Z., 1984. An analytical model of upwelling induced by tidal current past a peninsula. In ICHIYE, T. (ed.): "Ocean hydrodynamics of the Japan and East China Seas". Elsevier Science Publ. Co., New York, pp. 123-142.

HAMNER, W.M. and HAURI, I.R., 1977. Fine-scale currents in the Whitsunday Islands, Queensland, Australia: Effect of tide and topography. Aust J. Mar. Freshwater Res., 28, pp. 233-259.

HAMNER, W.M. and HAURI, I.R., 1981. Effects of island mass: water flow and plankton pattern around a reef in the Central Great Barrier Reef lagoon, Australia. Limnol. Oceanogr., 26 (6), pp. 1084-1102.

IMBERGER, J., 1983. Tidal jet frontogenesis. Inst. of Eng. (Australia), Mechanical Eng. Trans., ME8, pp. 171-180.

JOSHI, P.B. and Taylor, R.B., 1983. Circulation induced by tidal jets. ASCE, J. Waterway, Port., Coastal and Ocean Eng., 109, No.4, pp. 445-464.

LEIS J.M., 1982. Hawaiian creediid fishes (Crystallodytes Cookei and Limnichtys Donaldsoni): development of eggs and larvae and use of pelagic eggs to trace coastal water movement. Bull. Mar. Sci., 31 (1), pp. 166-180.

LINDEN, P.F. and HEIJST, G.F.J., 1982. Two-layer spin-up and frontogenesis. J. Fluid Mech., 143, pp. 69-94.

OKUBO, A., 1973. Effect of shoreline irregularities on streamline dispersion in estuaries and other embayments. Neth. J. Sea. Res., 6, pp. 213-224.

ONISHI, S., 1984. Study of vortex structure in water surface jets by means of remote sensing. In Nihoul, J.C.J. (ed.): "Remote sensing of shelf sea hydrodynamics". Elsevier Science Publ., Amsterdam, pp. 107-132.

PINGREE, R.D., and MADDOCK, L., 1979. The tidal physics of headlands flows and offshore tidal bank formation. Mar. Geol., 32, pp. 269-289.

SIMPSON, J.H., TETT, P.B., ARGOTE-ESPINOZA, M.L., EDWARDS, A., JONES, K.J. and SAVIDGE, G., 1982. Mixing and phytoplankton growth around an island in a stratified sea. Continental Shelf Res., 1, pp. 15-31.

THOMPSON, R.O.R.Y. and GOLDING, T.J., 1981. Tidally induced upwelling by the Great Barrier Reef. J. Geophys. Res., 86, No. C7, pp. 6517-6521.

THOMSON, R.E. and WOLANSKI, E., 1984, Tidal period upwelling within Raine Island entrance, Great Barrier Reef. J. Mar. Res., 42, pp. 787-808.

UDA, M. and ISHINO, M. 1958. Enrichment pattern resulting from eddy systems in relation to fishing grounds. J. Tokyo Univ. Fish., 44, pp. 105-129.

UNCLES, R.J., 1982. Residual currents in the Severn estuary and their effects on dispersion. Oceanologica Acta, pp.403-410.

WILKINSON, D.L., 1978. Periodic flows from tidal inlets. Proc. 16th Int. Coastal Eng. Conf., ASCE, Hamburg, pp. 1336-1346.

WOLANSKI, E., PICKARD, G.L. and JUPP, D.L.B., 1984a. River plumes, coral reefs and mixing in the Gulf of Papua and northern Great Barrier Reef. Estu. Coastal Shelf Sci., 18, pp. 291-314.

WOLANSKI, E., IMBERGER, J. and HERON, M.L., 1984b. Island wakes in shallow coastal waters. J. Geophys. Res., 86 (C6), pp. 10553-10569.

ZIMMERMANN, J.T.F., 1978. Dispersion by tide-induced residual vortices. In NIHOUL, J.C.J. (ed.): "Hydrodynamics of estuaries and fjords". Elsevier Oceanogr. Ser., Amsterdam, pp. 207-216.

Roles of Large Scale Eddies in Mass Exchange between Coastal and Oceanic Zones

Sotoaki Onishi

Science University of Tokyo, Japan

ABSTRACT

Roles of eddies in mass exchange between coastal and oceanic zones are discussed by using Landsat data as well as field measurements. As examples of the study, the flow dynamics at the Naruto strait and the Tohoku district off Japan are selected. In the Naturo strait, the Coriolis effect is not important, and the exchanging process in the strait can be described on the basis of simple vortex pair model. In the Tohoku district, the movements of the coastal water are geostrophic in nature and are affected by the oceanic currents. The large scale eddies produced by interactions between various currents may entrain occasionally the coastal water mass in it and may maintain it for a long period, say, several months.

1. Introduction

In the analysis of diffusion of various effluents from the land into the coastal water zone, interaction between th coastal water mass and the oceanic currents may make up one of the important boundary conditions. Few studies of that interaction have been reported so far, probably because the spatial scales of the relating phenomena are generally too large to be derived from ship-borne observation. Remote sensing techniques, however, are a tool to overcome this difficulty. In this paper, using Landsat data as well as field measurements, I will show some examples indicating how large scale eddies influence the mixing between the coastal and oceanic water masses. As examples of the study, the flow dynamics at Naruto strait and Tohoku district off in Japan will be selected.

2. The Naruto strait

The bathymetry of the Naruto strait produces strong tidal current accompanied by a series of vortices along free boundary layers. These vortices of organized structure amalgamate into a large scale vortex pair. ONISHI (1984) discussed some of the dynamic features of these vortices. In the present paper, the role of these vortices in the mass exchange between both sides of the strait will be discussed.

The Naruto strait is a narrow water course between the Shikoku- and Awaji island as shown in Fig. 1. The width of the strait (about 1000 m) is limited by the shoals and the maximum water depth reaches about 80 m as shown in Fig. 2.

The tide is semi-diurnal as shown Fig. 3 and at a spring tide the difference of the tidal levels across the strait reaches about 1.5 meter and causes the current velocity of about 10 knots. The current flowing to the Kii channel from the inland sea is a *"southward current"*, while the reverse is a *"northward current"*.

Landsat data for northward tidal currents at maximum velocity (A), slack tide (change from northward to southward currents (B) and for southward currents (C) were analyzed.

In Fig. 4, the water mass patterns obtained from Landsat data as well as the

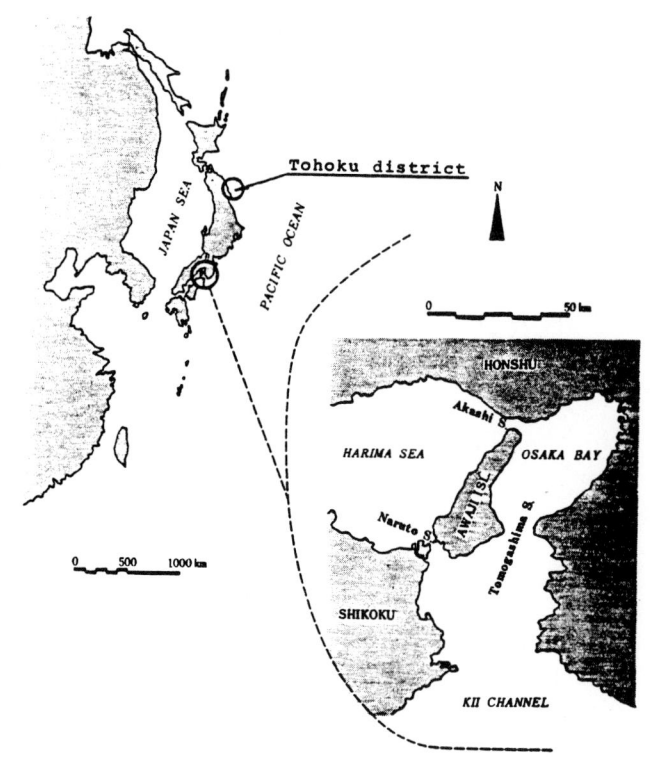

FIGURE 1. Locations of the Naruto strait and the Tohoku district

corresponding phases of the tidal currents are indicated.

PLATE 1 is an aerial photograph at southward currents taken at an altitude of 1000 meter, showing close views of the tidal jets (MARUYASU, ONISHI and NISHIMURA, 1981). One can see that the current patterns represents that of surface water jet and that the eddy size increases with distance downstream.

The dynamic features of these vortices have been reported (ONISHI, 1984). Suffice it to say here that the strength of these range from $60 \, m^2 \, sec^{-1}$ and $390 m^2 \, sec^{-1}$. In the near field, the order of the magnitude of the velocity (u) is $5 \, m \, sec^{-1}$ and the length (L) of the jet is 1000 m. Therefore, orders of the inertia effect and Coriolis effect are estimated as follows:

$$u \frac{\partial u}{\partial x} \rightarrow U^2 / L = 0.025 \, m \, sec^{-1}$$

$$2v\Omega \sin \theta = 2 \times 5 m \, s^{-1} \times (7.27 \times 10^{-5} / s) \times \sin 34° = 4 \times 10^{-4} m \, s^{-2}$$

This implies that the Coriolis effect can be neglected. The flow patterns in Fig. 4 shows larger scale vortex pairs formed in far field of the jet flow and that in Plate 1 shows the vortices in the turbulent region. Distribution of these eddies in the strait are schematically illustrated in Fig. 5.

Circulation of the vortex is defined by

$$\Gamma = \int v_s ds \tag{1}$$

where v_s is tangential velocity component and ds the element of the closed curve outside the vortex. ONISHI and NISHIMURA (1980) estimated the circulation of the vortex pair in the strait to be about $1.1 \times 10^4 \, m^2 \, sec^{-1}$. The vortex shape depends on the phase of the

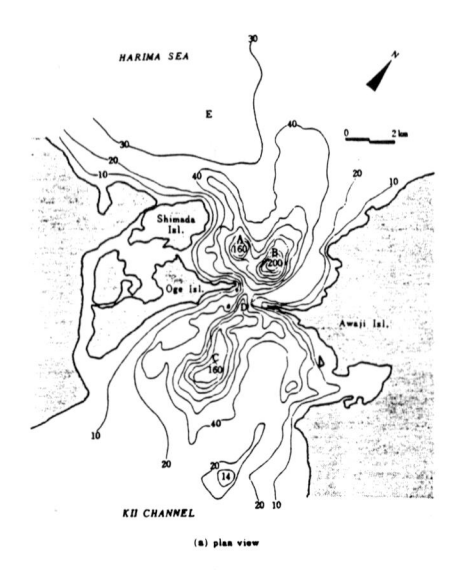

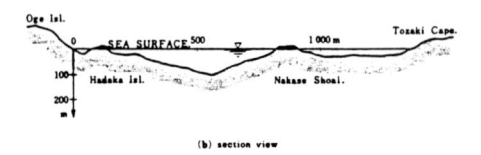

FIGURE 2. Geographical configuration of Naruto strait

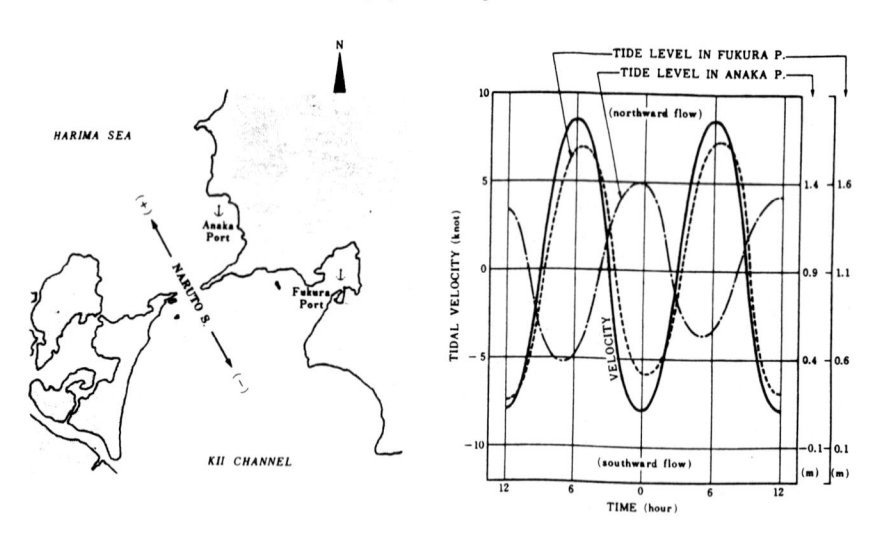

FIGURE 3. Relation among tidal currents and tidal levels at spring tide

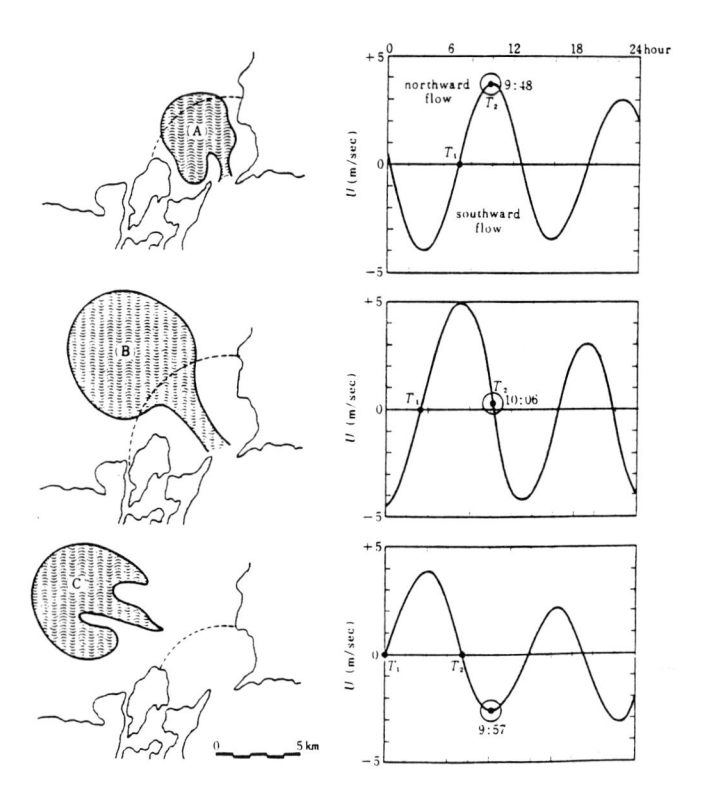

FIGURE 4. Water mass patterns in the different phases of the tidal currents obtained from the Landsat images

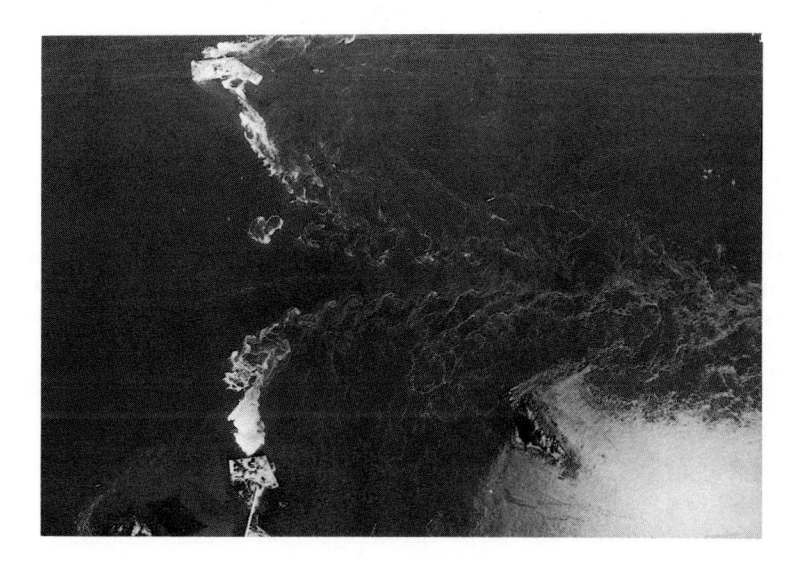

PLATE 1. Southward current observed at 1000 m height.

tidal cycle as sketched in Fig. 6.

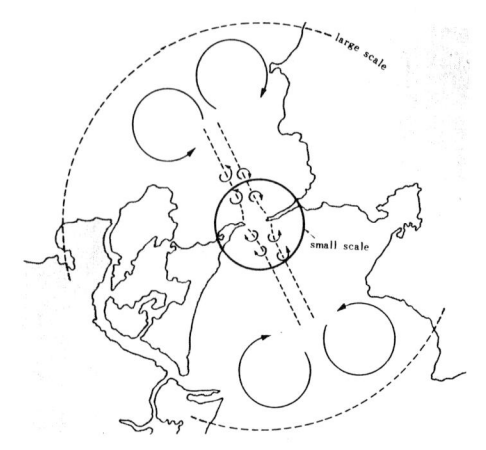

FIGURE 5. Modelling of the vortex information in Naruto strait.

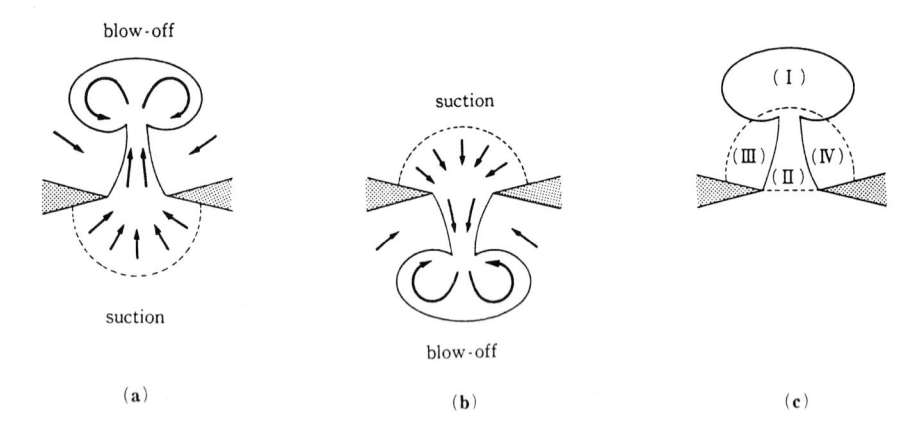

FIGURE 6. The effect of the shape of vortex pair on the mass exchange.

Figure 6(a) shows the situation at slack tide condition. Ambient sea water enclosed by the dotted line in the figure enters the strait and blows off into the wide sea area in the downstream side forming a narrow water course bounded by the two free boundary layers. Figure 6(b) shows the situation at the next slack condition (6 hours later). The figure indicates that the difference in the flow patterns between the sucking and the blowing sides causes the effective tidal exchange through the strait when the full tidal cycle of 12 hours has been completed.

The water volume (Q) passing through the strait during the first half of the tidal cycle is

$$Q = V_1 + V_2 \qquad (2)$$

where V_1 and V_2 are the water volumes contained in the regions (I) and (II) in the figure. In the above equation, turbulent entrainment of the ambient water into the vortex pair is neglected. In the same way, the volume flux of opposite sign during the next half of the tidal cycle is

$$Q' = V_2 + V_3 + V_4 \tag{3}$$

The water volume of (I) remains in the area during the next half tidal cycle and does not return. Therefore, the net volume flux through the strait during the full tidal cycle is V_1 and the tidal exchange ratio becomes

$$r = \frac{V_1}{V_1 + V_2} \tag{4}$$

From the Landsat images shown in Fig. 4, the values of $V_1 = 0.94 \times 10^9 \, m^3$, $V_1 + V_2 = 1.42 \times 10^9 \, m^3$ and $r = 0.66$.

3. Large scale eddies in the Tohoku district

In the Tohoku district, the Kuroshio current and the Oyashio current interact with each other and produce a confused water circulation containing many large scale eddies (KAWAI, 1955). Those eddies may play important roles in the mixing of the coastal water with the ocean water.

Fig. 7 shows the horizontal distributions of the water temperatures and velocity at the water surface measured in Apr.13, 1982.

The distributions of the water temperature and salinity along the lines $A-A$, $B-B$ and $C-C$ in Fig. 7, are represented in Figs. 8 to 10, respectively. Landsat data in Apr.26, 1982 were analyzed for current patterns. The result is sketched in Fig. 11. One can see that the coastal water mass moving offshore terminates in anti-clockwise eddies with a diameter of about 30 km and also that the eddy exists nearby section $B-B$ in Fig. 7. The order of the magnitude of the current velocity is $10 \, cm \, sec^{-1}$ and

$$2u\Omega \sin\theta = 2 \times 0.1 \times (7.27 \times 10^{-5}) \sin 30° = 10^{-5} \, m \, sec^{-1}$$

The length scale of the current is estimated approximately to be 1000 m for the inertia and Coriolis terms to be of equal order. The length scale of the current in Fig. 11 is about 30 km, i.e. much larger than 1000 m. Therefore, the currents are in geostrophic balance and the inertia effects can be neglected.

In Fig. 8, one can see that the eddy has a diameter of about 30 km and reaches the sea bottom at approximately 200 m water depth. The configurations of the contour lines of the salinity as well as the water temperature near the sea bottom suggest existence of upwelling from the bottom boundary layer toward the inside of the eddy.

The time scales of the eddies are controlled by Ekman pumping, i.e. the entrainment of the bottom boundary layer. The vorticity of the eddy and thickness of Ekman boundary layer are presented by the following equations, respectively.

$$\zeta = \partial v / \partial x - \partial u / \partial y \tag{5}$$

$$\delta = \sqrt{2\nu / f} \tag{6}$$

where, u and v are the components of velocity in horizontal directions (x and y directions), ζ is the vorticity, δ the thickness of the boundary layer, ν the coefficient of kinematic eddy viscosity and f the Coriolis parameter.

The velocity of upward flow into the inside of the eddy from the boundary layer, w_0, can be represented by the following equation (GREENSPAN, 1968).

$$w_0 = \frac{1}{2} \delta \cdot \zeta \tag{7}$$

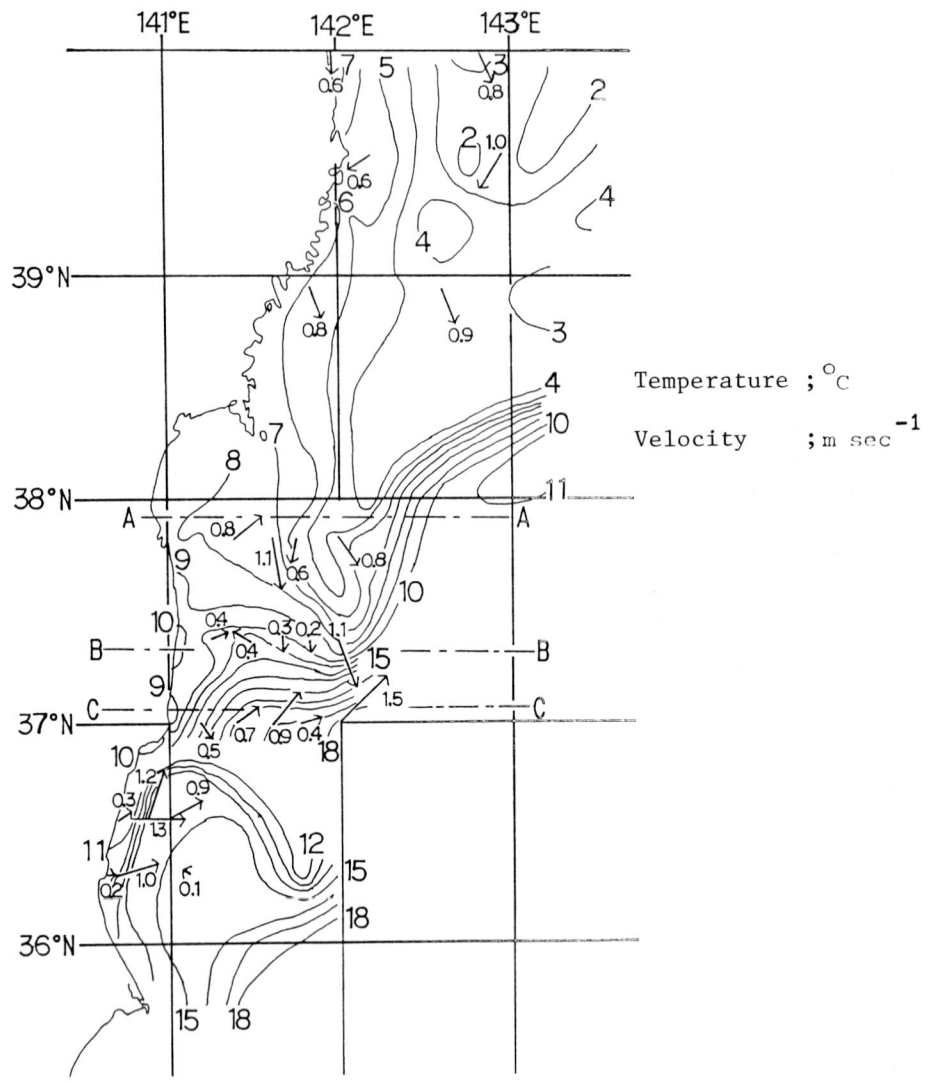

FIGURE 7. Horizontal distribution of water temperature and current velocity at sea surface in Apr. 1982.

Spin-down time t is presented by

$$t' = \frac{2H}{\delta \zeta}\left(\frac{\Delta\omega}{\omega}\right) \tag{8}$$

where H is the water depth, ω an angular velocity of the eddy, $\Delta\omega$ an increment in the angular velocity during a time interval t. Considering that $\zeta = 2\omega$, one obtains the following relation

$$t = \frac{\Delta\omega H}{\delta\omega^2} \tag{9}$$

In the present case, it is estimated that $H = 200\,m$, $\omega = v/R$ (R; the radius of eddy) $= 0.5\,ms^{-1}/10^4(m) = 5 \times 10^{-5}s^{-1}$ and $\nu = 10^{-4}m^2s^{-1}$. So that $\delta = 1.5\,m$ and the time scale of the existence of the eddy is of order of 15 days. The real time scale is likely to

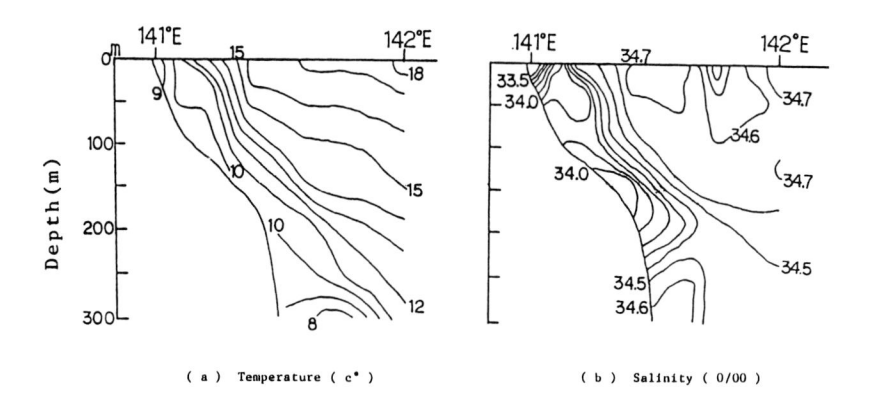

FIGURE 8. Vertical distributions along Section $A-A$ (Apr. 17 to 23, 1982).

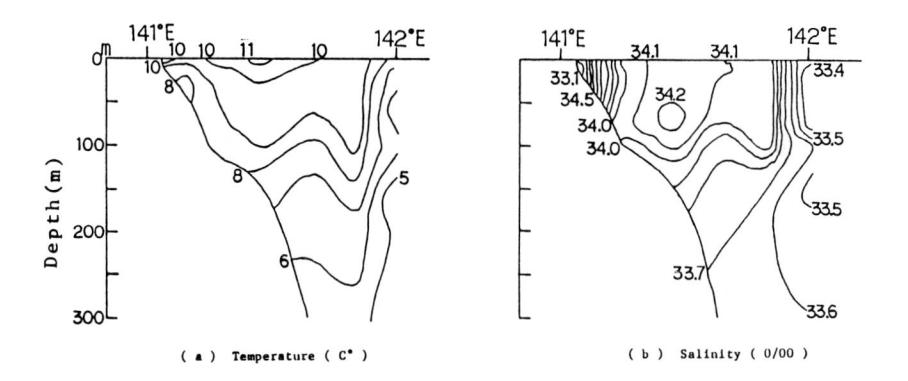

FIGURE 9. Vertical distributions along Section $B-B$ (Apr. 17 to 23, 1982).

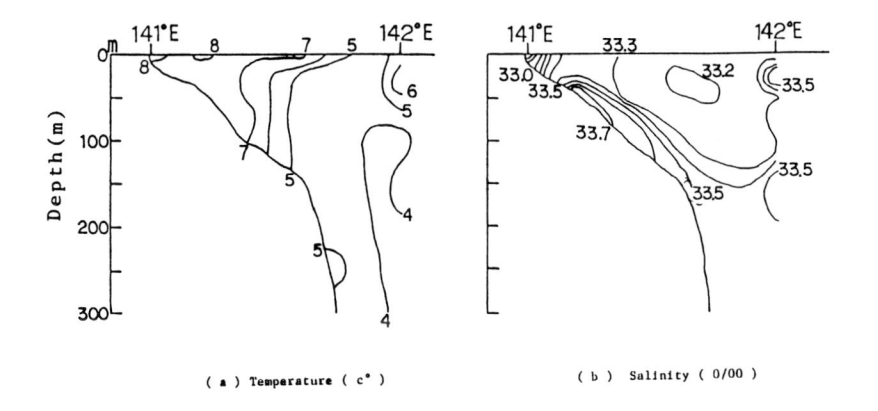

FIGURE 10. Vertical distributions along Section $C-C$ (Apr. 17 to 23, 1982).

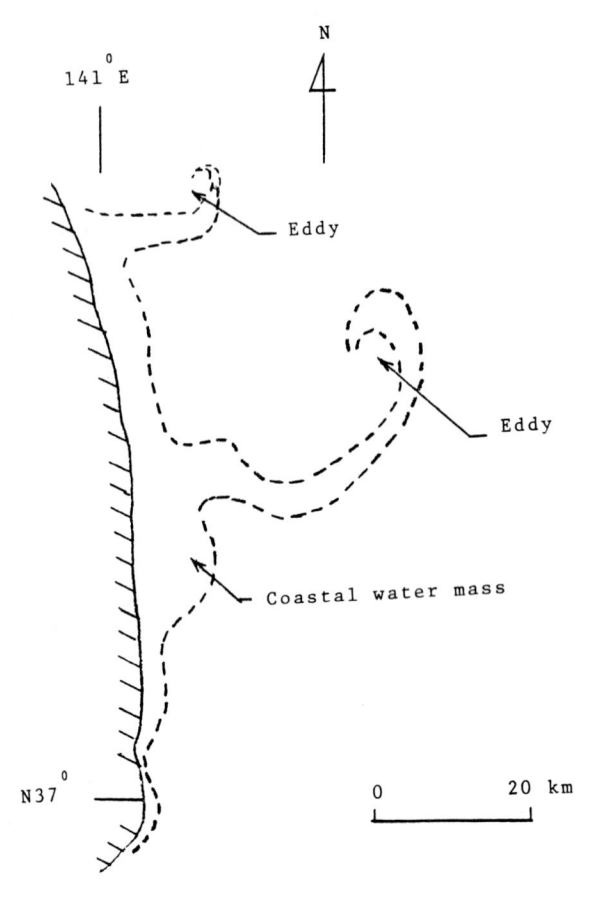

FIGURE 11. Large scale eddies obtained by Landsat in Apr.26, 1982

be much longer. Indeed, SAITO et al. (1983) measured life periods of eddies in deeper water (about 1000 m) and reported the meso scale eddies in this water are to exist for 9 months and more. Hence, in the coastal water region of this district, there exist meso scale eddies with the diameter of several tens of kilometers. Those eddies entrain the coastal water mass. (See also the vast literature on Gulf Stream Eddies.) Further, the water mass in the bottom boundary layer is upwelled into the inside of the eddies. The coastal water mass entrained in the inside of the eddies may remain for very long periods, of orders of weeks to months.

4. Conclusions

When the volume of the effluent in coastal water is large, or when the effluent contains pollutants of long life period, interaction between the coastal water and the oceanic water must be studied separately, as the oceanic and coastal flow dynamics may interface to produce complex flow circulation, including trapping in eddies. Although the studies on the diffusion phenomena in the coastal water zone are numerous, the interaction between the coastal and oceanic waters has not been studied in great detail so far. In this paper, the complexity of the resulting flow is illustrated by the examples, in Japan, of the water zones around the Naruto strait and the Tohoku district. The roles of large scale eddies in the mixing of coastal water with oceanic water seems to be very important.

In the Naruto strait, the Coriolis effect is not important. In this case the tidal residual flow plays an important role in the mass exchange between both sides of the strait. In this paper the author described the exchange processes in the strait on the basis of a simple vortex pair model.

In the Tohoku district, the movements of the coastal water are geostrophic in nature and are affected by the oceanic currents. In the Tohoku district, the coastal water mass may be entrained occasionally into the meso scale eddies with diameters on the order of tens of kilometers, located approximately 30 km offshore from the coastal line and having life times of several months. The large eddies in the Tohoku district are produced by the interactions between various oceanic currents, among which the effects of the Kuroshio current and the Oyashio current are considered to be most dominant. Therefore, meandering of the Kuroshio current may not be ignored in the analysis of the diffusion of the flow with the spatial scale more than several tens of kilometers.

References

GREENSPAN, H.P. 1968. The theory of rotating fluid, p.36 (Cambridge University Press).

KAWAI, H. 1955. On the polar frontal zone and its fluctuation in water in the north-east of Japan, Bulletin of Tohoku District Fish Research Laboratory.

MARUYASU, M., S. ONISHI and T. NISHIMURA 1980. Study on the tidal vortex in the Naruto strait through remote sensing, Bulletin of the Remote Sensing Laboratory, remote sensing series No.1, Science University of Tokyo.

ONISHI S. and T. NISHIMURA 1981. Study on tidal exchange phenomena at straits applying remote sensing from Landsat, Journal of Japan Society of Civil Engineering,, Vol. 298, pp.63-75.

ONISHI, S. 1984. Study of vortex structure in water surface jets by means of remote sensing, Remote sensing of shelf area hydrodynamics (edited by J.C.J. NIHOUL, Elsevier Science Publishers), pp. 107-132.

SAITO, S., S. MISHIMA and J. IIZUKA 1983. Observation of oceanic fronts in the water region of north-east zone of Japan, Proceedings of annual meeting of Japan Society of Oceanography, pp. 46-47 (in Japanese).

Coastal Circulation in the Key Largo Coral Reef Marine Sanctuary

Thomas N. Lee

Rosenstiel School of Marine and Atmospheric Science
University of Miami, USA

ABSTRACT

Analysis of current meter data obtained in the Key Largo Coral Reef Marine Sanctuary over a 3-year period from 1981 to 1983 indicates that the Sanctuary can be separated into two flow regimes with greatly different flow properties: the deep outer shelf where current and temperature variability is dominated by the Florida Current; and the shallower parts of the Sanctuary including Hawk Channel and the reef tracks, where tidal and atmospheric forcing controls current and temperature variability.

Large amplitude current fluctuations and strong northward mean flows that occur along the outer shelf are a direct consequence of the meandering and eddy shedding of the Florida Current within a broad period band of 2 to 15 days. The influence of the Florida current appears to be greater in the northern part of the Sanctuary than in the southern part, presumably due to the tendency for the Florida Current to follow the isobaths that converge shoreward toward the north.

Currents in Hawk Channel and shallower parts of the Sanctuary are strongly influenced by tides and local winds. Tides account for approximately 80% of the total cross-shelf current variance and 50% of the along-shelf. The remaining variance is most likely due to wind forcing which is coherent over spatial scales much greater than the dimensions of the Sanctuary

Mean flows in the shallower part of the Sanctuary show a seasonal cycle that appears to follow seasonal wind patterns; toward the south in fall and winter at 2 to 5 cm s^{-1} and toward the north in spring and summer at 2 to 4 cm s^{-1}. There is also a tendency for the mean flow to diverge, with northward flow occurring more often in the northern part of the Sanctuary and southward flow in the southern part. This flow divergence may be supported by a mean onshore flow at mid-depth from the outer shelf.

Coral reefs within the Sanctuary appear to be protected from harmful exposure to chilled Florida Bay waters following winter cold-air outbreaks by the solid coastal boundary of Key Largo, and the generation of southward coastal currents by winds from the north after cold front passages that advect cold Florida Bay waters southward, away from the Sanctuary.

1. Introduction

Current and temperature variability in the narrow coastal band off the upper Florida Keys is investigated. Current meter data were obtained in the Key Largo Coral Reef Marine Sanctuary (Pennekamp Park) during a three year period from November 1980 to November 1983 by General Oceanics, Inc. as part of the NOAA National Marine Sanctuaries Program.

Key Largo Coral Reef Marine Sanctuary consists of about 260 km² of reef track located east of Key Largo. The dimensions of the Sanctuary are approximately 32 km in length and 5-9 km in width, extending from near the shoreline to the 92 m isobath on the continental slope. The northern boundary lies approximately due east of the Angelfish Creek/Broad Creek inlet complex connecting Card Sound and South Biscayne Bay to the coastal waters. The southern boundary is situated east of Rodriguez Key and is about 13 km north of Tavernier Creek, which is the nearest waterway connecting Florida Bay to the coastal waters. Along the eastern boundary of the Sanctuary lies the strong northward flowing Florida Current.

Shelf circulation off Miami and Key Biscayne, 45 km to the north, results from the combined effects of tide, wind and Florida Current forcing (LEE and MAYER, 1977, MAYER and HANSEN, 1975). Low frequency currents tend to align with the local isobath orientation resulting in preferred northeast-southwest flow directions. LEE and MAYER (1977) found that the shelf off Miami Beach could be separated into characteristic flow regimes. Flow variability in the outer-shelf, in the vicinity of the shelf break, was primarily controlled by the Florida Current in the form of northward propagating wave-like meanders of the Florida Current front and random formation of spin-off eddies. Currents in the inner- to mid-shelf were mainly the result of tide and wind forcing. MAYER and HANSEN (1975) reported similar flow properties for the shelf off Virginia Key. In addition they found that the nearby Bear Cut and Norris Cut inlets produced a cross-shelf tidal current component due to the waters jetting through the inlets.

Tidal fluctuations of sea level in the Sanctuary are primarily semi-diurnal with a range of 0.6 to 1.2 m. Tidal currents in the Florida Straits have both semi-diurnal and diurnal components with amplitudes of about 4 and 6 cm s^{-1} respectively (SMITH, ZETLER and BROIDA, 1969; ZETLER and HANSEN, 1970; KIELMAN and DUING, 1974).

Current response to wind forcing is both direct and indirect. Direct response results from wind drag on the water surface producing a surface current with a magnitude of about 3% of the wind speed and in the direction of the wind in the shallow depths of the Sanctuary. The indirect response comes from wind-induced set-up and set-down of coastal sea-level, which produces a cross-shelf pressure gradient that is in geostrophic balance with an along-shelf barotropic flow (BEARDSLEY and BUTMAN, 1974; SCOTT and CSANADY, 1976; LEE, HO, KOURAFALOU and WANG, 1984). Northward winds (to the north) which occur with the onset of a winter cold front can set-down coastal sea level and drive a northward current in the nearshore waters. Southward winds behind the cold front can cause the opposite effect. During the winter months (November-April) cold fronts occur on the average of one per week. During the summer winds are light and variable, primarily from the southeast direction and the effects of wind forcing should be reduced.

Onshore/offshore meanders of the Florida Current western boundary and the formation of spin-off eddies occur on time scales of 2 days to 2 weeks throughout the year (LEE, 1975; LEE and MAYER, 1977). An offshore meander can produce a decline in northward current speed coupled to a rapid decrease in near bottom temperature brought about by upwelling of deeper, cooler Florida Current water. An onshore meander can reverse the process, resulting in strong northward flow and warmer temperatures. The passage of a spin-off eddy can produce a cyclonic current reversal and strong cooling of near bottom waters due to upwelling in the eddy's cold-core. LEE, ATKINSON and LEGECKIS (1981), and LEE and ATKINSON (1983) have found that off north Florida and Georgia, eddy induced upwelling is the major source of nutrients to the outer-shelf.

The cyclonic circulation in spin-off eddies can enhance the exchange of coastal and Florida Current waters. LEE and MAYER (1977) estimated a one week residence time for the shelf waters off Miami Beach as a direct consequence of the eddy induced flushing action, which occurred on the average of one per week.

Sudden cooling of the waters overlying the reefs may have a harmful effect, resulting in loss of living reef area (ROBERTS, ROUSE, WALKER and HUDSON, 1982). One aspect of this research is to determine if potentially harmful cooling events are due to Florida

Current induced upwelling, wind induced upwelling, direct cooling due to air-sea exchange during cold-front passages, or advection.

2. Measurements

Current and temperature data were collected in the Marine Sanctuary over a three year period from Nov. 1980 to Nov. 1983. The current meter array is shown in Fig.1; relevant information regarding mooring and instrument locations is given in Table 1.

TABLE 1: Relevant current meter information

Current Meter I.D.	Record Rotation	Mooring	Bottom Depth(m)	Inst. Depth(m)	Location
A11	50°	A	24	12	24°54.0'N 80°31.5'W
A12	50°		24	23	24°54.0'N 80°31.5'W
B11	50°	B	6	5	24°56.0'N 80°34.0'N
D11	35°	D	29	15	25°08.5'N 80°15.0'W
D12	35°		29	27	25°08.5'N 80°15.0'W
E11	35°	E	3	2	25°10.5'N 80°19.0'W
F11	15°	F	32	17	25°22.0'N 80°08.0'W
F12	15°			31	25°22.0'N 80°08.0'W
G11	10°	G	6	5	25°23.0'N 80°11.5'W
I11	10°	I	6	5	25°30.5'N 80°08.0'W
H11	10°	H	6	5	80°14.0'W

* Data Record Label example, A12:
- A is the mooring identification;
- 1 is the first mooring deployment period;
- 2 lower current meter.

Niskin Winged Current Meters (NWCM's) model 6011 were used at all moorings. On

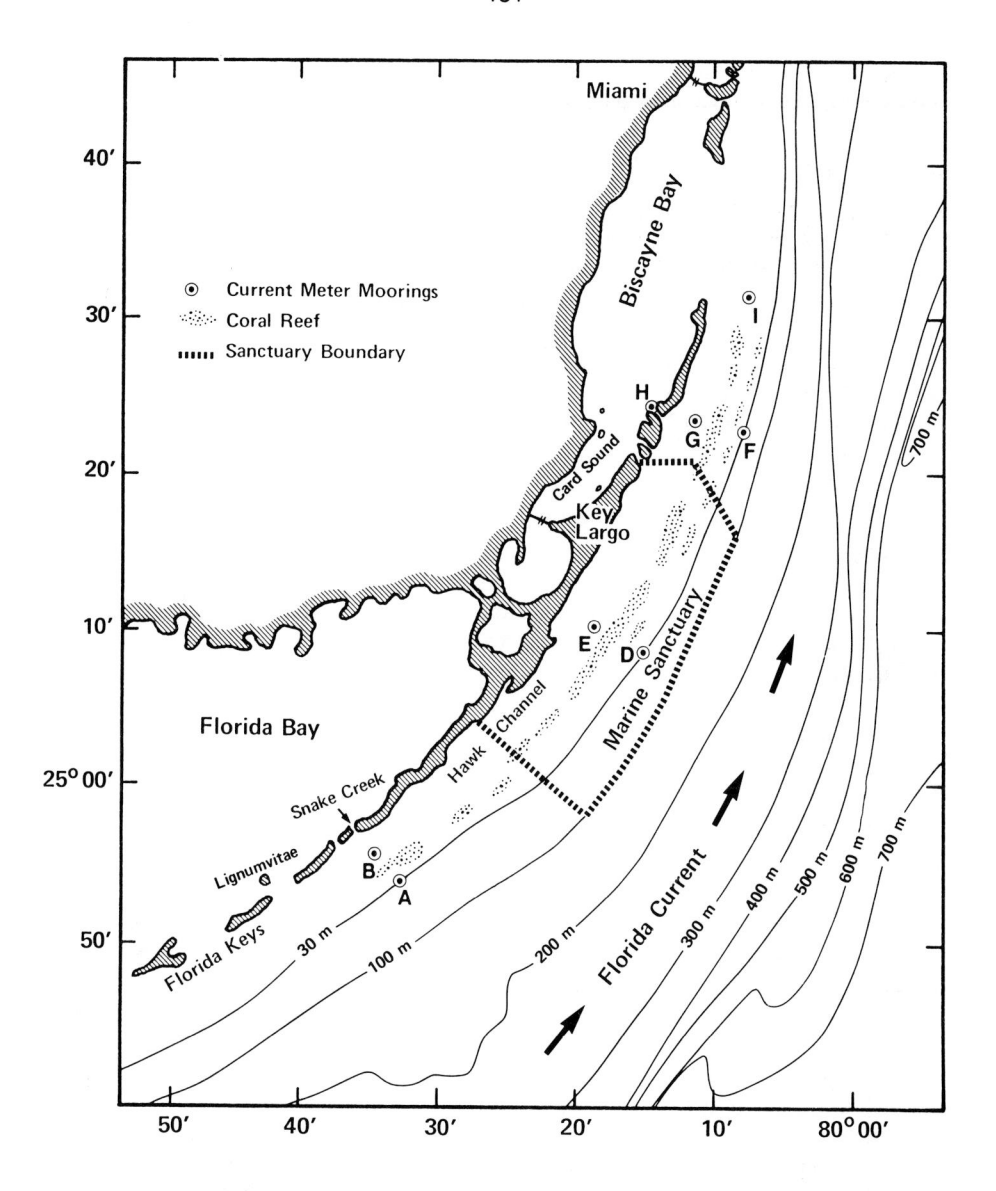

FIGURE 1: Key Largo Coral Reef Marine Sanctuary and curent meter array.

the shallow moorings the current meters were set in a *burst sampling mode* (8 samples spaced 1.32 seconds apart) to record current and temperature data every half hour in order to remove surface wave *noise*. During the first year moorings *B*, *D*, *E* and *G* were deployed with an additional mooring located between *D* and *E* (mooring #3) for the periods: Nov. 1980 to March 1981, and June to Oct 1981. Unfortunately during the first deployment the only usable data was obtained from mooring *E* due to lost moorings and equipment failures. The full eight mooring array was first deployed in July 1982 and maintained with minimum data loss until Nov. 1983.

The current meter array spans the along- and cross-shelf dimensions of the Sanctuary. Three moorings were positioned along the shelf edge (moorings *A*, *D* and *F* near the 30 m isobath) with current meters positioned at mid-depth and near bottom. Another three moorings were located in Hawk. Channel with current meters located near bottom (moorings *B*, *E*, *G*).

Mooring *I* was positioned at the northern boundary of the Sanctuary at mid-shelf and was equipped with a near bottom current meter. Mooring *H* was located in Ceasar's Creek with a near bottom current meter.

The data record labels are set up in sequence to define the mooring location (*A* to *I*); deployment period (1 to 5); and location of current meter on the mooring (1 = top, 2 = bottom). Wind data were obtained from the Miami Airport and Key West. All current data were rotated to transform current vectors into an isobath coordinate system with *v* corresponding to along-shelf flow (+ towards the north) and *u* cross-shelf flow (+ toward the east); and filtered with 3 and 40 hour low pass filters to smooth the data and remove tidal fluctuations, respectively.

3. Results

3.1. Outer shelf

Subtidal time series of currents and temperature for typical summer and winter periods are shown in Figs. 2-5. Mid-water current speeds along the shelf edge reach speeds >60 cm s^{-1} and were typically much stronger than in Hawk Channel, presumably due to the closer proximity of the Florida Current. Significant vertical shears were on the order of $10^{-2} s^{-1}$ for both the mean and fluctuating parts of the flow along the shelf break, which is also an indication of the presence of baroclinic Florida Current. Current speeds in Hawk Channel were generally <20 cm s^{-1}.

Outer shelf locations *D* and *F* show a significant Florida Current influence during all seasons. Mean along-shelf flows at mid-depth in the outer shelf were strong northward, ranging from about 15 to 30 cm s^{-1}, with standard deviations nearly as large as the mean ± 12 to 25 cm s^{-1}. Mean along-shelf currents near the bottom were northward ranging 5 to 15 cm s^{-1}. Mean cross-shelf flow in the outer shelf was onshore at mid-depth and offshore near bottom during all seasons with speeds ranging from 2-4 cm s^{-1}. Standard deviations of the cross-shelf flows were generally around 1 $\pm$ to 2 cm s^{-1}, which is considerably less than the along-shelf fluctuations, and shows the tendency for both the mean and fluctuating currents to align with the local topography.

Maximum currents of 50 to 60 cm s^{-1} northward were observed at station *F* at mid-depth during all seasons. The northward mean flow was also greatest at *F* (29 cm s^{-1}) and decreased with distance to the south reaching a minimum at station *A* of 1 to 3 cm s^{-1}. This trend for weaker currents in the southern part of the Sanctuary along the outer shelf was observed in each deployment period from the start of the 8 mooring array (July, 1982) and appears to be a real feature, most likely related to the proximity of the Florida Current front. Fig.1 shows that the topography of the Florida Straits converges shoreward near the northern region of the Sanctuary, which would bring the western edge of the Florida Current closer to the shelf break at station F than at station A on the average, and thus cause an increase in northward current speeds at F.

The percent of the total current or temperature variability occurring with periods longer than 40 hours (low-frequency) and at tidal periods was determined from the ratio of the variance of the 40 HLP filtered records to the variance of the 3 HLP filtered records. These results indicate that subtidal frequency fluctuations accounted for about 30% to 85% of the total along-shelf current variance in the outer shelf and 25% to 40% of the cross-shelf with no appreciable difference between mid-depth and near bottom locations. Tidal and inertial fluctuations accounted for the remaining variance, which indicates that cross-shelf current variations are mostly tidal. The percent of the total along-shelf variance due to low-frequency motions was highest at the northernmost station, F, where 70 to 85% of the variance was due to subtidal motion, and was lowest at A, where 25% to 65% was accounted for by low-frequency variations. The low-frequency part of the variance is made up of Florida Current and wind induced fluctuations, which again suggest a greater Florida Current influence in the northern part of the Sanctuary.

Temperature in the outer-shelf ranged from a minimum of about 24°C during the winter period to a maximum of about 30°C during summer/fall. Mean temperatures were quite uniform throughout the outer shelf.

3.2. Hawk channel

Mean along-shelf flows in the shallow Hawk Channel area were toward the north at 1 to 4 cm s^{-1} during the summer and southward at -2 to -5 cm s^{-1} during the winter. At times a divergence in the along-shelf mean flow occurred with southward flow in the southern part of the Sanctuary and northward flow in the northern region. This mean flow divergence may have been compensated for by mean onshore flow observed at mid-depth in the outer shelf. Mean cross-shelf currents in Hawk Channel were also onshore at -1 to -2 cm s^{-1}.

Low-frequency current fluctuations in Hawk Channel had standard deviations of ± 6 to 8 cm s^{-1} in the along-shelf and ± 1 to 2 cm s^{-1} in the cross-shelf, which is considerably less than occurred in the outer-shelf. Maximum subtidal currents were about 15 cm s^{-1} in the along-shelf direction.

Current variability in Hawk Channel is strongly influenced by tidal/inertial motions, with 30 to 60% of the total along-shelf current variance accounted for by tidal/inertial motions and about 85% of the cross-shelf current variance. Site H is located within Ceasar's Creek, which is tidally dominated as expected with about 98% of the current variance produced by the tides. The mean flow at site H was mostly cross-shelf, since the inlet is oriented in an east/west direction, westward during winter/spring and eastward during summer/fall. These mean flow directions are consistent with the seasonal evaporation/precipitation cycle of the region. During the winter/spring dry season evaporation exceeds precipitation in Biscayne Bay and Card Sound and a net flow into the Bay should result. During the summer/fall wet season the opposite occurs and a net flow out of the Bay should result (LEE and ROOTH, 1976).

Temperature fluctuations throughout the Sanctuary are primarily produced by processes with periods longer than the tidal/inertial periods. Approximately 60 to 98% of the total temperature variance at all stations occurred as low-frequency fluctuations, the cause of which is most likely atmospheric air-sea heat exchange in Hawk Channel in the shallow part of the Sancutary, and Florida Current forcing in the outer shelf.

3.3. Summer

Large amplitude current fluctuations, that are coherent over the lower half of the water column, occurred along the shelf edge in the 2-day/2-week period band (Fig. 2). At times the current reversed for short periods toward the south and occasionally strong southward flow occurred following a cyclonic current reversal coincident with a drop in temperature. The current and temperature signature of this type event is very similar to that described for cyclonic spin-off eddies (LEE, 1975, LEE and MAYER, 1977). These eddies form along the Florida Current cyclonic front during an offshore meander of the Current. The drop in temperature is produced by upwelling of deeper, nutrient rich Florida Current water within the cold-core of the cyclonic eddy. These events appear at times during the summer to influence the flow field in Hawk Channel as well as the shelf edge.

Subtidal current fluctuations at stations D and F appear to be visually coherent for most of the record, indicating a coherent length scale of greater than 28 km. Current reversals occurred about 18 hours later at station F than at D, indicating a northward propagation of about 40 cm s^{-1}. LEE, BROOKS and DUING (1977) found the downstream coherence length scale of 9-12 day period northward propagating along-stream current fluctuations in the western side of the Florida Straits off Miami to be about 60 km. From mid-August to mid-October the visual coherence of several day period current fluctuations appears to be significant between all stations except A. This indicates that either the influence of the Florida Current fluctuations are extending into the shallow Hawk Channel region during this period or that wind forcing is driving the flow events, which become amplified at the shelf break. However, wind forcing is typically not very strong during summer and it is unlikely that wind forcing could produce the large amplitude, vertically sheared, northward propagating disturbances at the shelf break.

Low frequency temperature time series (Fig. 3) show high visual coherence over the lower half of the water column at the shelf break, with larger amplitude fluctuations occurring at the near bottom locations. Vertical stratification was generally weak, about .03°C/m, over the 11 and 12 m separation distances at A and D, and .06°C/m at F. Since the upper meters were positioned at depths of 12 to 17 m they may have been below the summer seasonal thermocline so that the vertical gradients may not reflect that of the total water column. Temperature fluctuations were generally visually coherent between stations along the shelf edge as were stations along Hawk Channel, but correlation was poor between shelf break sites and Hawk Channel.

Typical energy spectra for the along-shelf and cross-shelf current components from the shelf edge and Hawk Channel for the summer are shown in Figs. 6 and 7, respectively. The energy levels of along-shelf current fluctuations were about an order of magnitude higher than that of the cross-shelf for all periods longer than tidal at both sites. Well-defined coherent spectral peaks occurred at periods of the semi-diurnal tide (12.4 hrs), diurnal tide (24 hr.), 2-3 days and 4-10 days. Generally the current components were about 180° out of phase, or rectilinear. Energy levels of the along-shelf component were higher at the shelf edge than in Hawk Channel, but the cross-shelf components were about the same, except that Hawk Channel site was much more energetic at the semi-diurnal tidal period.

3.4 Winter

Subtidal time series of currents and temperature from the winter period are shown in Figs. 4 and 5. At the shelf break energetic current fluctuations in the 2-day to 2-week period band show significant visual correlation over the lower half of the water column

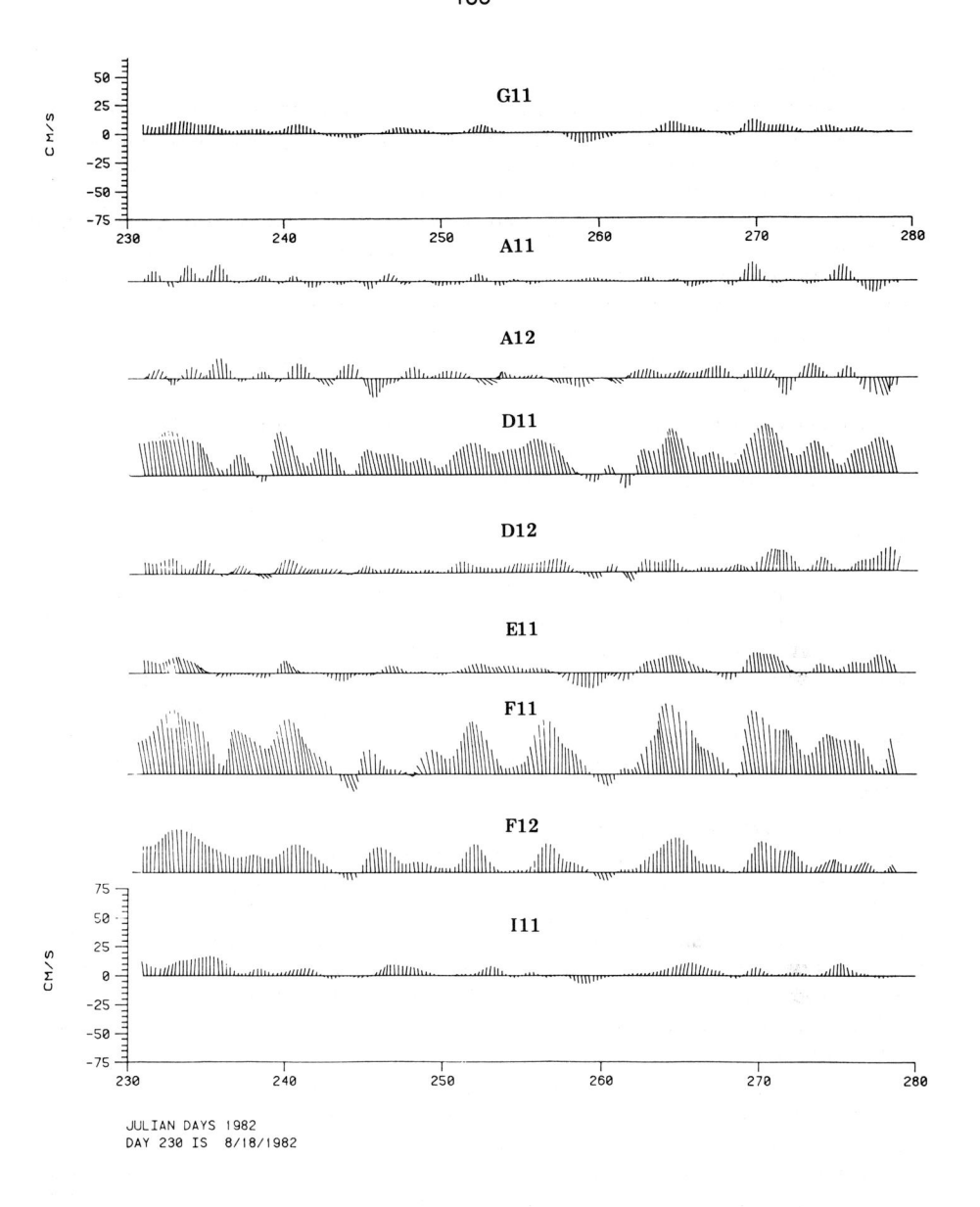

FIGURE 2: Subtidal current vectors during summer 1982.

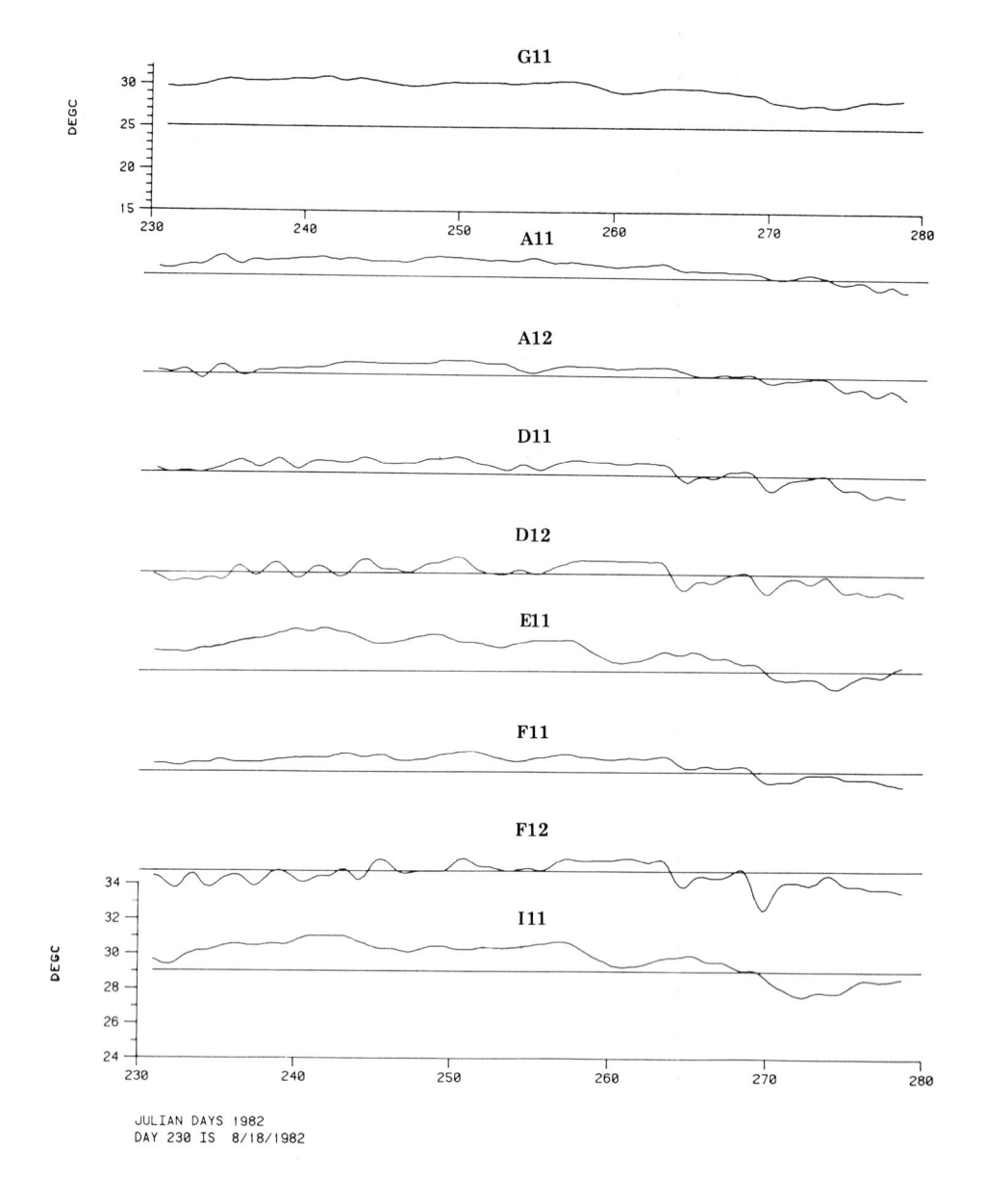

FIGURE 3: Subtidal temperature time series during summer 1982.

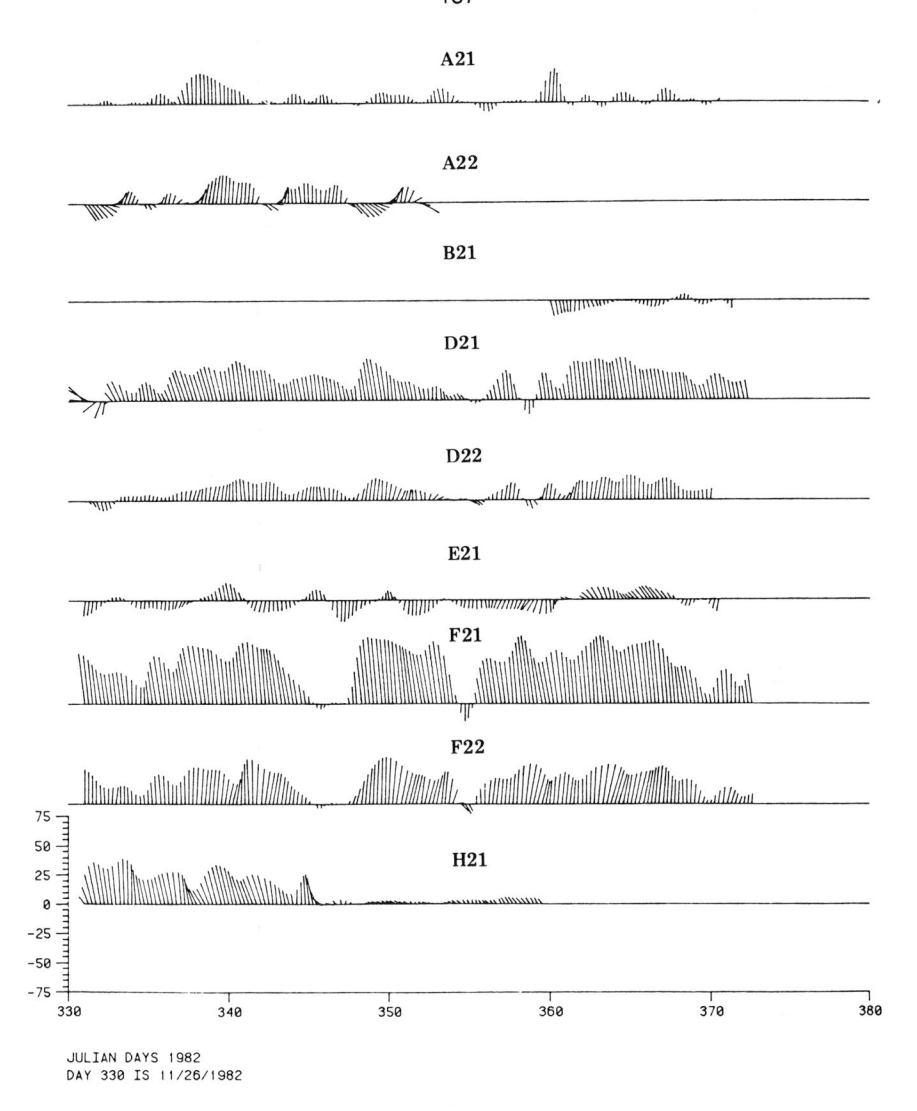

FIGURE 4: Subtidal current vectors during winter 1982.

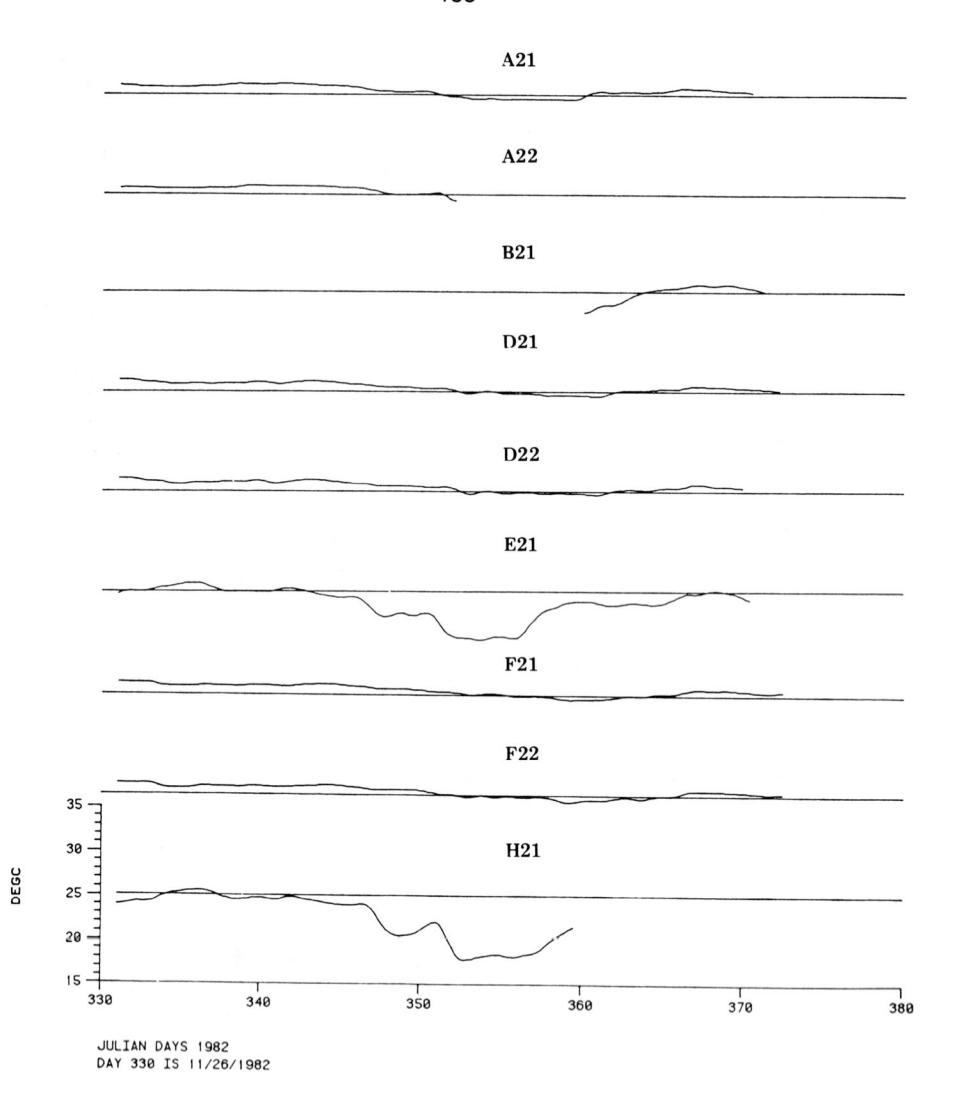

FIGURE 5: Subtidal temperature time series during winter 1982.

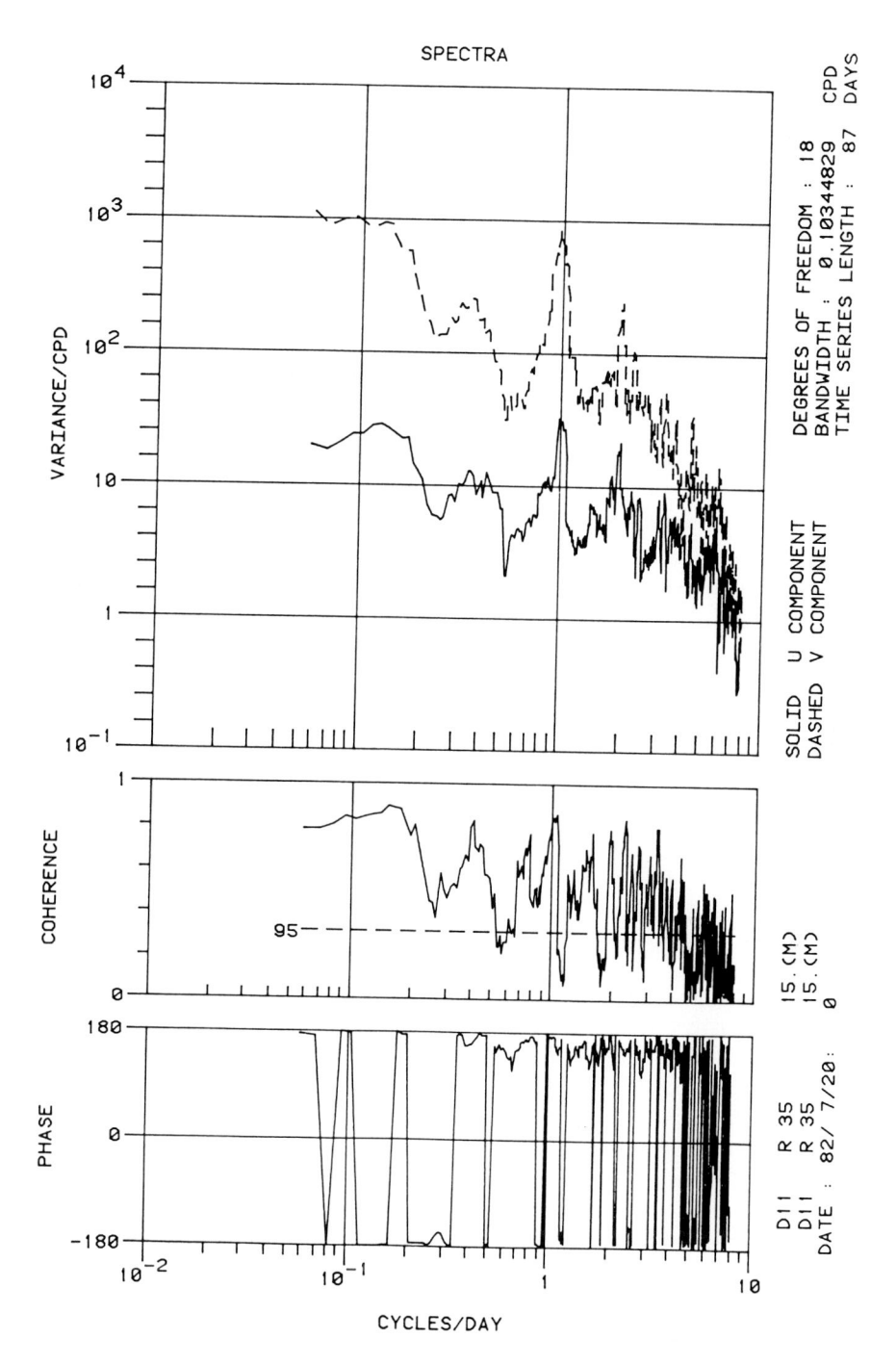

FIGURE 6: Spectra, coherence squared and phase of cross-shelf (U) and along-shelf (V) velocity components al D11 during summer 1982.

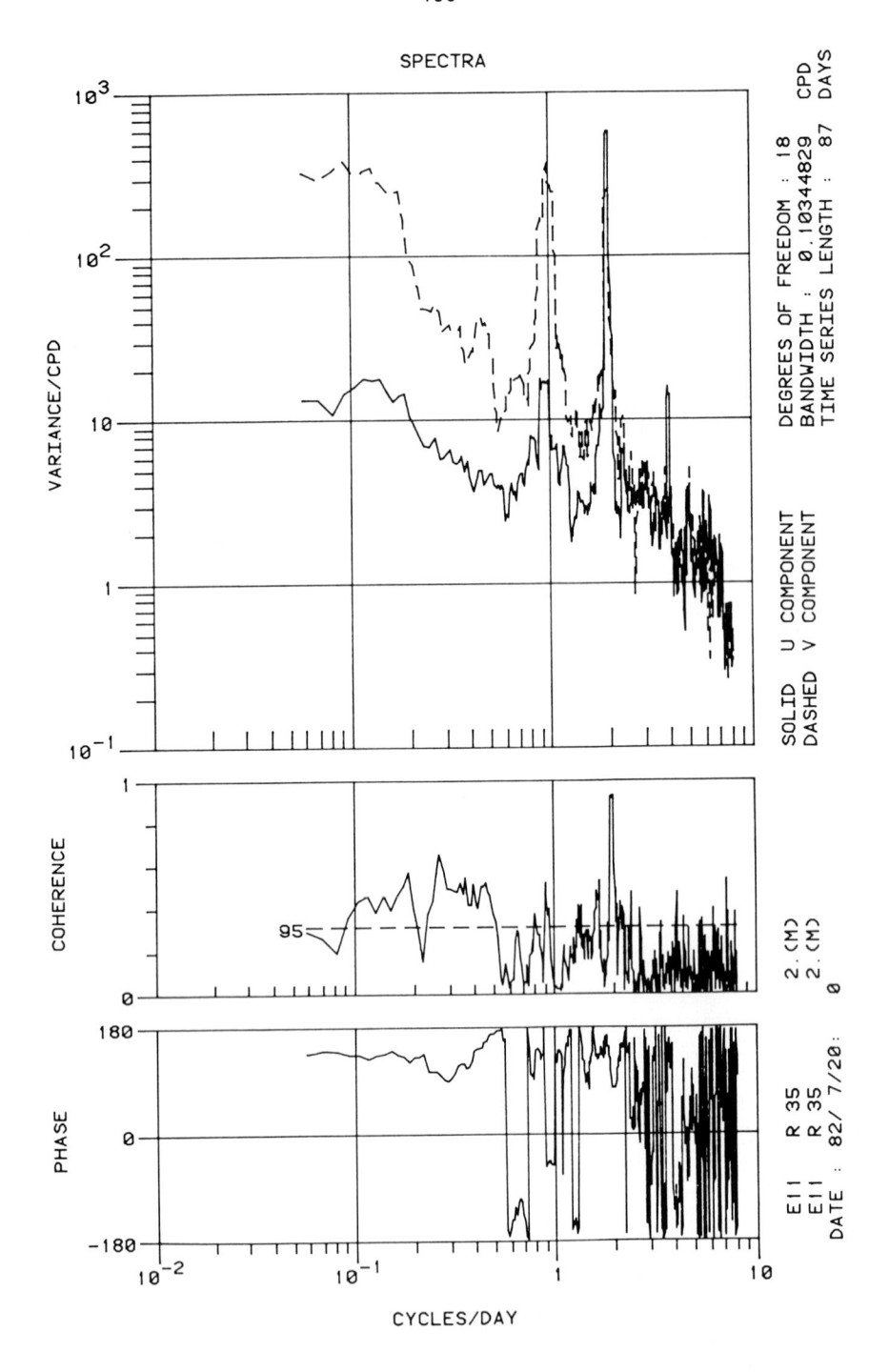

FIGURE 7: Spectra, coherence squared and phase of cross-shelf (u) and alsong-shelf (v) velocity compoents at E11 during summer 1982.

and between *D* and *F*. There is little visual coherence between the outer shelf stations and Hawk Channel. Spectra, coherence squared and phase of cross-shelf and along-shelf velocity components from the outer shelf and Hawk Channel during Winter 1982 are shown in Figs. 8 and 9, respectively. Along shelf current fluctuations were about an order of magnitude more energetic than cross-shelf variations as was found for the summer/fall period. Also, the level of energy and shape of the spectra are not significantly different than the summer/fall period. The primary difference between the two seasons is in the coherence of the velocity components. During winter *u* and *v* were coherent in the tidal and 2-3 day period band in the outer shelf, but only the semi-diurnal tidal components were above the 95% significance level of coherence in Hawk Channel. During the summer/fall the several day period band was coherent between *u* and *v* as well.

LEE and MAYER (1977) have shown that water motions in the shallow portion of the shelf off Miami are primarily driven by the local winds in winter, whereas the outer shelf currents are mainly influenced by the Florida Current. A similar condition appears to prevail in the Sanctuary during winter. As wind forcing intensifies during winter, with the weekly progression of cold fronts, the shallow waters of the Sanctuary may respond more readily to the wind forcing than to the Florida Current influence, which still prevails over the outer shelf. Also, atmospheric cooling of the shallow waters will increase the density relative to the warmer Florida Current and thus produce a horizontal density gradient which inhibits Florida Current water from penetrating across the shelf.

Temperature time series show the Hawk Channel site to be normally 2-3°C cooler then the outer shelf during winter. Temperature and current appear to be strongly correlated in Hawk Channel during the winter, opposite to the summer period when no correlation was evident. Temperature declined during periods of southward currents. A clear example of this occurred on days 344 to 360 where temperature dropped 6°C to a low of 19°C following a series of southward current episodes. At this time the water in Hawk Channel was 6°C colder than at the outer shelf. These wintertime cooling events in Hawk Channel appear to be atmospherically induced through air-sea heat exchange following cold front passages.

An example of the shelf's response to cold fronts is clearly shown in Fig. 10. Vector time series of 40 HLP currents from Hawk Channel (*E*) and Miami airport winds are shown together with 40 HLP temperature from station *E* and 3-hourly air-temperature from Miami airport for Jan. and Feb. 1971. Low frequency currents appear to be strongly related to local wind forcing. Northward (southward) wind events were followed by northward (southward) currents with small phase lags. Low-frequency currents were primarily in the along-shelf direction with speeds of about ± 20 to $30 \, cm \, s^{-1}$. Cross-spectra, coherence squared and phase between the 40 HLP along-shelf component of the current at station *E* and Miami airport winds (not shown) show that the current fluctuations in the 2-3 and 7-10 day period bands were highly coherent with the local wind, with phase lags of less than one day.

Water temperatures in Hawk Channel were strongly influenced by atmospheric forcing. Diurnal variations of water temeprature of about 1°C are visually coherent with daily changes in Miami airport air temperature.

Minimum air and water temperatures follow cold front passages with prolonged winds from the northwest to north (examples are shown by the vertical event lines in Fig. 10). Minimum water temperature of 13.9°C occurred on Jan. 14 and remained below 16°C (temperature reported to be detrimental to reef-building coral: ROBERTS et al., 1982) for about one day. This was the only period that water temperatures less than 16°C were observed in the Sanctuary. Cold winds from the north behind a front will

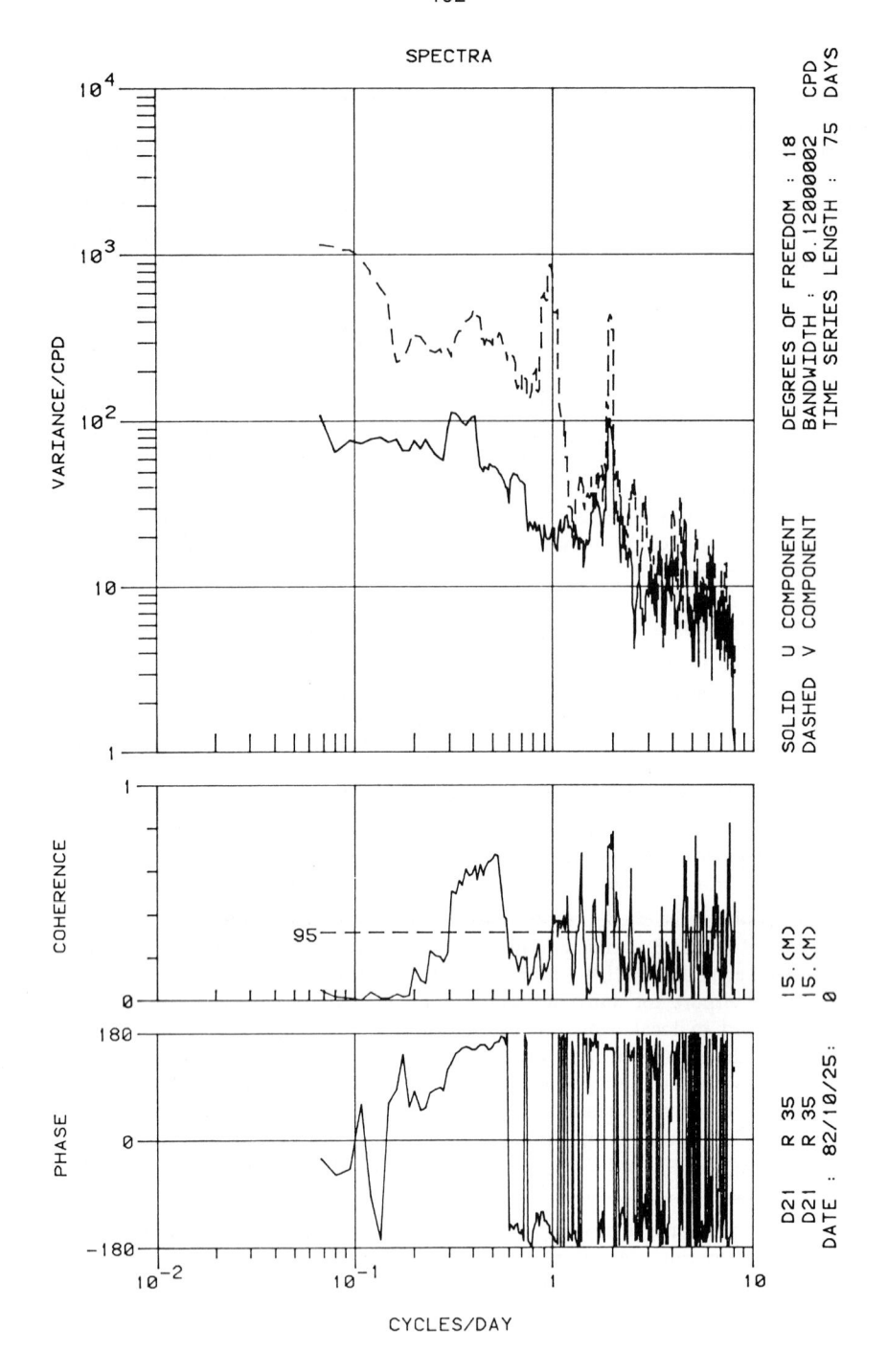

FIGURE 8: Spectra, coherence squared and phase of cross-shelf (u) and along-shelf (v) velocity components at D21 during winter 1982.

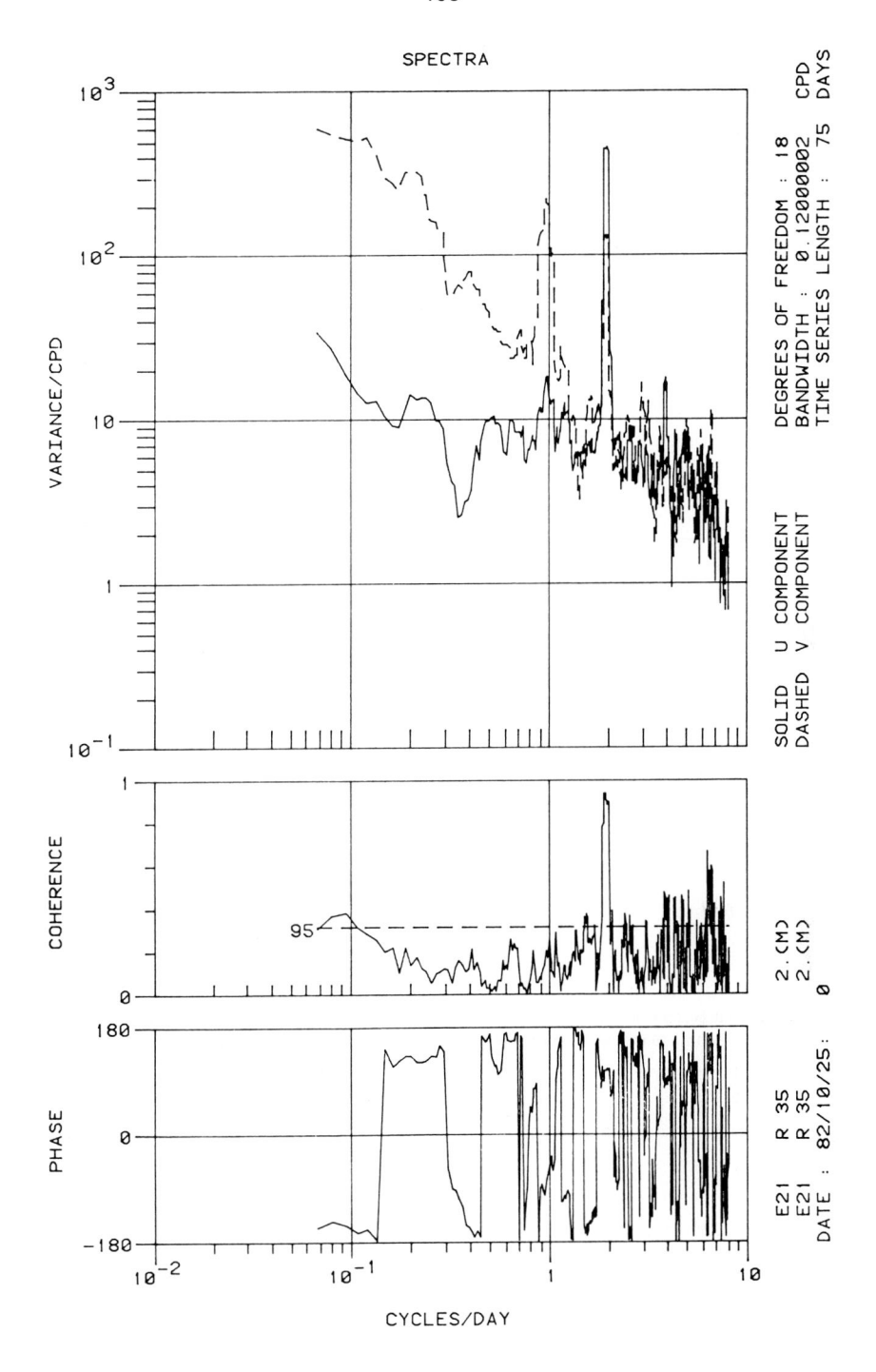

FIGURE 9: Spectra, coherence squared and phase of cross-shelf (u) and along-shelf (v) velocity components at E21 during winter 1982.

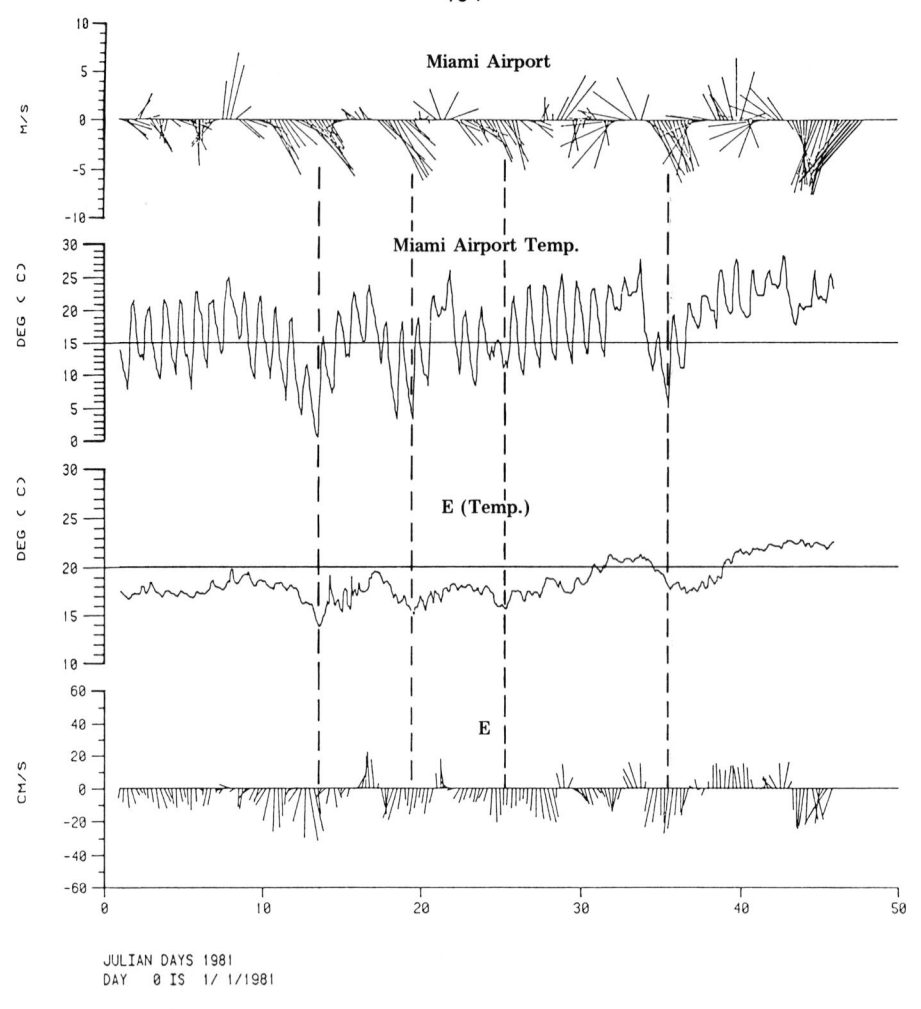

FIGURE 10: Subtidal current and temperature time series from E11 and Miami-airport 40 HLP filtered wind vectors with 3-hourly air temperature from the airport during winter 1981.

generate southward coastal currents and increase vertical mixing, which will enhance cooling of the shallow waters.

TABLE 2: Seasonal Variation of Mean Flow Conditions in Hawk Channel and Outer Shelf from 14 Day Average of 3 Hr LP Data

Current Meter ID	Location	Water Depth (m)	Instr. Depth (m)	Time Period	No. Days	Season	$v(cms^{-1})$	$u(cms^{-1})$	$T(°C)$
\multicolumn{10}{c}{Hawk Channel}
E	25°10.4'N 80°18.3'W	5	3	12/09/80 3/03/81	85	Winter 1980/81	-5.4	-0.3	19.9
E	25°10.4'N 80°18.3'W	5	3	6/19/81 9/18/81	92	Summer 1981	2.0	-1.2	32.3
E11	25°10.5'N 80°19.0'W	3	2	7/17/82 10/01/82	77	Summer 1982	4..3	-1.8	30.4
E11 21/31	25°10.5'N 80°19.0'W	3	2	10/02/82 2/27/83	54	Fall 1982/83	-5.8	-1.7	25.9
G 31/41	25°23.0'N 80°11.5'W	6	5	2/28/83 6/05/83	99	Spring 1983	2.7	-0.5	24.0
G41	25°23.0'N 80°11.5'W	6	5	6/06/83 7/17/83	42	Summer 1983	2.8	-0.4	29.4
B51	24°56'N 80°34'W	6	5	8/31/83 11/09/83	70	Fall 1983	-3.9	+0.3	21.5
\multicolumn{10}{c}{Outer Shelf}
D	25°08.5'N 80°15.0'W	30	15	7/03/81 10/07/81	97	Summer 1981	19.9	2.0	31.1
D11	25°08.5'N 80°15.0'W	29	15	7/17/82 10/01/82	77	Summer 1982	23.5	-4.6	29.2
D11	25°08.5'N 80°15.0'W	29	15	10/02/82 11/24/82	54	Fall 1982	5.7	-4.0	27.4
F 21/31	25°22.0'N 80°08.0'W	32	17	11/25/82 2/27/83	64	Winter 1982/83	24.7	-4.0	24.4
F 31/41	25°22.0'N 80°08.0'W	32 32	17 17	2/28/83 6/04/83	98	Spring 1983	25.0	-1.7	23.4
D41	25°08.5'N 80°15.0'W	29	15	6/02/83 8/10/83	69	Summer 1983	13.5	-2.3	28.2
D51	25°08.5'N	29	15	8/31/83 11/09/83	70	Fall	13.0	-1.2	28.5

3.5. Seasonal variations

Table 2 gives seasonal mean flows and temperatures in the outer shelf and Hawk Channel for central locations in the Sanctuary. These data were obtained from 2-week averages of 3 hour low passed filtered data from the entire data set.

There appears to be a seasonal variation of mean along-shelf flow in Hawk Channel. During the fall and winter mean flows were toward the south (along the isobaths) at -2 to -5 cm s^{-1}. During the spring and summer the mean along-shelf flow was toward the north at 2 to 4 cm s^{-1}. There was an onshore component during all seasons that ranged from about -1 to -2 cm s^{-1} with no clear seasonal trend. This seasonal change in along-shelf flow appears to result primarily from a change in atmospheric forcing. During the short fall season of Sept. and Oct. (*"Mariners Fall":* as described by WEBER and BLANTON, 1980, for the South Atlantic Bight), it is normal for a series of southward wind events to occur that drive the southward currents. During winter, the passage of cold fronts appear responsible for the southward mean flows, for winds are normally stronger from the north (southward) behind the front than the northward winds that occur before frontal passage. The spring and summer periods are characterized by winds from the southeast, which would tend to drive a northward mean coastal current

4. Conclusions

Analysis of current meter data obtained in the Key Largo Coral Reef Marine Sanctuary over a 3-year period from 1981 to 1983 indicates that the Sanctuary can be separated into two regions with greatly different flow properties: the deeper outer shelf region where current and temperature variability is dominated by the Florida Current; and the shallower parts of the Sanctuary including Hawk Channel and the reef tracks where tidal and atmospheric forcing controls current and temperature variability.

Large amplitude current fluctuations and strong northward mean flow that occur along the outer shelf are a direct consequence of the meandering and eddy shedding of the Florida Current within a broad period band of 2 to 15 days. The influence of the Florida Current appears to be greater in the northern part of the Sanctuary than in the southern part, presumably due to the tendency for the Florida Current to follow the isobaths that converge onshore toward the north.

Currents in Hawk Channel and shallower parts of the Sanctuary are strongly influenced by tides and local winds. Tides account for approximately 80% of the total cross-shelf current variance and 50% of the along-shelf. The remaining variance is due primarily to local wind forcing in the winter and a combination of wind and Gulf Stream forcing during the summer. Wind forcing occurs through two mechanisms: direct forcing due to the frictional drag of the surface wind stress; and indirect forcing due to an Ekman response where along-shelf winds generate cross-shelf slopes in the sea surface that are in geostrophic balance with along-shelf barotropic currents.

Mean flows in Hawk Channel show a seasonal cycle; toward the south in fall and winter at -2 to -5 cm s^{-1} and toward the north in spring and summer at 2 to 4 cm s^{-1}, which appears to be due to seasonal changes in local wind forcing. There is also a tendency for the mean flow in Hawk Channel to diverge, with northward mean flow occurring more often in the northern part of the Sanctuary and southward mean flow in the southern part. This flow divergence may be supported by an observed mean onshore flow at mid-depth from the outer shelf.

The time a water particle resides within the Sanctuary or *residence time* is relatively short. From seasonal mean flow considerations the residence time in the Hawk Channel area could range from 10 to 25 days. However, the amplitude of typical flow events in

Hawk Channel is about 10 cm s^{-1}, which would take 5 days to transport a particle from one end of the Sanctuary to the other. Residence time for waters in the outer shelf is extremely short due to the strong northward mean flow from the Florida Current influence. Residence times in the outer shelf are estimated at 2 to 5 days for seasonal mean flows, and possibly 1 day for events when the Florida Current is close to the outer shelf.

Temperature in the shallow water reef regions of the Sanctuary is primarily controlled by air-sea heat exchange. Minimum water temperatures follow the passage of cold fronts with winds from northwest to north that last several days. As the wind shifts to northeast the air mass is modified by heat exchange with the warm waters of the Florida Current, causing air temperatures and the shallow water temperatures in the Sanctuary to increase. Therefore a rapidly moving cold front will have less effect on temperature reduction in the Sanctuary than a slowly moving front that maintains a cold, southward air flow over land for several days.

Studies on the effects of lower temperatures on coral reef survival indicate that prolonged exposure to temperatures of 16°C or less is stressful to reef building corals (ROBERTS et al., 1982). The coldest waters in the vicinity of the Sanctuary occur in the shallow, exposed Florida Bay (Fig. 1) following winter cold-air outbreaks. Water temperatures of 8.7°C were recorded on Jan. 13, 1981, at Lignumvitae Key in Florida Bay (ROBERTS et al. 1982). Fortunately the coral reefs within the Sanctuary are protected from direct contact with the cooled Florida Bay waters, due partly to their location seaward of a solid coastal boundary (Key Largo), and partly to the generation of southward coastal currents by winds from the north following cold-air outbreaks. The nearest inlet connecting Florida Bay to the coastal waters is Snake Creek (Fig. 1) located approximately 10 km south of the Sanctuary. Therefore the chilled Florida Bay waters will enter the coastal waters at Snake Creek and other inlets to the south following a cold front passage. Winds are typically from the northwest to north during this time, which will generate a southward along-shelf coastal current and advect the cold waters offshore and toward the south-southwest. Northward currents, which would advect the cold waters into the Sanctuary, do not occur until the winds have shifted to be from the east and warmed by flow over the Florida Current. Thus the coool Florida bay waters will undergo substantial warming before entry into the Sanctuary. This may explain why the coral reefs within the Sanctuary are better developed and in a healthier state than the reefs to the south, such as Hen and Chickens and Alligator reefs.

Acknowledgements

I thank Bob Calvert and John Halas for their help in the data collection part of the study, Jere Green and Lila Ptak for typing the manuscript and Jean Carpenter for drafting the figures. This work was supported by the NOAA National Marine Sanctuaries Program through Contract No. N0-80-SAC-00832 to General Oceanics of Miami, Florida. The data products were supplied by Science Applications Inc. of Raleigh, N.C.

References

BEARDSLEY, R.C. and B. BUTMAN, 1974. Circulation on the New England continental shelf: response to strong winter storms. *Geophys. Res. Lett.,* 1: 181-184.

KIELMAN, J. and W. DUING, 1974. Tidal and sub-inertial fluctuations in the Florida Current. *J. Phys. Oceanogr.,* 4: 227-236.

LEE, T.N., 1975. Florida current spin-off eddies. *Deep-Sea Res.,* 22: 753-763.

LEE, T.N., and C. ROOTH, 1976. Circulation and exchange processes in southeast Florida's coastal lagoons. University of Miami: Sea Grant Program, Biscayne Bay Symposium I Proceedings: 51-63.

LEE, T.N., and D.A. MAYER, 1977. Low-frequency current variability and spin-off eddies on the shelf off Southeast Florida, *J. Mar. Res.*, 35 (1): 193-220.

LEE, T.N., I. BROOKS and W. DUING, 1977. The Florida Current — Its Structure and Variability. University of Miami Technical Rerport UM-RSMAS 77003: 275pp.

LEE, T.N., L.P. ATKINSON and R. LEGECKIS, 1981. Observations of a Gulf Stream frontal eddy on the Georgia continental shelf, April 1977. *Deep-Sea Res.*, 28A (4): 347-378.

LEE, T.N. and L.P. ATKINSON, 1983. Low-frequency current and temperature variability from Gulf Stream frontal eddies and atmospheric forcing along the southeast U.S. Outer Continental Shelf. *J. Geophys. Res.*, 88 (C8): 4541-4568.

LEE, T.N., W-J HO, V. KOURAFALOU and J.D. WANG, 1984. Circulation on the Southeast U.S. continental shelf, Part I — subtidal response to wind and Gulf Stream forcing. *J. Phys. Oceanogr.*, 14 (6): 1001-1012.

MAYOR, D.A. and D.V. HANSEN, 1975. Observations of currents and temperatures in the southeast Florida coastal zone during 1971-1972. NOAA Tech. Report ERL AOML ERL 346-AOML 21, Atlantic Oceanographic and Meteorological Laboratories, Miami, FL 36pp.

ROBERTS, H.H., L.J. ROUSE, JR, N.D. WALKER and J.H. HUDSON, 1982. Cold-water stress in Florida Bay and northern Bahamas: A product of winter cold-air outbreaks. *J. Sed. petrol,* 42: 145-155.

SCOTT, J.T. and G.T. CSANADY, 1976. Nearshore currents off Long Island. *J. Geophys. Res.*, 81: 5401-5409.

SMITH, J.A., B.A. ZETLER and S. BROIDA, 1969. Tidal modulation of the Florida current surface flow. *Mar. Tech. Soc. Jr.,* 3 (3): 41-46.

WEBER, A.H. and J.O. BLANTON, 1980. Monthly mean wind fields for the South Atlantic Bight. *J. Phys. Oceanogr.,* 10, 1256-1263.

ZETLER, B.D. and D.V. HANSEN, 1970. Tides in the Gulf of Mexico — a review and proposed program. *Bull Marine Soc.,* 20: 57-69.

Effects Of Coastal Boundary Layers On The Wind-Driven Circulation In Shallow Sea

Su Jilan

Second Institute of Oceanography
People's Republic of China

ABSTRACT

Through analyzing the asymptotic solutions of several examples, it is shown that, if the nearshore water depth is not large, the interior flow of a wind-driven shallow sea can be obtained without considering the coastal boundary layers. However, in case of deep nearshore water depth that varies over a narrow region the effects of the coastal boundary layers can be significant.

1. Introduction

Knowledge of circulation patterns is important for the study of sediment transport, pollutant dispersion, and the marine ecosystems. The current field is usually complicated and variable because of the complex topography and changeable driving forces. In a shallow sea with weak stratification one often employs a set of vertically integrated equations to investigate the wind-driven currents. In these equations the horizontal friction terms are usually neglected and the transport at the coast is required only to be parallel to the boundary, i.e., the effects of the coastal boundary layers are neglected. It is generally believed that such a method of obtaining the circulation will yield the correct lowest order solution in the interior away from the coasts. As far as we know, whether this conjecture is a valid one has not been systematically explored, although TEE (1976) has demonstrated numerically the importance of the coastal boundary layer on the tidally driven residual flow field in the interior.

In the following, wind-driven circulation in shallow seas is studied. Through asymptotic analyses of several simple examples it is concluded that the above method of finding the circulation is in general valid, but the method becomes questionable when there is a narrow deep trough near the coast.

2. Formulation of problem

For simplicity, we consider a shallow homogeneous water with constant horizontal and vertical eddy viscosities, subjected to a steady wind field. Since the inertial terms are usually small except during storms, they will also be neglected. Assuming hydrostatic approximation the linearized dimensionless equations can be shown to be

$$\vec{e}_3 \times \vec{q} = -\nabla \zeta + E_H \nabla^2 \vec{q} + E_V \vec{q}_{zz}, \tag{1}$$

$$\nabla \cdot \vec{q} + w_z = 0, \tag{2}$$

where ζ represents the free surface elevation and $\vec{q}, \nabla, \nabla^2$ are the horizontal component of the velocity, del operator, and Laplacian operator, respectively. w is the vertical component of the velocity and z is the vertical coordinate with unit vector $\vec{e}_3$, positive upward. E_H and E_V are, respectively, the horizontal and the vertical Ekman numbers. In deriving Eq.(1) the f-plane approximation has been used and the atmospheric pressure was taken to be constant in both space and time. The linearized boundary conditions at the sea surface are

$$\vec{q} = \vec{\tau} \text{ and } w = 0, \quad \text{at } z = 0, \tag{3}$$

where $\vec{\tau}$ is the wind stress vector. Velocities at both the sea bottom and coast are required to vanish identically. If there are open boundaries other conditions have to be specified there.

In this study we concentrate our interests in shallow seas only, i.e., where E_V is of order 1. However, E_H is in general a small quantity. In the following we will take, for simplicity, E_V to be unity and denote E_H by a small parameter ϵ, i.e., $\epsilon \ll 1$. It is easy to see that the lateral friction terms are important only inside a coastal boundary layer across which the horizontal velocity component is rapidly brought to zero. If these terms are dropped from the equations one usually relaxes the coastal boundary conditions such that only the horizontal velocity component normal to the coast vanishes there. The question is then whether such a formulation wil yield the right lowest order solution compared with the correct solution of the complete equation.

To investigate the structure of coastal boundary layers JANOWITZ (1972) proposed a scheme to find the local solution of Eqs. (1) and (2) for a basin of uniform depth. However, his results cannot provide an answer to the question posed above since, in arriving at his solution, the normal gradient of the free surface elevation at the coast was assumed to be known. Considering the fact that finding analytical solutions to Eqs. (1)-(2) is in general difficult, we take the alternate approach by studying the transport equations, i.e., the vertically integrated equations of Eqs. (1) and (2),

$$\vec{e}_3 \times (h\vec{Q}) = -h\nabla\zeta + \epsilon\nabla^2(h\vec{Q}) + \vec{\tau} - \vec{\tau}_b, \tag{4}$$

$$\nabla \cdot (h\vec{Q}) = 0, \tag{5}$$

where $h(x,y)$ is the water depth, $\vec{Q}$ is the vertical average of $\vec{q}$, and $\vec{\tau}_b$ is the bottom stress vector. In deriving Eq. (4) the term $\epsilon\nabla h \cdot \nabla \vec{q}|_{z=-h}$ was neglected. This is probably a valid assumption since $\vec{q}$ is identically zero along the bottom.

To keep the equations linear we take

$$\vec{\tau}_b = k\vec{Q}, \tag{6}$$

where k is a constant coefficient. From Eq. (5) a stream function, ψ, can be introduced such that

$$h\vec{Q} = -\vec{e}_3 \times \nabla\psi. \tag{7}$$

Substituting the above two equations into Eq. (4) and then eliminating ζ we obtain the vorticity equation

$$\epsilon(\nabla^4\psi - \nabla\ln h \cdot \nabla\nabla^2\psi) - (k/h^2)(h\nabla^2\psi - 2\nabla h \cdot \nabla\psi)$$
$$-(\nabla\ln h \times \nabla\psi)\vec{e}_3 = \vec{e}_3 \cdot h\nabla \times (\vec{\tau}/h). \tag{8}$$

The boundary condition for ψ along a coast with a vertical shoreface is

$$\psi = \text{constant and } \vec{n} \cdot \nabla\psi = 0, \tag{9}$$

where $\vec{n}$ is the unit normal vector at the coast.

Because ϵ is a small quantity the standard way to solve for the wind-driven circulation is to drop the lateral friction term, i.e.,

$$(k/h^2)(h\nabla^2\psi - 2\nabla h \cdot \nabla\psi) - (\nabla \ln h \times \nabla\psi)\vec{e}_3 \tag{10}$$

$$= -\vec{e}_3 \cdot h \nabla \times (\vec{\tau}/h)$$

and to simplify the boundary condition at the coast to just

$$\psi = 0. \tag{11}$$

The loss of higher order derivative terms implies that Eq. (10) does not hold inside a coastal boundary layer. It is always a tacit assumption held among researchers that Eqs. (10) and (11), though invalid inside the coastal boundary layer, will nevertheless yield the correct lowest order solution in the interior. This is the question we will address through the following examples.

3. Examples

It is evident from Eq. (8) that the structure of the coastal boundary layer depends on the local topography. Since many coastlines are locally straight and the nearshore isobaths are usually parallel to the coast, local solutions may be modelled by considering a straight coast along the y-axis with the water depth varying only in the x-direction, i.e., $h = h(x)$. If the wind field is taken as uniform Eq. (8) then becomes

$$\epsilon(\underbrace{h\nabla^4\psi - h_x\nabla^2\psi_x}_{(A)}) - \underbrace{k(\nabla^2\psi - 2(\ln h)_x\psi_x)}_{(B)}$$

$$-\underbrace{h_x\psi_y}_{(C)} = -\underbrace{\tau h_x}_{(D)}, \tag{12}$$

where τ now represents the wind stress component in the y-direction. The boundary condition, Eq. (9), becomes simply

$$\psi = \psi_x = 0, \text{ at } x = 0. \tag{13}$$

In Example 3 we will consider a rectangular basin where similar boundary conditions are to be imposed at the other three coastal sections of the basin. To simplify discussions the symbols (A), (B), (C), and (D) are used to denote the lateral friction, bottom friction, Coriolis, and wind stress terms, respectively.

When the nearshore water depth is of order 1 or smaller (B) will be important everywhere from the interior up to the coast. (C) will also be important everywhere except in the case of small nearshore water depth when it becomes important only in the interior. However, (A) is important only inside the coastal boundary layer. Therefore, Eq. (10) is valid outside the coastal boundary layer. The validity of Eq. (11) is not obvious and has to be inferred from analysis of examples.

If the nearshore water depth is deep, i.e., much greater than 1, the Coriolis term (C) will now be important everywhere. (A) is still only significant inside the coastal boundary layer, while (B) dominates only away from the shore. There are two possible combinations, one where the zones of dominance of (A) and (B) are separated (Fig. 1a) and the other where the two zones overlap or are next to each other (Fig. 1b). In Fig. 1 the coastal boundary layer due to the action of bottom friction are also depicted. This kind of coastal boundary layer acts to bring the interior geostrophic flow to one that has a zero normal velocity component at the coast and it only comes into being when the lateral friction is ignored. In Fig. 1 they are shown to be narrower than the corresponding normal coastal boundary layers. In fact, the width of a bottom-friction coastal boundary layer depends on both the value of k and the local topography, and it can be wider than the normal coastal boundary layer.

If we disregard the part of the interior region where (B) becomes important, the situation in Fig. 1a is similar to the classical wind-driven ocean circulation (STOMMEL, 1965, p. 83), except that the topographic effect takes place of the β-effect. From the

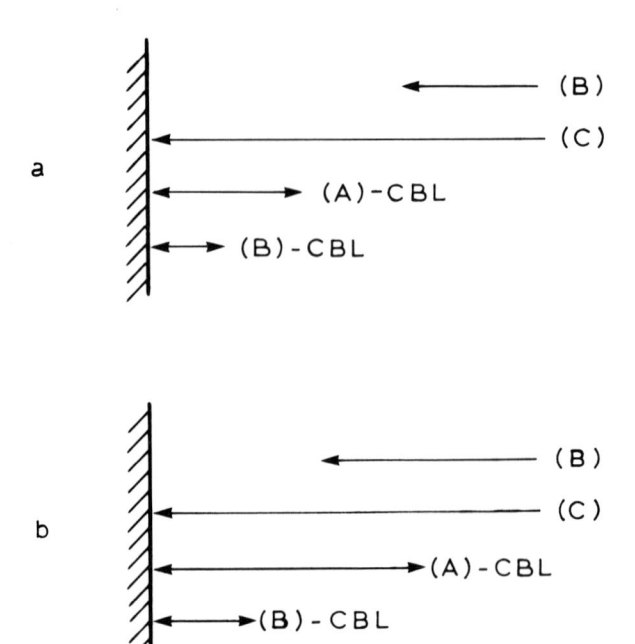

FIGURE 1. Zones of dominance (Eq. (12));
a) (A) and (B) separated;
b) (A) and (B) overlapped or next to each other

solution of the classical wind-driven circulation we infer that the solution of Eqs. (10) and (11) will be valid outside the coastal boundary layer. The situation in Fig. 1b is quite different. As mentioned before, Eq. (10) is valid in regions where (B) is important but (A) is not important. Now that the zones of dominance of (A) and (B) overlap or are next to each other, from the point of view of asymptotic matching the boundary condition of Eq. (10) must depend on whether lateral friction is considered in the coastal boundary layer. In other words, the boundary condition approximation, Eq. (11), may not be valid. Again, this conjecture is based on the assumption that the pseudo coastal boundary layer is narrower than the normal boundary layer.

In the following we will take three examples for further discussions. For convenience of asymptotic analysis the water depth functions are assumed to depend on the small parameter ϵ as well, i.e., $h = h(x;\epsilon)$.

(a) Example 1. $h = \theta + 2 - 2\exp(-x/\theta)$, $\theta = \epsilon^{-\frac{1}{4}}$

In this example the water depth is very shallow near the coast but increases rapidly to unity offshore. It is recalled that the interior region was previously defined as where the horizontal friction is not important. According to this definition the interior region can be further divided into two subregions. One subregion is the inner interior region with $x = 0(1)$, inside which the lowest order equation is simply

$$k\nabla^2 \psi_0 = 0. \tag{14}$$

This is well-known equation for steady wind-driven circulation in a constant water depth region. The other subregion is the outer interior region where $x = 0(\theta)$. Choosing the stretched coordinate $\eta = x/\theta$ the lowest-two order solution in this region can be shown to be

$$\overline{\psi}(\eta,y) \approx \theta \overline{\psi}_0(\eta,y) + \theta^2 \overline{\psi}_1(\eta,y), \tag{15}$$

where

$$\overline{\psi}_0 = A_0(y) + B_0(y)\{2\eta + 4\exp(-\eta) - \exp(-2\eta)\}$$
$$\quad - 2(\tau/k)\{\eta + \exp(-\eta)\},$$

$$\overline{\psi}_1 = A_1(y) + B_1(y)\{2\eta + 4\exp(-\eta) - \exp(-2\eta)\} - \tau\eta/k$$
$$\quad + 2(B_0 + A_0'/k)\{\eta + \exp(-\eta)\} + B_0' P(\eta).$$

$P(\eta)$ is a complicated but known function of η. A_0, B_0, A_1, and B_1 are all function of y to be determined. We mention that $\overline{\psi}_0$ and $\overline{\psi}_1$ are also the asymptotic solutions of Eq. (10) but the next order solution $\overline{\psi}_3$ will include effects of the lateral friction.

For the coastal boundary layer the horizontal friction term is balanced with the bottom friction term and the stretched coordinate becomes $\xi = x/\theta^{5/2}$. The lowest-two order solution can be shown to be

$$\tilde{\psi}(\xi,y) \approx \theta^{7/2} F_0(y) G(\xi)$$
$$\quad + \theta^4 \{F_1(y) G + (-2sF_0 + \tau/k)\xi^2 + F_0 H(\xi)\exp(-s\xi)\}, \tag{16}$$

where $s = k^{1/2}, G = 1 - s\xi - \exp(-s\xi)$, and H is a known polynomial of ξ. Both F_0 and F_1 are functions of y to be determined.

In deriving the above expressions the asymptotic matching principle (VAN DYKE, 1964) has been used. The unknown functions of the solutions are found to be related through

$$A_0 + 3B_0 - 2\tau/k = 0, \tag{17}$$

$$A_1 + 3B_1 + 2A_0'/k + 2B_0 + B_0' P(0) = 0, \tag{18}$$

$$F_0 = \tau/s^3. \tag{19}$$

However, these functions cannot be completely determined within the local analysis. In other words, global analysis considering other boundary conditions along both y-directions and at large x are needed.

From Eq. (15) and (17)-(19) it is seen that

$$\lim_{\eta \to 0}(\theta \overline{\psi}_0 + \theta^2 \overline{\psi}_1) = 0. \tag{20}$$

Next consider the case when the lateral friction term (A) is neglected. From Eqs. (10) and (11) we find that the lowest-two order solution in the $x = 0(\theta)$ domain can be written as

$$\hat{\psi}(\eta,y) \approx \theta \hat{\psi}_0 + \theta^2 \hat{\psi}_1, \tag{21}$$

where

$$\hat{\psi}_0 = D_0(y)\{2\eta - 3 + 4\exp(-\eta) - \exp(-2\eta)\}$$
$$+ 2(\tau/k)\{1 - \eta - \exp(-\eta)\},$$
$$\hat{\psi}_1 = D_1(y)\{2\eta - 3 + 4\exp(-\eta) - \exp(-2\eta)\} - \tau\eta/k$$
$$+ 2(D_0 - 3D_0'/k)\{\eta - 1 + \exp(-\eta)\} + D_0'\{P(\eta) - P(0)\}.$$

It is seen that the above solution has the same form as Eq. (14). In fact, they will be exactly the same if D_0 and D_1 are equal to B_0 and B_1, respectively. However, this question cannot be answered here since it relies on the global solution. Consequently, we can only conclude that, for this example, the solutions of Eqs. (10) and (11) is probably valid outside the coastal boundary layer.

(b) Example 2. $h = 1 + x$

In this example the water depth is of $O(1)$ everywhere. The lowest order equation in the interior is

$$k(\nabla^2\psi_0 - \frac{2}{1+x}\psi_{0_x}) + \psi_{0_y} = \tau. \tag{21}$$

which is just the wind-driven equation normally used, i.e., Eq. (10). Its solution is bounded and for $x \ll 1$ it can be expanded in power series

$$\psi_0(x,y) = C_0(y) + C_1(y)x + O(x^2). \tag{22}$$

The stretched coordinate for the coastal boundary layer is $\xi = x\epsilon^{-1/2}$ and the lowest-two order solution can be shown to be

$$\tilde{\psi}(\xi,y) \approx \epsilon^{1/2}F_0(y)G(\xi)$$
$$+ \epsilon\{F_1(y)G + (\tau\xi^2/2k) + F_0R(\xi)\exp(-s\xi)\}, \tag{23}$$

where s and G have the same definitions as before. $R(\xi)$ is a known polynomial of ξ, and both F_0 and F_1 are still unknown functions of y.

Asymptotic matching yields $C_0(y) \equiv 0$. Therefore,

$$\lim_{x \to 0} \psi_0(x,y) = C_0(y) = 0. \tag{24}$$

Hence, similar to the conclusion reached in Example 1, the solution of Eqs. (10) and (11) for the finite water depth case is probably also valid outside the coastal boundary layer.

(c) Example 3. $h = 1 + \beta(\epsilon)\exp(-\alpha(\epsilon)x), \alpha \gg 1, \beta \gg 1$

For this example a rectangular basin is used so that the explicit solutions, Eqs. (27) and (28), can be found. The water depth is now very deep nearshore at the west boundary, but it decreases rapidly to $O(1)$ away from the coast. It can be shown from analyzing Eq. (12) that if $\alpha \geqslant \epsilon^{-1/3}$ and $\beta \geqslant \alpha$, the zones of dominance of (A) and (B) are next to each other and if $\alpha \geqslant \epsilon^{-1/2}$ then these two zones overlap. Either category belongs to the case of Fig. 1b, and both have coastal boundary layer widths wider than the width over which the water depth varies significantly. To be able to find explicit asymptotic solutions we choose $\alpha = \epsilon^{-1}$ and $\beta = \epsilon^{-2}$. In the following the notations $\theta = \epsilon^{1/2}$ and $s = k^{1/2}$ are used.

The coastal boundary layer can be divided into two subregions, the $O(\theta)$ region and the $O(\theta^2)$ region. In the farshore subregion the lowest two-order solution can be shown to be, with $\xi = x/\theta$,

$$\tilde{\psi}(\xi,y) \approx B_0(y)$$

$$+\theta\{B_1(y)\xi - B_0 + (\tau/s^3)(1-\exp(-s\xi))\}, \quad (25)$$

where $B_0 = s\tau y + A_0$. A_0 is an unknown constant and $B_1(y)$ is an unknown function. In the nearshore subregion, by letting $\zeta = x/\theta^2$ the lowest two order solution is found to be

$$\tilde{\psi}(\zeta,y) \approx \{\theta B_0 + \theta^2(B_1 + \tau/k)\}(\zeta - 1 + \exp(-\zeta)). \quad (26)$$

It should be pointed out that the explicit functional dependence of $B_0(y)$ is obtained only after matching the solutions of the two subregions up to the fifth asymptotic order.

If the lateral friction is ignored the pseudo coastal boundary layer, i.e., the bottom-friction coastal boundary layer, can also be divided into two subregions, now the $O(\theta^2)$ region and the $O(\theta^3)$ region. In the farshore subregion, with the stretched coordinate $\zeta = x/\theta^2$ the lowest order solution is found to be

$$\hat{\psi}(\zeta,y) \approx \tau y + D_0(1 - \exp(-2\zeta)). \quad (27)$$

where D_0 is an unknown constant. In the nearshore subregion if we introduce $\eta = x/\theta^3$ the lowest order solution is found to be

$$\hat{\hat{\psi}}(\eta,y) \approx \tau y\, erf(\lambda) + \{erf(\lambda) - 1\}(\tau/2k)\eta^2$$

$$+ (k\tau/3\pi)(\lambda\exp(-\lambda^2))y, \quad (28)$$

where $\lambda = \eta/(4ky)^{1/2}$. To obtain the above two solutions higher order equations have also been used, in addition to matching between the two solutions themselves. We have also made the assumption that $\hat{\psi} \to 0$ as $y \to 0$.

In the interior region where $x = O(1)$, the lowest order governing equation remains to be the Laplace equation whether the lateral friction is considered or not. As in Example 2 the lowest order interior solution has a power series expansion for $x \ll 1$, i.e.,

$$\psi(x,y) \approx \mu(\theta)\psi_0(x,y) \quad (29)$$

$$\psi_0(x,y) = C_0(y) + C_1(y)x + O(x^2) \text{ for } x \ll 1. \quad (30)$$

The ordering parameter $\mu(\theta)$ in Eq. (29) has to be obtained from matching. If lateral friction is considered matching between Eqs. (25) and (29) yields

$$\mu(\theta) = \theta^{-1},\ C_0 = 0,\ C_1 = \tau sy + A_0. \quad (31)$$

On the other hand, if the lateral friction is neglected, matching has to be done between Eqs. (27) and (29) and the following result is obtained

$$\mu(\theta) = 1,\ C_0 = \tau y + D_0. \quad (32)$$

It is obvious that the two sets of asymptotic results, Eqs. (31) and (32), are completely different in terms of both the ordering parameter and the functional dependence.

Eq. (32) shows that the ordering parameter is much greater than one, which seems to indicate that there is a very large volume flow rate in the interior. The following arguments show that this is probably not so, although to find asymptotic solutions at the corner layers prevents us from giving a rigorous proof. When the lateral friction is considered, then along each of the three sides other than the west there is also a coast boundary layer with a boundary layer structure similar to the function G in Eq. (23). The asymptotic behaviour of these boundary layer solutions at the edge of the coastal boundary layer indicate that the interior solution should have the same ordering parameter and the same asymptotic functional property as those in Eq. (32). Consequently, as far as the lowest order interior solution is concerned the stream function assumes zero value along all four sides of the boundary. Since the lowest order interior solution is governed by the Laplace equation this can only mean that the solution is identically zero. Evidently, because the coastal boundary layer is wider than the width over which the depth varies, it effectively blocked the depth variation information from transmitting to the interior region, i.e., the region where Eq. (10) holds. The topographically induced wind-driven volume flux only goes around the basin, confined inside the coastal boundary layers at the four sides of the coast. The nontrivial lowest order interior solution is of 0(1). Its value is still zero at the west boundary but finite at the other three sides. Based on the above discussions we sketched in Fig. 2a the streamline pattern for the case where the lateral friction term is in effect.

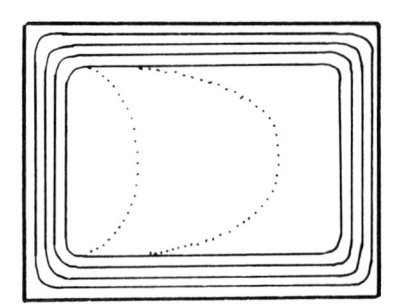

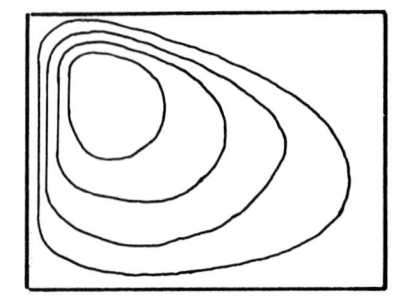

FIGURE 2. Streamline pattern for Example 3; a. lateral friction considered; b. lateral friction neglected

When lateral friction is ignored Eq. (10) is then applicable to the whole basin. In deep water region away from the coast the Coriolis term (C) dominates over the bottom friction (B) and Eq. (26) is the lowest order expansion of its solution. As the west boundary is approached the bottom friction resumes its dominance because of the high flow rate, and the lowest order expansion of the solution there is given by Eq. (27). This pseudo coastal boundary layer is induced by the bottom friction, and it is similar to the Stommel layer (STOMMEL, 1965) in the classical ocean circulation model except that here the topographic effect has taken the role of the β-effect. Other than the west shore there is no coastal boundary layer along any of the other three boundaries. Eq. (32) indicates that the variation of the bathemetry has significant influences on the interior solution. If the y-component of the wind stress vector is directed northward, the total volume flux directs southward inside the pseudo coastal boundary layer along the west coast, where the water is also the deepest. As the flow moves south the streamlines gradually leave the coastal boundary layer and branch into the interior along the way. They go around the interior region in a counterclockwise fashion and reenter the pseudo coastal boundary layer at the northwest corner of the basin, inducing a corner layer with strong singularity (Fig. 2b).

4. Discussions and conclusion

Eq. (8) was the basic equation used for the above analysis. Its derivation was based on several assumptions, among which we will examine the steady state and the linear bottom friction law assumptions.

If the flow field is unsteady, the basic equation will be similar to Eq. (8) except for an extra time derivative term

$$-h\frac{\partial}{\partial t}\nabla\cdot(\frac{1}{h}\nabla\psi).$$

As far as the spatial differentiation is concerned this term is of the same order as the bottom friction term. Since the boundary layer analysis is principally based on the spatial derivatives the asymptotic results presented above should still hold. As for the relation between the bottom friction stress vector and the total flow vector, it is known that the linear law is a valid assumption for shallow seas with strong tidal flow (CSANADY, 1976). In other situations a quadratic law is more appropriate, and then the corresponding governing equation will be nonlinear. The nonlinear term is the product of a second order derivative and a first order derivative. Since the lateral friction term is a fourth order derivative, i.e., one order higher than the nonlinear term, from the point of view of boundary layer analysis the asymptotic results described previously should prove still valid. However, the scale parameters for each region will be different.

In summary, if the nearshore water depth is of the same order as or smaller than the Ekman depth, Eqs. (10) and (11) are valid for wind-driven circulation problems. However, if the nearshore water depth is larger than the Ekman depth and if the coastal boundary layer width is wider than the width over which the depth varies significantly, although Eq. (10) is still valid sufficiently far from the shore Eq. (11) is no longer a valid boundary condition to Eq. (10). The correct boundary condition to Eq. (10) has to be found through matching with the coastal boundary layer condition.

Finally, take the Hangzhou Bay for illustration purpose. In general the water depth in the Bay is less than 10 meters, but there are narrow deep troughs along the northern shore with maximum depth reaching 50 meters. If we regard the troughs as having the depth variation as that in Example 3, then the effects of the coastal boundary layer may have to be considered if the water depth nearshore is greater than $\epsilon^{-1/3}H$ and the width over which the depth varies is narrower than $\epsilon^{+1/3}L$. H and L are, respectively, the characteristic depth and horizontal length of the Bay, taken to be $H=0$ m and $L=100$ km. In summer the wind over the Bay is mild, the vertical and the horizontal eddy viscosities may be chosen to be $10^{-2}m^2/s$ and $10^2m^2/s$, respectively. Thus $\epsilon=10^{-3}$ and

$$\epsilon^{-1/3}H=100m \quad \text{and} \quad \epsilon^{+1/3}L=10km.$$

Since the largest trough has characteristic dimensions of

$$\beta H=30m \quad \text{and} \quad \alpha^{-1}L=20km,$$

it is seen that in summer the coastal boundary layer is not important to the interior flow field of the wind-driven circulation. During winters or other times of stoms the horizontal eddy viscosity may be chosen to be $10^4 m^2/s$, following the work in Pohai Sea by QIN and FENG (1975). Now $\epsilon=10^{-2}$ and thus

$$\epsilon^{-1/3}H=45m \quad \text{and} \quad \epsilon^{+1/3}L=22km.$$

The critical depth now is slightly larger than the trough depth, but the critical width is almost the same as the trough width. Therefore, when discussing the wind-driven circulation in Hangzhou Bay in winter or during storms, the lateral friction effects may be important.

References

CSANADY, G.T., 1976. Mean circulation in shallow seas. J. Geophys. Res., 81, pp. 5389-5399.

JANOWITZ, G.S., 1972. The effects of finite vertical Ekman number on the coastal boundary layers of a lake. Tellus, 24, pp. 414-420.

QIN ZHENGHAO & FENG SHIZUO, 1975. Preliminary studies on the dynamics of storm surge in shallow seas, (in Chinese). Scienca Sinica, pp. 64-78.

STOMMEL, H., 1965. The Gulf Stream. Cambridge Univ. Press, pp. 248.

TEE, K.T., 1976. Tide-induced residual current, a two-dimensional numerical tidal model. J. Marine Res., 34, pp. 603-628.

VAN DYKE, M., 1964. Perturbation methods in fluid mechanics. Academic Press, 299 pp.

VI

SEDIMENT TRANSPORT

Suspended Sediment Transport in Rivers and Estuaries

M. Markofsky, G. Lang & R. Schubert

Institut für Strömungsmechanik,
Universität Hannover.
Federal Republic of Germany

ABSTRACT

Results of the numerical simulation of vertically discretized suspended sediment transport are presented in which the influence of the following parameters on the turbidity maximum region were investigated:
- flow field
- settling velocity
- vertical dispersion coefficient
- ocean and upstream boundary conditions for salinity and sediment
- bottom boundary condition (erosion, sedimentation)

Comparisons are made with analytical and numerical solutions for both steady and unsteady nonstratified conditions.
Additional calculations for the stratified case illustrate basic physical phenomena associated with the suspended sediment
transport process.

1. Introduction

The deposition of suspended sediment in german shipping channels and harbours results in annual dredging costs in excess of 25 million dollars. Additional environmental problems are associated with the heavy metals absorbed on the suspended sediment, namely the resuspension of pollutants during dredging and the nonrecyclability of dredged spoil which might have been used for agricultural purpose.

The transport of suspended sediment is governed by the interaction between the current and the particle characteristics such as size, specific weight, organic content, flocculation properties etc. The unsteady, in part density driven flow in estuary salinity intrusion regions plays a dominant role in this process through the generation of dead water zones, net upstream bottom currents and downstream surface currents, vertical velocities and oscillating flows resulting in alternating erosion and sedimentation phases.

A common phenomenon in salinity intrusion regions is the occurence of a turbidity maximum, i.e. a concentration maximum of suspended particles, resulting in increased deposition. This has been reported for example for the Thames (OWEN and ODD, 1972), the Elbe (CHRISTIANSEN, 1974) and the Seine (SALOMON, 1982). In order to better understand this phenomenon, research using a two-dimensional numerical model is presently being conducted at the Institut f. Strömungsmechanik (Universität Hannover) through the support of the Deutsche Forschungsgemeinschaft.

Three main hypotheses for the formation of a turbidity maximum are reported in the literature:

1) biological reactions: the death of organic substances due to brackish water (CASPERS and SCHULTZ, 1964; NOETHLICH, 1972);

2) physiochemical reactions: electrolitical forces resulting in flocculation (WHITEHOUSE

et al. 1959);

3) flow induced trapping due to density induced upstream bottom currents, downstream surface currents and vertical flows (POSTMA and KALLE, 1955; FESTA and HANSEN, 1977).

Both biological and physiochemical reactions effectively change the settling characteristics of the suspended matter. As such this paper is restricted to flow induced trapping. Algorithms for 1 and 2 will be not treated.

2. Qualitative description

The basic characteristics of a turbidity maximum can be seen in turbidity contours measured in the Weser Estuary (Northern Germany). Fig. 1 shows that the maximum turbidity region in this case lies between the 4‰ and 16‰ isohalines. The primary deposition in the turbidity maximum region is in the form of silt. Further upstream and downstream fine sand predominates. In other words, the turbidity maximum, governed by the salinity distribution tends to act as a filter within the estuary.

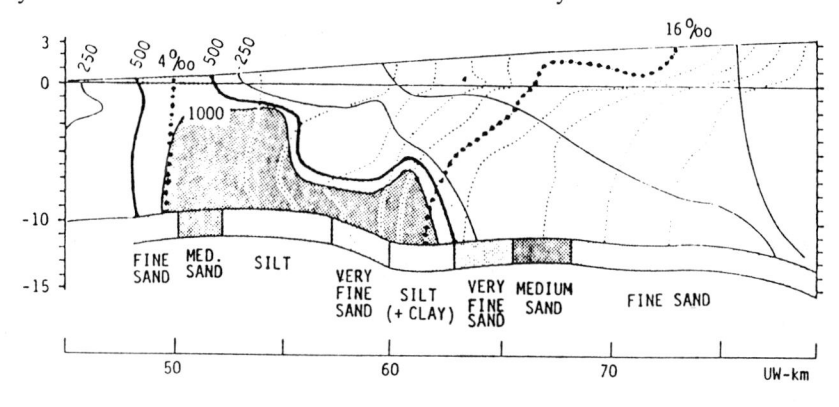

FIGURE 1. Turbidity measurements in the Weser (FRG) (after WELLERSHAUS, 1981). Contour lines for salinity (...) are shown together with those for turbidity (---).

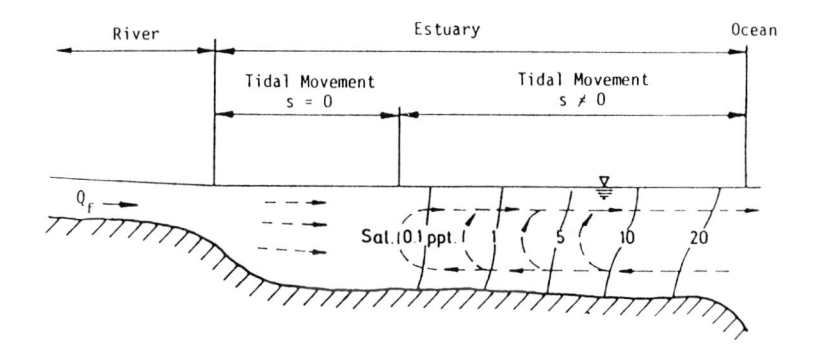

FIGURE 2. Schematic representation of tidal averaged flow and salinity fields in a partially mixed estuary.

A mechanism leading to the generation of a turbidity maximum in estuaries can be found in the temporal mean tidal flow field in the salinity intrusion region. This circulation pattern depends primarily on the fresh water flow rate Q_f, the tidal range, salinity gradients and the estuary geometry. In stratified and partially stratified estuaries this can result in a net downstream surface current and an upstream bottom current (Fig. 2). This implies a vertical velocity distribution based on continuity considerations. Sediment particles entering this region from both the river and ocean reaches can become trapped in this circulation pattern and concentrated in a given salinity zone. The concentration distribution will depend on the interaction between the particle settling velocity (which may increase significantly due to flocculation, in the case of cohesive particles), the flow field and the flow turbulence (which may be reduced due to vertical density gradients) tending to keep the particles in suspension.

These phenomena are further influenced by the alternating periods of sedimentation (during slack tides) and erosion (during strength of tides) in the tidal zone.

3. Governing equations

A mathematical formulation of this problem is given by the three-dimensional momentum- continuity- and transport-equations (Eqs. 1-5) in which the following assumptions are made:

- convective terms with vertical velocities w are negligible
- viscous stresses $\ll$ turbulent stresses
- hydrostatic pressure distribution.

x-momentum:

$$u_{,t} + uu_{,x} + vu_{,y} - fv = -\frac{1}{\rho}p_{,x} + (A_x u_{,x})_{,x}$$
$$+ (A_y u_{,y})_{,y} + (A_z u_{,z})_{,z} \quad (1)$$

y-momentum:

$$v_{,t} + uv_{,x} + vv_{,y} + fu = -\frac{1}{\rho}p_{,y} + (A_x v_{,x})_{,x}$$
$$+ (A_y v_{,y})_{,y} + (A_z v_{,z})_{,z} \quad (2)$$

hydrostatic pressure:

$$P_{(x,y,z)} = -g \int_{\eta_{(x,y)}}^{z} \rho_{(x,y,z')} dz' \quad (3)$$

continuity:

$$w_{,z} = -u_{,x} - v_{,y} \quad (4)$$

transport:

$$c_{,t} + (uc)_{,x} + (vc)_{,y} + ((w+w_s)c)_{,z} =$$
$$(K_x c_{,x})_{,x} + (K_y c_{,y})_{,y} + (K_z c_{,z})_{,z} + Q \quad (5)$$

The dependent and independent parameters are defined in Appendix 1. It can be seen from Eq. 1,2 and 5, that the flow turbulence has been parameterized using *dispersion coefficients* rather than a turbulence model (e.g. $k - \epsilon$ model). These dispersion coefficients can, however, be functions of space and time, and coupled to the density stratification through, for example, a parameterization using the Richardson number. The reason for choosing this approach at this phase of the research is to simplify the sensitivity analysis for investigating the basic parameters influencing the development of the turbidity maximum. An elaborate turbulance model would complicate an interpretation of the results and increase the computation time unnecessarily.

The flow field is coupled to the salinity field through the pressure term (Eq. 3), which also includes the horizontal density (salinity) gradients. The transport surface boundary condition is that of a non-zero concentration gradient in which an equilibrium exists between the bottom-directed flux resulting from particle settling and the surface-directed flux due to turbulence. This results in no net transport across the water surface.

a) SURFACE (no net transport):
$$q_{z_s} = -(K_z \frac{\partial c}{\partial z})\Big|_{z_s} + w_s c \Big|_{z_s} = 0$$

b) BOTTOM

1) No erosion/sedimentation or equilibrium between erosion and sedimentation (no net transport)

$$q_{bot} = -(K_z \frac{\partial c}{\partial z})\Big|_{bot} + w_s c \Big|_{bot} = 0$$

2) Erosion and sedimentation:

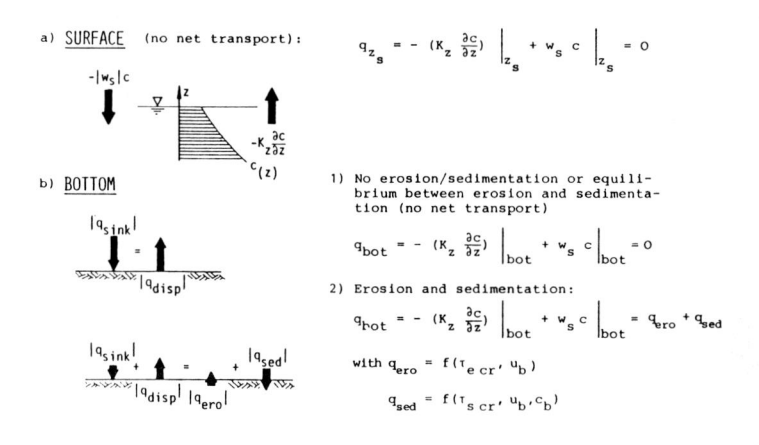

$$q_{bot} = -(K_z \frac{\partial c}{\partial z})\Big|_{bot} + w_s c \Big|_{bot} = q_{ero} + q_{sed}$$

with $q_{ero} = f(\tau_{e\,cr}, u_b)$

$q_{sed} = f(\tau_{s\,cr}, u_b, c_b)$

FIGURE 3. Boundary conditions used to solve the transport equation.

The bottom transport boundary condition is analogous to that at the surface for the case of no erosion or sedimentation. The consideration of the last two phenomena results in two additional terms as shown in Fig. 4. There are many possible functions for mathematically parameterizing these phenomena. Since we are dealing with small particles on the order < 100 μm, with corresponding low settling velocities and cohesive characteristics, erosion and sedimentation have been simulated based on the parameterization given in Fig. 4 (see also KANDIAH, 1974). This involves the specification of a critical shear stress at which erosion begins, and a critical shear stress for the onset of sedimentation. Once these shear stresses have been reached, the respective phenomena varies directly proportional to the square of the near bottom velocity.

4. Numerical model

The solution of Eqs. 1-5 is achieved using a three-dimensional finite difference model, schematized in Fig. 5. In its present form the horizontal and transverse length steps are constant; the vertical length is variable. A staggered numerical grid is used, where volumetric quantities (density, temperature, concentration ...) are computed at the element middle and velocities at the element boundaries.

a) Erosion

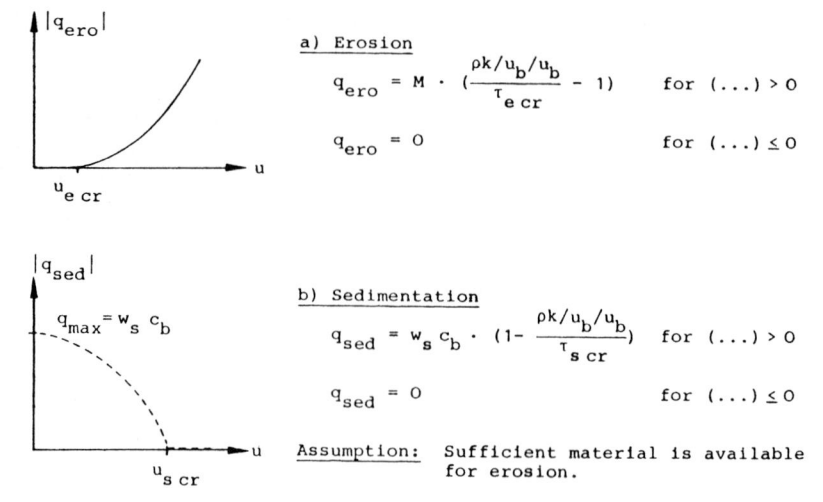

$$q_{ero} = M \cdot \left(\frac{\rho k /u_b/u_b}{\tau_{e\,cr}} - 1\right) \quad \text{for } (\ldots) > 0$$

$$q_{ero} = 0 \quad \text{for } (\ldots) \leq 0$$

b) Sedimentation

$$q_{sed} = w_s c_b \cdot \left(1 - \frac{\rho k /u_b/u_b}{\tau_{s\,cr}}\right) \quad \text{for } (\ldots) > 0$$

$$q_{sed} = 0 \quad \text{for } (\ldots) \leq 0$$

Assumption: Sufficient material is available for erosion.

FIGURE 4. Parameterization of erosion and sedimentation (after KRONE (1962) and PARTHENIADES (1962)).

a) MODEL GEOMETRY

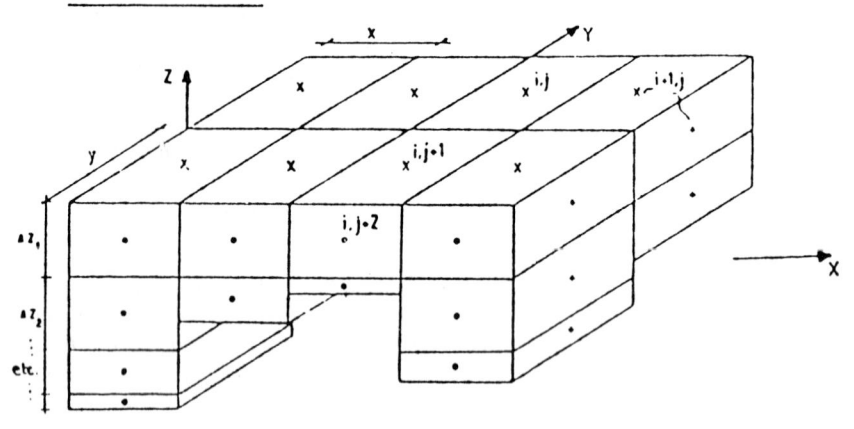

FIGURE 5. Numerical schematization. +, o, x velocities.

Horizontal advective and diffusive terms are represented explicitely, while the vertical convective terms are implicitly computed. The degree of impliciteness can be controlled by the choice of a weighting factor. This mixed explicit - implicit technique offers sufficient accuracy in an economic way (e.g. PERRELS and KARELSE, 1981).

Upstream differences are used for the advective terms and central differences for the convective ones. Upstream differences were chosen because explicit central difference schemes tend to generate numerical instability for small values of the eddy diffusivity. In such a situation stability could only be attained for uneconomically small time-steps (e.g. PEYRET and TAYLOR, 1983, p. 41) a problem which does not occur for upstream differences. On the other hand, upstream differences have only first order accuracy, while central differences are of order two in space (e.g. MITCHEL and GRIFFITHS, 1980). The resulting numerical diffusion has been corrected using the antidiffusion scheme given by SMOLARKIEWICZ (1983).

For the vertical terms, the use of an implicit procedure - like Crank-Nicolson-, Galerkin- or a fully-implicit-scheme - is favourable, because of its stability-properties (e.g. HAMILTON, 1984). The Crank-Nicolson-scheme has second order accuracy in time, while the other schemes are only of first order accuracy.

The numerical solution is generated using a time variable time step for the solution of the transport equation, which is normally significantly larger than the time step for the momentum equation. This reduces both the computer time and the occurance of numerical dispersion. Details of this procedure are presented in MARKOFSKY (1984).

5. Calculations for idealized flow situations

Two dimensional (laterally averaged) steady and unsteady suspended sediment transport calculations help to illustrate basic transport phenomena as well as aspects associated with their numerical simulation. The flow field was calculated using either a constant or parabolic eddy diffusivity (Fig. 6) which have the following velocity profiles for the steady state case: constant eddy diffusivity

$$u(z) = \frac{u_*}{k^{\frac{1}{2}}} + \frac{u_*^2 D}{2A_z}(1-(1-z/D)^2) \tag{6}$$

parabolic eddy diffusivity

$$u(z) = \bar{u} + \frac{u_*}{\kappa}(\ln(z/D)+1) \tag{7a}$$

with

$$\bar{u} = \frac{u_*}{(f/8)^{\frac{1}{2}}} \tag{7b}$$

Fig. 6 shows, that the numerical solution is in good agreement with the theory.

5.1 Steady State Concentration Profiles

5.1.1 Constant Eddy Diffusivity

The following vertical concentration profile results for the case of a constant eddy diffusivity, no flux of suspended matter across the surface and bottom boundaries and a constant settling velocity (e.g. ROUSE, 1938):

$$c_{(z)} = c_a \exp(\frac{-|w_s|}{K_z}z) \tag{8}$$

The non-dimensional concentration profile then depends on the ratio w_s/K_z. For high values of this ratio (e.g. 2 : 1 m^{-1}), which is equivalent to high settling velocities or low turbulence, the sediment is concentrated at the channel bed. For low values of this ratio (e.g. 1 : 100 m^{-1}) a near uniform concentration profile results. As can be seen in Fig. 7, agreement between numerical and analytical results is quite good. Only for large settling velocity: diffusivity ratios (e.g. 1:1 or 2:1) is there some deviation close to the bottom. This is due to the insufficient resolution of the numerical grid used in this case (18 elements, constant thickness of 0.76m). Resolution can be improved by adding supplementary computational points close to the bottom, where concentration gradients are large.

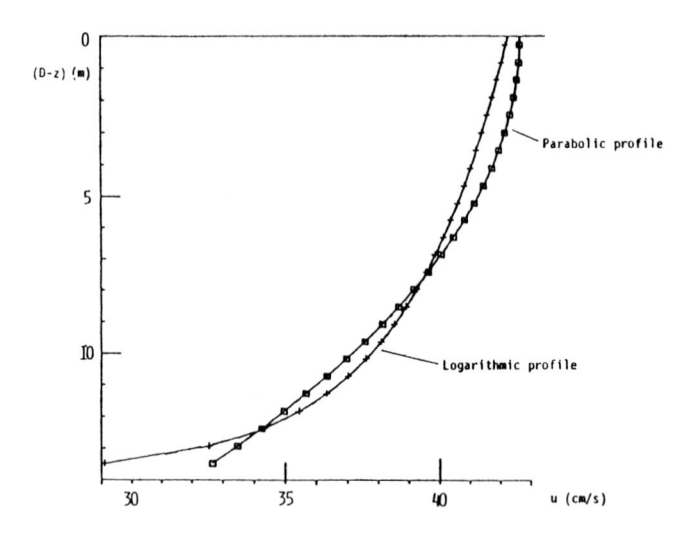

FIGURE 6. Parabolic and logarithmic velocity profiles. Comparison between numerical (...) and analytical (x) results (Eq. 6 and 7a). 25 vertical elements, $\Delta z = .55m$.

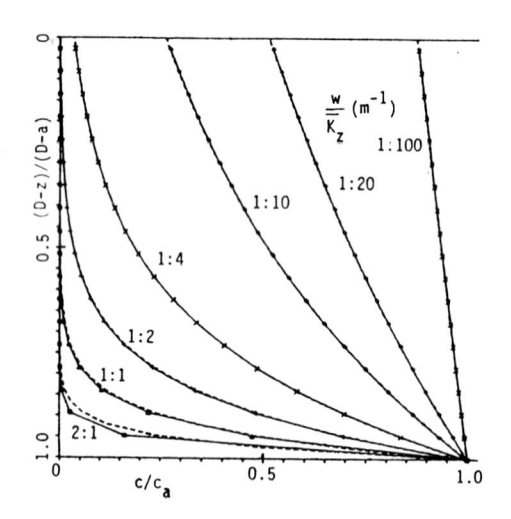

FIGURE 7. Vertical concentration profiles for constant diffusivity. Comparison between numerical (---) and analytical (...) results (Eq. 8).

5.1.2 Parabolic eddy diffusivity

If the vertical distribution of eddy-diffusivity is a parabolic function of depth, and no flux across the upper and lower boundaries occurs, then the following analytical solution is obtained for the case of a constant settling velocity (e.g. ROUSE, 1937):

$$c_{(z)} = c_a \left(\frac{D-z}{z} \frac{a}{D-a} \right)^{\frac{|w_s|}{\kappa u_*}} \tag{9}$$

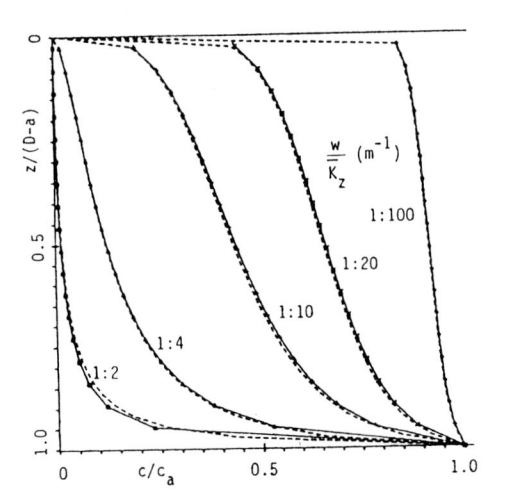

FIGURE 8. Vertical concentration profiles for a parabolic diffusivity. Comparison between numerical (---) and analytical (...) results (Eq. 9). $\bar{K}_z$ is the depth averaged eddy diffusivity.

It should be noted that for the same depth-averaged concentration (see Fig. 9), the bottom concentration is higher for a parabolic eddy diffusivity distribution when compared with the constant diffusivity case. This difference increases for large settling velocities or low turbulance (e.g. $w_s:K_z=1:10$).

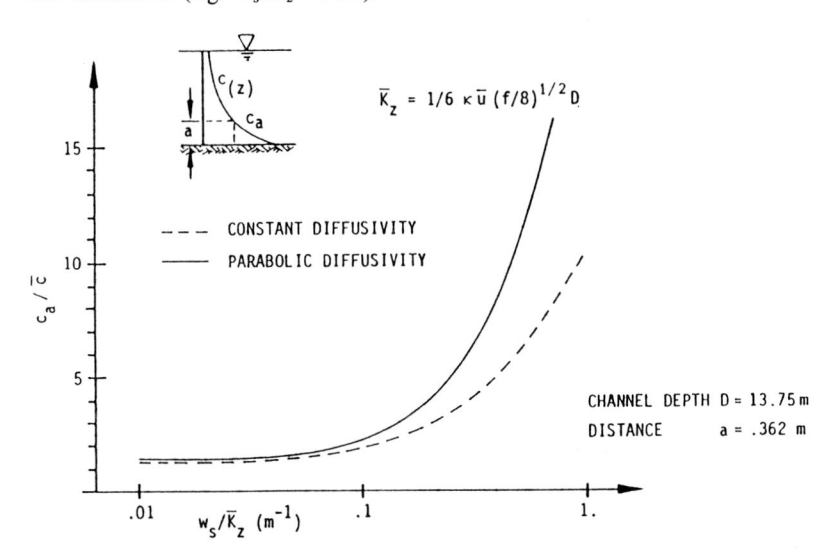

FIGURE 9. Concentration of suspended load close to the bottom for constant total suspended mass. The concentration is normalized with the mean concentration $\bar{c}$.

5.2 Transient concentration profiles

Transient behavior of suspended sediment in a uniform turbulence field was experimentally investigated by DOBBINS (1943). He derived analytical solutions for the time dependent behaviour of vertical concentration profiles including consideration of erosion and sedimentation.

5.2.1 Sedimentation without resuspension

Starting with an exponential equilibrium profile at time $t=0$, the sediment flux $w_s c_b$ is imposed at the bottom boundary. The flux across the water surface is zero. For this case - at times $t>0$ - the concentration adjusts to a zero-gradient at the bottom. Fig. 10 shows vertical concentration profiles for 6 different times after the onset of sedimentation (numerical, analytical and experimental results). The time history of the sedimentation process illustrates the ability of the model to simulate unsteady suspended sediment transport in idealized situations.

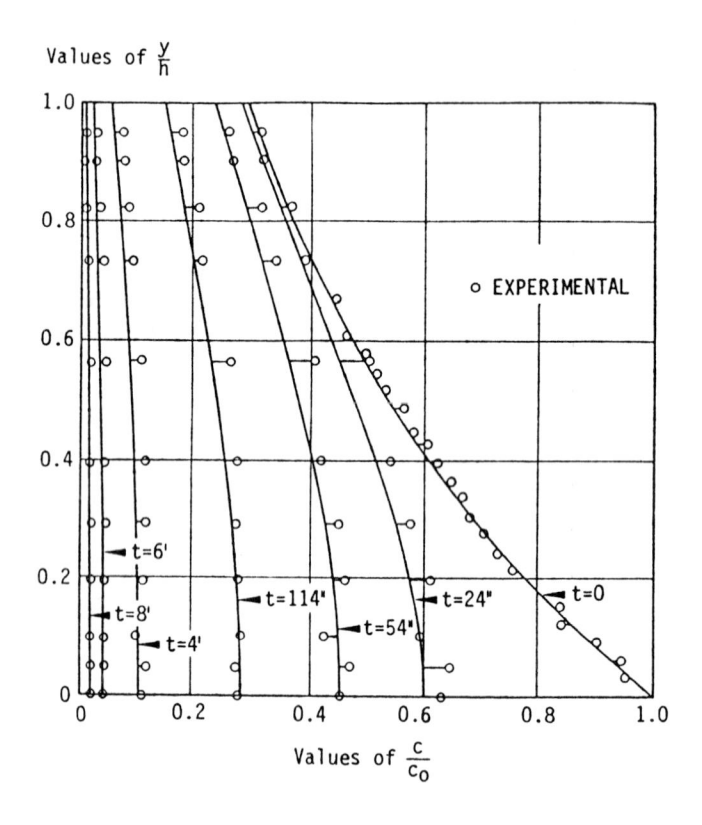

FIGURE 10. Transient vertical concentration profiles for sedimentation without resuspension. Comparison between numerical (---), analytical (---) and experimental results (0) (after DOBBINS, 1943). 11 vertical elements.

Vertical profiles calculated with a constant diffusivity K_z are compared in Fig. 11 with those using a parabolic diffusivity. The mean value $\overline{K_z}$ is the same for both calculations. Starting with a constant concentration at $t=0$, sedimentation takes place at the rate $w_s c_b$ at subsequent times without resuspension of material from the bottom. The following differences can be observed:

- for a parabolic K_z, the concentration at the surface drops to zero immediately after the onset of sedimentation, while for constant K_z the surface concentration always has a finite value;
- at intermediate depths, the vertical concentration gradient is steeper for a constant than for parabolic K_z, while close to the bottom and near the surface the contrary can be observed.

These differences result from the higher values of K_z at intermediate depths in the parabolic case and the associated lower values close to the upper and lower boundaries, when compared with a uniform distribution of K_z.

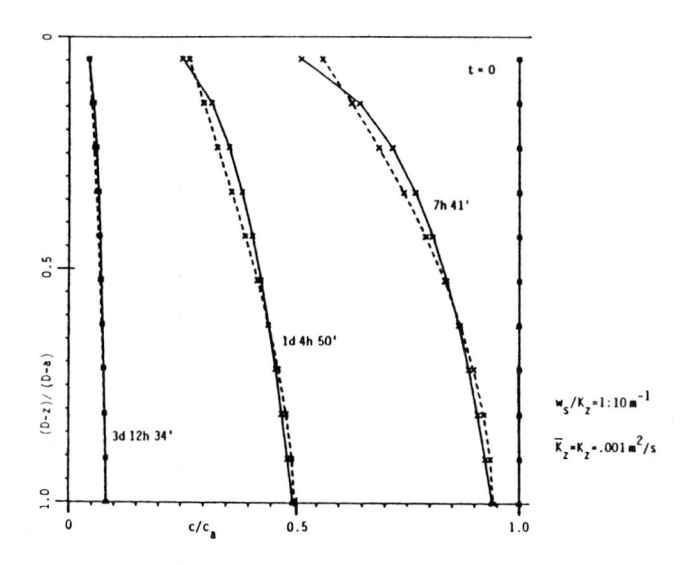

FIGURE 11. Comparison between transient concentration profiles for constant (--- ---) and parabolic (---) diffusivities. 11 vertical elements.

On the other hand, both distributions have the same sedimentation rate across the lower boundary, so that for large times (which depends only on the mean value of K_z) the concentration profiles approach each other. The absolute value of this time depends also on the mean value for K_z.

5.2.2 Comparison between transient concentration profiles under stratified and unstratified conditions

Fig. 12 shows transient concentration profiles for a vertical one-dimensional case for unstratified and stratified conditions. K_z is assumed to be only dependent on the Richardson number Ri, in which the functional relationship between K_z and Ri is given by (MUNK and ANDERSON, 1948):

$$K(z) = K_0(1+\frac{10}{3}Ri)^{-3/2} \text{ with } Ri = -\frac{g}{\rho}\frac{\rho_{,z}}{u_{,z}^2+v_{,z}^2} \qquad (10)$$

The stratified case was simulated as a two-layer-model, for which Ri is zero everywhere except in a small zone at mid depth (one element thick), where the local value of Ri

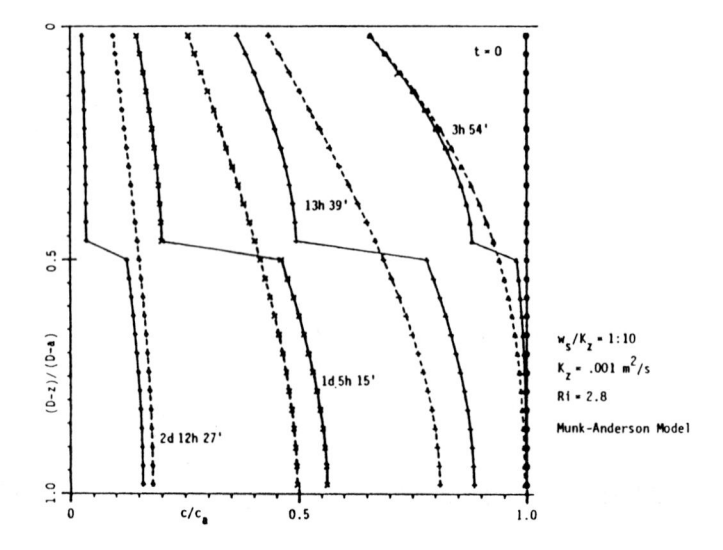

FIGURE 12. Comparison between transient concentration profiles for stratified (---) and unstratified (--- ---) conditions. Concentrations are normalized by c_a, which corresponds to the concentration a distance a above the bottom at time $t=0$. 25 vertical elements.

equals 2.8. At $t=0$ the concentration was assumed to be constant over the depth. Subsequently, sedimentation with $w_s c_b$ takes place at the bottom. No erosion was simulated. The main differences between the two calculation results are:

- for the stratified case, the concentrations in the upper layer are lower and in the lower layer greater than those for the nonstratified case. This is due to the reduced value of K_z at mid-depth for the stratified case resulting in a reduction of the upward directed flux across the boundary;
- the suspended sediment settles out faster in stratified situations as a consequence of the higher concentration in the lower layer and the associated sedimentation flux $w_s c_b$.

5.3 Sedimentation with periodic erosion

5.3.1 Critical bottom shear stress τ_{ecr} and transient concentration profiles

The influence of a critical shear stress τ_{ecr} at the bottom boundary is shown in Fig. 13. The erosive flux is given by the following formula (KRONE, 1962; see also Fig. 3):

$$q_{ero} = m(\frac{\tau_b}{\tau_{e\,cr}} - 1) \quad \text{if } \tau_b > \tau_{e\,cr} \tag{11a}$$

$$q_{ero} = 0 \quad \text{if } \tau_b < \tau_{e\,cr} \tag{11b}$$

The bottom shear stress can be computed from the bottom velocity and the Taylor friction factor. In the example shown, a sinusoidally varying u_b and different values for $\tau_{e\,cr}$ have been assumed. Erosion occurs during the shaded period on the time axis in Fig. 13.

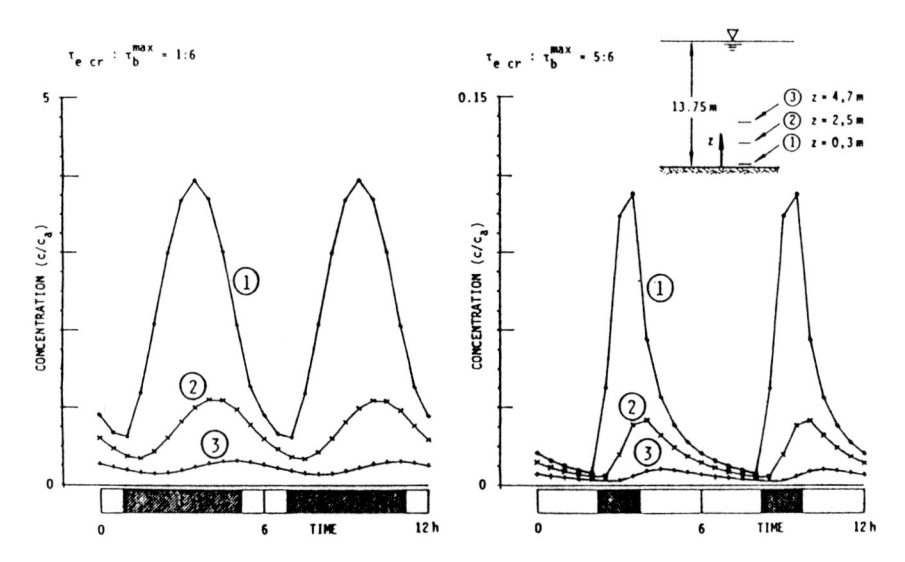

FIGURE 13. Time-dependent concentrations at different depths for periodic erosion. Constant diffusivity, Concentrations are normalized by the respective concentration at the bottom, given by Eq. 8 (equilibrium profile). $\tau_b^{max} / \tau_{e\,cr} = 6.0$ (left), $\tau_b^{max} / \tau_{e\,cr} = 1.2$ (right), 25 vertical elements.

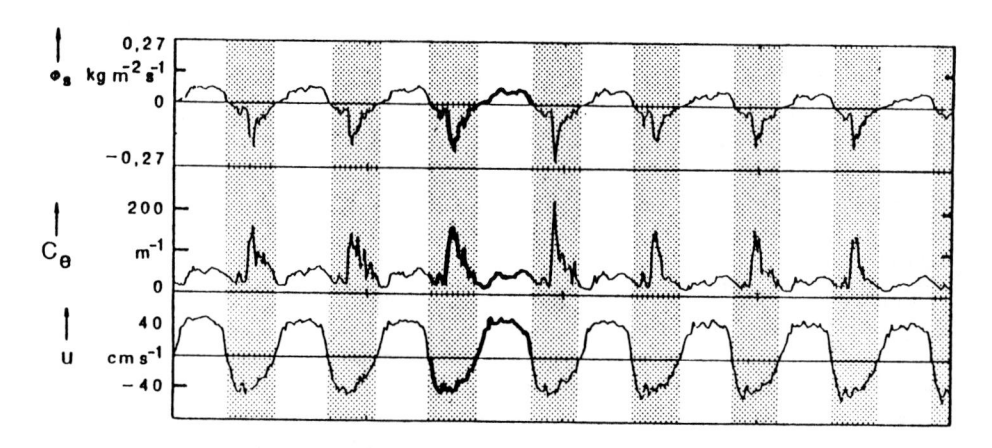

FIGURE 14. Measurements at Blexen in the Weser (FRG) 1 m above bottom.(After Inst. f. Meeresforschung, Bremerhaven (FRG).)

The sedimentation rate was simulated as $(q_{sed} = w_s c_b)$. The following results were obtained:

- the lower the value of $\tau_b^{max} / \tau_{e\,cr}$ (e.g. 1.2) the more asymmetric the curves of concentration versus time become, since the period of erosion is short; they indicate that a sharp increase in concentration occurs during the erosion phase and a long (exponential like) period of sedimentation after erosion ceases;

- at greater distances from the bottom, the curves show more and more a sinusoidal shape, because the higher frequencies are damped by diffusion;

- when $\tau_b^{max} / \tau_{e\,cr}$ is larger (e.g. 6.0), the concentration varies sinusoidally even close to the bottom.

Results of measurements 1 m above the bed in the turbidity maximum region of the Weser-estuary are shown in Fig. 14. The turbidity C_e shows the typical characteristics of the numerical simulation, namely a sharp increase at the onset of the flood tide followed by an exponential decay. Whether these observations can be explained by erosion-sedimentation or by consideration of advection is presently under investigation.

6. Turbidity maximum in a stratified estuary
6.1 Steady state salinity and suspended sediment distributions

Calculations were made of steady-state salt wedge and suspended sediment distributions for the case of a constant vertical eddy diffusivity. The value of K_z was assumed to be the same for the (turbulent) diffusion of salt and suspended sediment. The investigation included two distinct values of K_z ($10^{-4} m^2/s$ and $10^{-5} m^2/s$) and several w_s/K_z-ratios (1:1, 1:2, 1:10, 1:20 and 1:100). The simulations were made with the following boundary-conditions:

- fixed Q_f (5.1 $m^3/s/m$; mean velocity = .37 m/s) and no salinity at the upstream boundary (km 30);
- at the downstream boundary the salinity was zero in the upper 6.8 m and 30 %, in the lower 6.95 m. The total depth was held constant at 13.75 m.

Two different turbidity boundary conditions were investigated:

- zero turbidity downstream and an equilibrium profile (Eq. 8) for suspended sediment upstream;
- zero turbidity upstream and an equilibrium profile (Eq. 8) for suspended sediment downstream

The grid-spacing was 3 km in the horizontal direction and varied between 1.38 m and .69 m in the vertical direction (12 elements over the whole depth). The horizontal diffusivity was set at 0 m^2/s.

Contour lines for the horizontal and vertical velocity fields are shown in Fig. 15 along with the associated salinity distributions. The salt wedge penetrates deeper into the estuary for the smaller value of K_z. This generates an upstream shift of the region of maximum vertical velocities (located directly upstream of the tip of the salt wedge). Fig. 16 demonstrates, that there exists no pronounced turbidity maximum in the region at the tip of the salt wedge for a ratio of $w_s/K_z = 1:10$ m^{-1}, whereas for a ratio of 1:2 m^{-1} a clearly pronounced maximum can be observed. Additionally, the existence of this maximum does not depend on whether the suspended load comes from up- or downstream (see Fig. 16) whereas its strength does depend on the location of the sediment source. For the same vertical suspended sediment profile at one of the estuary boundaries, the maximum concentration within the turbidity maximum is the higher when the sediment comes from the ocean end ($w_s/K_z = 1:2$ m^{-1}; $(c/c_0)_{max} = 9.0$; c_0 = boundary concentration at the bottom).

The calculations indicate that not only the ratio w_s/K_z is important for the development of a turbidity maximum, but also the absolute value of K_z. Fig. 16 shows that for $w_s/K_z = 1:2$ m^{-1}, the maximum concentrations are lower for $K_z = 10^{-5}$ m^2/s than for $K_z = 10^{-4}$ m^2/s. This is attributed to the larger relative settling velocities ($w_s + w$) associated with the higher eddy diffusivity when w_s/K_z is kept constant. In other words, the settling velocity w_s, for $K_z = 10^{-4}$ m^2/s is a factor of 10 greater than that for $K_z = 10^{-5}$ m^2/s. The vertical velocities for $K_z = 10^{-4}$ m^2/s are at most a factor 2 greater than those

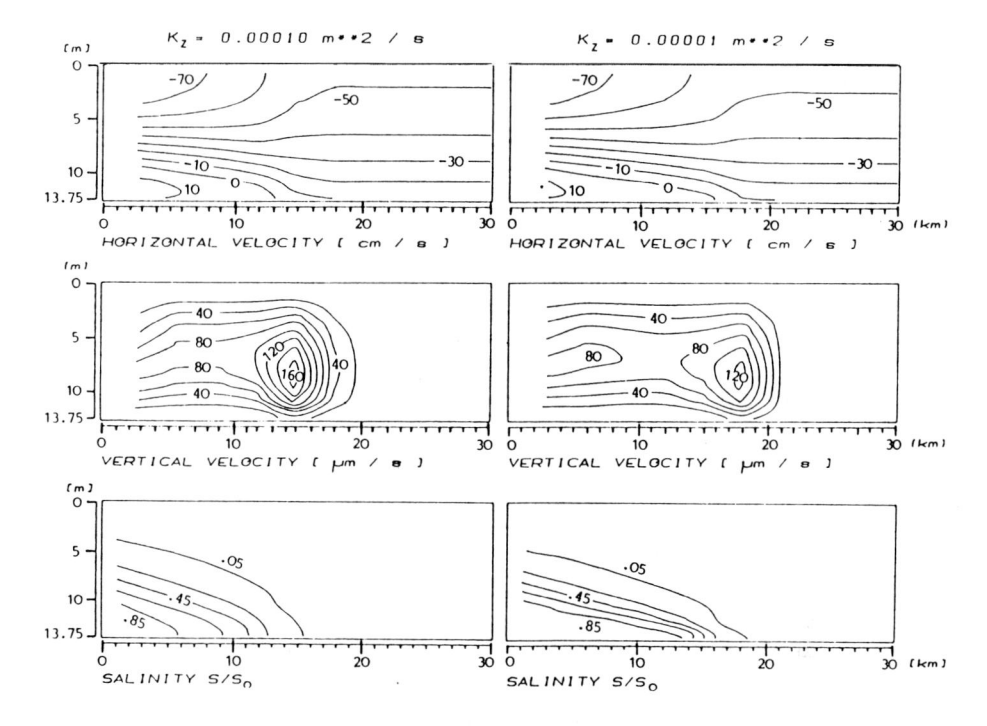

FIGURE 15. Steady state horizontal and vertical velocities and salinities for two values of K_z. Differences between contour lines are: 10 cm/s for u-velocities, 20 μm/s for w-velocities and .2 for (normalized) salinity, $s_{0=30}$ promile.

for $K_z = 10^{-5}$ m²/s.

7. Summary and conclusions

Results of numerical simulations of suspended sediment transport in stratified and non-stratified river and estuarine environments have been presented.

The validity of the numerical model was illustrated through comparisons with analytical solutions for steady state parabolic and logarithmic vertical velocity profiles, as well as steady state suspended sediment concentration profiles including erosion and sedimentation.

The assumed form of the vertical eddy diffusivity (constant vs. parabolic) did not significantly affect the calculated sedimentation rates. Reduced turbulence due to density (salinity) stratification can, however, be important under given situations.

Investigations were conducted on the simulation of erosion in a sinusoidal velocity field based on a critical bottom shear stress above which the erosion rate is proportional to the velocity squared. It was found that the amount of suspended material and the temporal behaviour of the suspended sediment concentration is strongly dependent on the ratio of the actual bottom shear stress to the critical shear stress. These results show some similarity to the temporal changes of turbidity measured in the Weser estuary.

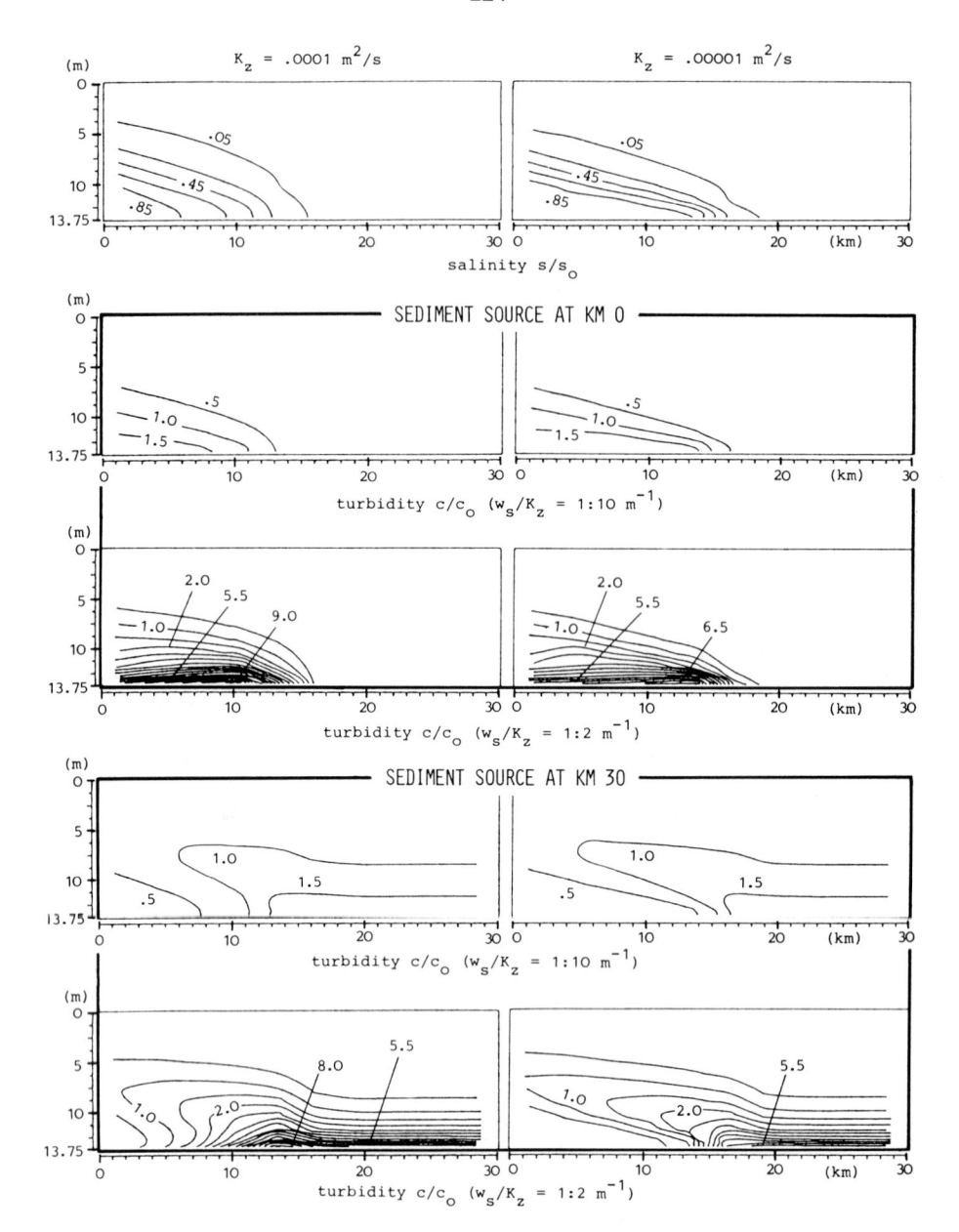

FIGURE 16. Steady state salt wedge. Concentration of suspended sediment for $w_s/K_z = 1:10$ m^{-1} and $w_s/K_z = 1:2$ m^{-1}. Constant diffusivity, $K_z = 10^{-4}$ m^2/s and 10^{-5} m^2/s. Differences between contour lines are: *turbidity*. Data have been normalized with c_0, the mean concentration at the boundary.

A sensitivity analysis to several parameters affecting the development of a turbidity maximum in the salinity intrusion region indicates that the strength and form of the turbidity maximum depends primarily on the interaction between the particle settling velocity w_s, turbulent diffusion K_z and the vertical velocity field as well as the location of the sediment source (ocean or river end).

The applicability of the model to field conditions in the Weser is presently under investigation.

It can be concluded that a numerical model is a valuable tool for studying the effects of multiple physical parameters involved in the generation of an estuarine turbidity maximum. For the near future a better representation of the geometry (bottom profile, three-dimensional effects) in the numerical model is needed, as well as algorithms for coagulation or salinity induced flocculation which may play an important role for the behaviour of cohesive sediments often found in the salinity intrusion region.

Acknowledgement

This work is being conducted through the support of the Deutsche Forschungsgemeinschaft (German Research Society) in the context of the interdisciplinary research group Sonderforschungsbereich 205 - Küsteningenieurwesen (Coastal Engineering) at Hannover University.

Field data is being obtained from: the Institut für Meereseforschung, Bremerhaven; Strom und Hafenbau, Hamburg and GKSS, Geesthacht and the Wasser und Schiffahrtsamt, Bremerhaven in the context of a research project of the Kuratorium für Forschung im Küsteningenieurwesen (KfKI) (Council for Coastal Engineering Research) through the support of Bundesministerium für Forschung und Technologie (Federal Ministry for Research and Technology). The support of these various organizations is greatfully acknowledged.

References

BLOSS, S., M. MARKOFSKY, & W. ZIELKE, 1982. Numerische Simulation von Ausbreitungsvorgängen in Astuarien: Fallstudie Untere Trave, Deutsche Gewässerkundliche Mitteilungen, 26. 1982, H 1/2: 6-13.

CASPERS, H. & H. SCHULZ, 1964. Die biologischen Verhältnisse in der Elbe bei Hamburg, Hydrobiol., 60: 53-88.

CHRISTIANSEN, H., 1974. Über den Transport suspendierter Feststoffe in Astuarien am Beispiel der Elbemündung bei Neuwerk. Theses, Universität Hannover, H. 28.

DOBBINS, W.E., 1943. Effect of Turbulence on Sedimentation. American Society of Civil Engineers, Transactions, 109: 629-656.

HAMILTON, P., 1984. Hydrodynamic Modelling of the Columbia River Estuary. Science Applications Inc., Raleigh, North Carolina.

KANDIAH, A., 1974. Fundamental Aspects of Surface Erosion of Cohesive Soils, Thesis, University of California.

KRONE, R.B., 1962. Flume Studies of the Transport of Sediment in Estuarial Shoaling Processes. Hydraulic Engineering Laboratory and Sanitary Engineering Laboratory, University of California, Berkeley.

FESTA, J.F. & D.V. HANSEN, 1977. Turbidity Maxima in Partially Mixed Estauries: A Two Dimensional Numerical Model. Estuaries and Coastal Marine Science, 7: 347-359.

MARKOFSKY, M., 1984. A Technique for Reducing the Computing Time of Two-Dimensional Thermal and Water Quality Explicit Finite Difference Scheme and Finite Element Numerical Models. Publ. in: Proc. 4th Congr. Asian and Pacific Regional Division International Association for Hydraulic Research, Chiang Mai, Thailand.

MITCHEL, A.R. & D.F. GRIFFITHS, 1980. The Finite Difference Method in Partial Differential Equations. Wiley, New York.

MUNK, W.H. & E.R. ANDERSON, 1948. Note on the Theory of the Thermocline. Journal of Marine Research, 7: 276-295.

NOETHLICH, I., 1972. Untersuchungen über den Schlickhaushalt in der Unterelbe mit besonderer Berücksichtigung der biologischen Komponenten. Wasser und Schiffahrts-Direktion Hamburg, No. 17, H. 1/2.

OWEN, M.W. & N.V.M. ODD, 1972. A Mathematical Model in the Effect of a Tidal Barrier on Siltation in an Estuary. Tidal Power (Ed. Gray, I.J. and Gashaus, O.K.), Plenum Press: 457-458.

PARTHENIADES, E., 1962. A Study of Erosion and Deposition of Cohesive Soils in Salt Water. Thesis, University of California, Berkeley.

PERRELS, P.A.J. & M. KARELSE, 1981. A Two-Dimensional, Laterally Averaged Model for Salt Intrusion in Estuaries. Publ. in: Transport Models for Inland and Coastal Waters (ed. by Fischer, H.B.). Academic Press, New York.

PEYRET, R. & T.D. TAYLOR, 1983. Computational Methods for Fluid Flow. Springer, New York.

POSTMA, H. & K. KALLE, 1955. Die Entstehung von Trübungszonen im Unterlauf von Flüssen, speziell im Hinblick auf die Verhältnisse in der Unterelbe, Deutsche Hydrograph. Zeitschrift, 8: 137-144.

ROUSE, H., 1937. Modern Concepts of Mechanics of Turbulence. Trans. Am. Soc. Civil Engrs., 102.

ROUSE, H., 1938. Experiments on the Mechanics of Sediment Suspension. Proc. 5th Int. Congr. for Applied Mechanics, Cambridge, Mass.

SALOMON, J.C., 1982. Modelling the Turbidity Maximum. Ecohydrodynmaics, Proc. 12th Int. Liege Colloq. in Ocean Hydrod.: 285-317.

SMOLARKIEWICZ, P.K., 1983. A Simple Positive Definite Advection Scheme with Small Implicit Diffusion, Monthly Weather Revue, 111: 479-486.

WELLERSHAUS, S., 1981. Turbidity Maximum and Mud Shoaling in the Weser Estuary. Arch. Hydrobiol., 92, 2: 161-198.

WHITEHOUSE, U.G., L.M. JEFFREY, & J.D. DEBBRECHT, 1959. Differential Settling Tendencies of Clay Minerals in Saline Waters. Clays and Clay Minerals, Proc. of the 7. National Conference on Clays and Clay Minerals, Washington 1958, Int. Ser. Mono. Earth Sciences.

Appendix 1: List of Terms

A_x, A_y, A_z	components of eddy-viscosity tensor
D	channel depth
K_x, K_y, K_z	components of eddy-diffusivity tensor
Q	sink/source term for sedimentation/erosion
Q_f	fresh water flow
Ri	Richardson number
c	concentration
c_a	c at distance a above the channel bottom
f	Coriolis parameter, Darcy friction factor
k	Taylor friction factor
m	erodibility
p	pressure
q_{ero}	erosion flux at the channel bottom
q_{sed}	sedimentation flux at the channel bottom
u	velocity in x-direction
v	velocity in y-direction
w	velocity in z-direction
w_s	settling velocity
x, y, z	spatial coordinates
κ	Karman constant
ρ	density
τ_b	bottom shear stress
$\tau_{e\ cr}$	critical stress for erosion
$\tau_{e\ cr}^{max}$	maximum erosion shear stress occuring during a tidal cycle
$\tau_{s\ cr}$	critical stress for sedimentation
η	water level

Tide-Induced Residual Transport of Fine Sediment

J. Dronkers

Rijkswaterstaat, Delta Department,
The Netherlands

ABSTRACT

A qualitative and quantitative analysis is presented of the net displacement of fine sediment in a homogeneous tidal basin. The transport dynamics, first pointed out by POSTMA some 30 years ago, is based on the existence of an ebb-flood asymmetry in the cycle of sedimentation, resuspension and subsequent displacement by the tidal current. This process is defined as *"tide-induced residual sediment transport"*. For a quantitative description of tide-induced residual sediment transport a one-dimensional tide-integrated analytical model is constructed. Some predictions based on the model are compared with experimental data.

1. Introduction

In many estuaries the observed import of fine marine sediment can be explained by gravitational circulation, driven by the longitudinal density gradient caused by fresh water inflow. Examples are the Thames (ODD & OWEN 1972), the Gironde (ALLEN et al. 1977), Columbia river (GELFENBAUM 1983) and Savannah river (ARIATHURAI, MACARTHUR & KRONE 1977). With this transport mechanism a turbidity maximum is found near the upstream boundary of the gravitational circulation (FESTA & HANSEN 1978). Even in vertically well mixed estuaries, where the contribution of gravitational circulation to the residual salt transport is of minor importance, the influence on the residual sediment transport may be considerable.

In tidal basins with no river inflow, or with a small river inflow as compared with the average tidal discharge (ratio in the order of 1% or less) no significant gravitational circulation is present. Yet a landward residual sediment transport and a landward increase of the suspended sediment concentration can be observed in this type of tidal basin (POSTMA 1967).

In an estuary, sediment may accumulate on the bottom in regions where the shear stress from tidal currents and wind waves does not exceed some critical value. The rate of accumulation depends on the capacity of the currents to induce a residual transport of sediment from the sea boundary to the sedimentation areas. The exchange of estuarine waters and sea water by horizontal circulation and shear diffusion is not responsible for this residual import of sediment. These exchange processes would result in an export rather than an import of sediment as a consequence of the higher suspended sediment concentration of the estuarine waters.

Some thirty year ago POSTMA (1954,1961) described a class of sediment transport mechanisms which may explain the import of marine sediment in homogeneous tidal basins. These mechanisms are based on the existence of an ebb-flood asymmetry in the cycle of sedimentation, resuspension and subsequent tidal displacement of sediment. An ebb-flood asymmetry is defined here as follows: certain tidal characteristics designated by $f(t)$ (for example the current velocity, or the bottom shear stress), evaluated in a frame moving with the cross-sectionally averaged tidal velocity, violate the equality

$f(t)+f(t+\frac{1}{2}T)=0$, where T is the length of the tidal period.

The essential assumption in Postma's theory is that the current speed necessary to resuspend fine settled material from the bottom is larger than the current speed necessary to keep this material in suspension and to prevent it from settling. In that case a sediment particle deposited in the period around HWS and subsequently eroded when ebb currents are strong enough is displaced in a landward direction relative to a neighbouring sediment particle which remained in suspension. Similarly a sediment particle which settles at LWS undergoes a relative seaward displacement. If the tidal and geometrical conditions experienced by sediment particles during ebb and flood are not symmetrical then (a) the sediment masses which are deposited at HWS and LWS; and (b) the resulting relative displacements of these sediment masses relative to average flow will be different. The net transport which results is called *"tide-induced residual sediment transport"*. In section 2 Postma's theory is generalized and a mathematical formulation of the residual sediment transport due to the above mechanism is derived.

In section 3 the assumption of critical shear stresses for sedimentation and resuspension is discussed on the basis of field observations. In section 4 some predictions of the model are compared with measurements of residual transport of fine sediment in a few homogeneous tidal basins in the Netherlands.

The role of wind waves is ignored in this study. It can be shown (DRONKERS 1984) that the residual transport of fine sediment through the sea inlet of a long tidal basin (length at least several times the tidal excursion) is influenced by wave action only during severe storms. However, the suspended sediment concentration in the interior part of a tidal basin can be strongly increased already by moderate waves.

2. Analytical expression of tide-induced residual sediment transport

This section deals with a mathematical formulation of the tide-induced residual sediment transport. Introducing certain simplifying assumptions which are not very restrictive an analytical solution is derived.

The cross-section of the tidal basin is divided in two parts: one part represents the main channel, with cross-section area A_s, average water-depth h and average current velocity U. The other part represents the shallow regions along the channel banks, where the current velocity is on the average much smaller than U. Consider a section of the main cannel $A_s dx$ moving with the average tidal velocity U. From the sediment budget of this moving section the following equation can be derived:

$$A_s \frac{dC}{dt} + \frac{\partial}{\partial x}(A_s \overline{u'c'}) - \frac{\partial}{\partial x}(A_s \epsilon_x \frac{\partial C}{\partial x}) = \dot{E} - \dot{S} \qquad (1)$$

Here the following conventions and notations are used: $d/dt = \partial/\partial t + U\partial/\partial x$ is the time derivative in the moving frame, C is the cross-sectionally averaged suspended sediment concentration, u', c' are deviations of the suspended sediment concentration and the current velocity from the averaged values, and ϵ_x is a diffusion coefficient. $\dot{E}$ and $\dot{S}$ are sediment fluxes per unit length resulting from respectively erosion and sedimentation of bottom and channel sides. The terms $\overline{u'c'}$ and $\epsilon_x \partial C/\partial x$ stand for sediment transport related to renewal of water inside the section $A_s dx$ by dispersion and diffusion processes, and will not be taken into account further on. The exchange of sediment with the shallow regions along the banks will also be neglected. The rate of change of the suspended sediment concentration in the moving section $A_s dx$ is then entirely due to local deposition and resuspension.

The exchange of sediment with the bottom depends on the shear stress τ exerted on the bottom by the current, the settling velocity and the degree of consolidation of the bottom sediment. Many investigations of the sedimentation and erosion functions have been conducted, yielding different formulations depending inter alia on the salinity, viscosity and turbulence of the water and on the chemical and physical properties of the sediment (see, for example, EINSTEIN & KRONE 1962, MIGNIOT 1968). A common feature is that for high bottom shear stress, i.e. τ larger than a critical value τ_e, erosion dominates over deposition, while for bottom shear stress smaller than a critical value τ_d deposition dominates. For $\tau_d < \tau < \tau_e$ erosion and deposition are in equilibrium: $E - S = 0$.

Due to the inhomogeneity of the sediment composition the values of τ_d and τ_e in natural estuaries vary during a tidal cycle, especially when a moving frame is considered. For the sake of simplicity, however, τ_d and τ_e will be assumed constant.

In the absence of wind waves the bottom shear stress is related to the bottom topography and the current velocity. It will be assumed that a unique relationship exists between the bottom shear stress and the cross-sectionally averaged current velocity U. This is equivalent to the assumption of a constant Chézy friction coefficient. In reality this coefficient depends on the water depth h, the bottom roughness and ebb-flood differences and lag effects in turbulent intensity.

Only estuaries are considered in which the depth-averaged concentration of fine sediment does not exceed a few hundred mg/l. Most tidal systems at moderate latitudes belong to this category. Higher concentrations are generally related to estuarine circulation. This phenomenon is represented in the transport equation by the term $\overline{u'c'}$, which has been assumed to be negligible.

In the less turbid estuaries the bottom layer formed by sedimentation in the period around slack water contains a few $kg.m^{-2}$ sediment or less. Erosion rates reported in literature (KRONE 1974, McDOWELL & O'CONNOR 1977) are on the order of $1 gm^{-2}s^{-1}$ at an applied shear stress of twice the critical erosion shear stress. This means that in moderately turbid estuaries the sediment deposited in the period around slack water is resuspended in a short lapse of time after the critical erosion shear stress is reached. In the derivation of an expression for the residual sediment transport it will also be assumed that resuspension of the recently deposited bottom sediment takes place instantaneously when $U > U_e$.

The evaluation of the tidally averaged transport is based on a simple bookkeeping method. A plane $X(t)$ moving with the cross-sectionally averaged velocity U is considered. The amounts of sediment passing through the plane during different stages of the tidal period are inventorized and summed up.

The plane $X(t)$ moves about the fixed plane x_0 between $x_0^- = x_0 - L/2$ at $t^- = LWS$ and $x_0^+ = x_0 + L/2$ at $t^+ = HWS$. As the velocity distribution is assumed to be homogeneous in the cross section, sediment in suspension cannot pass the moving plane. Only the sediment which has settled on the bottom passes the moving plane.

A cyclic tidal variation of water levels and sediment concentration is assumed. No residual discharge is present: the plane $X(t)$ returns to its initial position after a tidal period. In the range $[x_0^-, x_0^+]$ no net bottom erosion or bottom accretion takes place. The amount of sediment passing through the moving plane $X(t)$ during a tidal period thus equals the net amount passing through the fixed plane x_0, and is designated as $M(x_0)$.

The total amount of sediment M_F passing through the moving plane during the flood is composed of three terms:

- the amount of sediment settled on the bottom at $t^- = LWS$ that is, the sediment deposited during the last stage of the ebb during which $t > t_{1d}^-$, i.e. $|u(x,t)| < u_d$,

$$(a) = - \int_{t_{1d}^-}^{t^-} dt \int_{x_0^-}^{x_0^+} dx \theta(U_d - |U(x,t)|) \dot{S}(x,t).$$

In this equation U is the local, Eulerian velocity; U_d is the highest velocity at which sediment is deposited; t^- is the time of LWS at the location of the moving plane $X(t)$; t_{1d}^- is the time preceding LWS at which the velocity U in the range $[x_0^-, x_0^+]$ has reduced to U_d.

The Heavyside function θ ($\theta(u)=1$, $u>0$; $\theta(u)=0$, $u<0$) could be omitted, as by definition $\dot{S}=0$ if $U>U_d$. It is included, however, as it makes the algebra easier to understand.

- the amount of sediment settled between the moving plane $X(t)$ and the limit of the flood excursion x_0^+:

$$(b) = - \int_{t^-}^{t^+} dt \int_{X(t)}^{x_0^+} dx \theta(U_d - |U(x,t)|) \dot{S}(x,t).$$

- the amount of sediment eroded in front of the moving cross-section

$$(c) = \int_{t^-}^{t^+} dt \int_{X(t)}^{x_0^+} dx \dot{E}(x,t).$$

If it assumed that all the sediment deposited during the preceding LWS period is eroded instantaneously when $|U(x,t)| = U_e$, then

$$\dot{E}(x,t) = \delta(t - t_e^-) \int_{t_{1d}^-}^{t_{2d}^-} dt \theta(U_d - |U(x,t)|) \dot{S}(x,t).$$

Here δ is the Dirac delta function, t_{2d}^- delimited the first stage of the flood during which $U(x,t) < U_d$ and t_e^- is the subsequent time at which $U(x,t_e^-) = U_e$. Substitution in (c) yields:

$$(c) = \int_{t_{1d}^-}^{t_{2d}^-} dt \int_{X_e^-}^{x_0^+} dx \theta(U_d - |U(x,t)|) \dot{S}(x,t).$$

Here $X_e^- \equiv X(t_e^-)$. It has been assumed that the location $X_e(t)$ at which the local velocity U equals U_e, changes faster with t than the moving plane $X(t)$. The opposite, $|dX_e/dt| < |dX/dt|$, will occur in practice only exceptionally, for example, if the velocity is decreased locally at a widening of the estuary and remains there a longer time below the threshold of erosion than in neighbouring locations.

The amount of sediment M_F passing through the moving plane during the flood is given by the sum of the integrals (a), (b) and (c). For the amount of sediment M_E passing through the moving plane during ebb identical expressions can be written, replacing the superscript $-$ by $+$ and vice versa. Collecting all the terms, the following expression is found for the residual sediment transport during a tidal cycle, caused by tidal displacement in combination with sedimentation and resuspension (see appendix):

$$M(x) = \int_{t_{1d}^+}^{t_{2d}^+} dt \int_{X_e^+}^{X(t)} dx \, \dot{S}(x,t) - \int_{t_{1d}^-}^{t_{2d}^-} dt \int_{X(t)}^{X_e^-} dx \, \dot{S}(x,t). \qquad (2)$$

In the absence of a residual discharge, this quantity $M(x)$ represents the sediment transport through a fixed plane integrated over the period between two successive times of maximum flood or maximum ebb. The expression (2) applies to that part of the tidal basin where no net bottom erosion of accretion takes place. Sufficiently remote from the region of net bottom accretion Eq. (2) can be approximated by an expression with the following structure (see appendix):

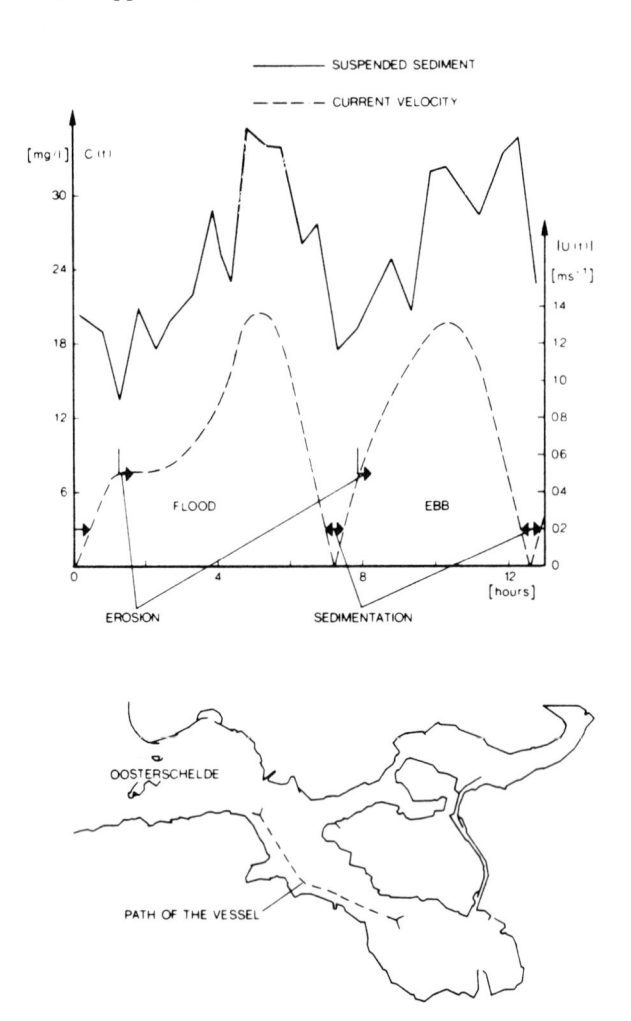

FIGURE 1. The tidal variation of the depth-averaged suspended sediment concentration in the Eastern Scheldt basin, measured from a vessel moving with the tidal current

$$M(x) = \mu^+ \lambda^+ - \mu^- \lambda^- \tag{3}$$

Here λ^+ (resp. λ^-) is the relative displacement with respect to the water body of the sediment mass μ^+ (resp. μ^-) which is deposited per unit length during the period around $t^+ = HWS$ (resp. $t^- = LWS$) in the landward limit $x^+ = x + L/2$ (resp. seaward limit $x^- = x - L/2$) of the tidal excursion. Estimates for λ and μ are given by (see appendix).

$$\mu^\pm(x) = \omega^\pm(x) \cdot C_{max}(x) \tag{4}$$

$$\omega^{\pm}(x) \cong A_s(x^{\pm}, t^{\pm})[1 - \exp(-w \frac{\Delta t_d^{\pm}(x^{\pm})}{h(x^{\pm}, t^{\pm})})]$$

$$\lambda^{\pm}(x) \simeq \frac{1}{2} U_e \cdot \Delta t_e^{\pm}(x^{\pm}) \tag{5}$$

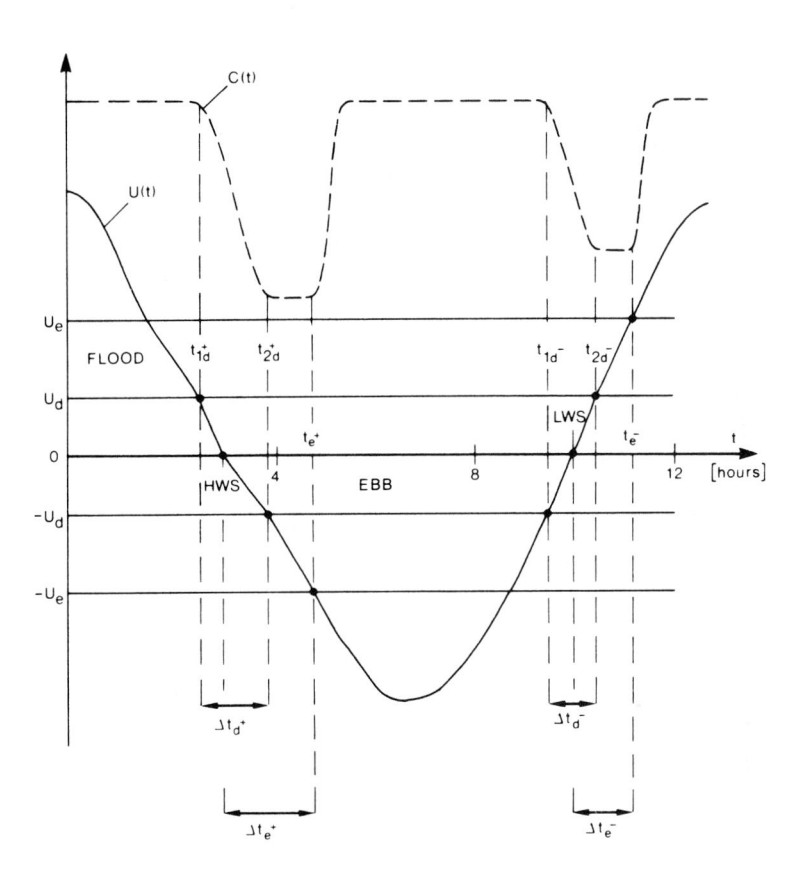

FIGURE 2. Definition sketch of critical depth averaged current velocities U_d and U_e corresponding to thresholds for deposition and erosion. Definition of the time intervals $\Delta t_d^{\pm}$ and $\Delta t_e^{\pm}$.

The suspended sediment concentrations at maximum flood and maximum ebb are assumed to be equal and are designated by C_{max}. The time interval during which sedimentation takes place and the time interval between slack water and resuspension are designated by Δt_d and Δt_e respectively (see Fig. 2). The superscripts $+$, $-$ refer to the HWS, resp. LWS periods. The duration of the time intervals Δt_d and Δt_e depends primarily on the time variation of the current velocity around slack water. In particular, if Δt_d, $\Delta t_e \ll T$ then

$$\Delta t_d^{\pm} \cong 2 U_d \cdot \left| \frac{dU}{dt}(x^{\pm}, t^{\pm}) \right|^{-1}, \quad \Delta t_e^{\pm} \cong U_e \cdot \left| \frac{dU}{dt}(x^{\pm}, t^{\pm}) \right|^{-1}.$$

In Eq. (4) the constant w represents the average near bottom settling velocity multiplied by the average ratio of near bottom and depth-averaged suspended concentrations. If the fraction of deposited sediment $[1-\exp(-w\Delta t_d/h)]$ is small, then the expression (4) can be further approximated by

$$\omega^{\pm}(x) \cong w \cdot b_s(x^{\pm}, t^{\pm}) \cdot \Delta t_d^{\pm}(x^{\pm}), \qquad (6)$$

where b_s designates the width A_s/h of the main channel.

From the Eqs (3)-(6) it is clear that a net import or export of suspended matter may exist in a tidal basin, independently of the existence of residual discharge, gravitational circulation of other dispersion mechanisms. An increasing width in the landward direction favours a landward tide-induced transport of sediment, as in that case in Eq. (6), $b(x^+, t^+) > b(x^-, t^-)$. However, if there is an ebb-flood asymmetry in the tidal variation of the current velocity U, then the time derivative $|dU/dt|$ at slack water is the main parameter controlling import or export. If, for example,

$$|\frac{dU}{dt}|_{x^+,t^+} < |\frac{dU}{dt}|_{x^-,t^-}, \quad \text{then} \quad \Delta t_d^+ > \Delta t_d^- \quad \text{and} \quad \Delta t_e^+ > \Delta t_e^-.$$

From the Eqs (2)-(6) it clearly follows that a landward tide-induced transport of fine sediment will occur in such cases.

3. In situ investigation of sedimentation and resuspension

In the Eastern Scheldt estuary (a dammed-off part of the Rhine-Meuse-Scheldt delta in The Netherlands) an attempt has been made to follow the tidal motion of a water volume and to determine the average suspended sediment concentration; samples were taken in the vertical from a vessel which followed a subsurface drogue.

If it is assumed that the same water volume is sampled during the entire tidal cycle (which of course is at best a rough approximation, due to variations in current speed and direction as a function of depth) then the variation of the average suspended sediment concentration can be ascribed to local erosion and sedimentation processes. The results are shown in Fig. 1. Rather large fluctuations in the suspended sediment concentrations are due to the smallness of the individual samples. Around slack water approximately 5% of the suspended sediment has a diameter larger than 50μ. At the highest current speed this fraction increases to 40% (during flood).

Fig.1 clearly shows that a time lag is present between the variation of the suspended sediment load and the variation of the current velocity. The sediment load only increases when the current speed has exceeded a critical value U_e ($U_e \cong 0.5$ ms^{-1}). A decrease of the suspended sediment load is observed in periods when the current speed has fallen below some critical value for deposition U_d ($U_d \cong 0.2$ ms^{-1}).

A more detailed inspection of Fig.1 shows that the increase of suspended load holds on as long as the current velocity increases, in particular during flood. This is partially explained by erosion of course bottom material. In addition, fine sediment trapped at the leeside of bottom ripples is probably removed and brought into suspension only when the course sediment is eroded. A few consequences for the application of Eq. (3) appear.

Firstly, the assumption of instantaneous erosion when $|U| > U_e$ does not hold. However, if the resuspension of sediment deposited during the previous slack water period is completed well before the maximum current speed is reached, then Eq. (5) still can be applied. The values of Δt_e and U_e, corresponding to the onset of the resuspension process, should be replaced by values corresponding approximately to the averages of Δt and U for the entire resuspension time interval.

Secondly, if the resuspension process goes on till the maximum current speed is

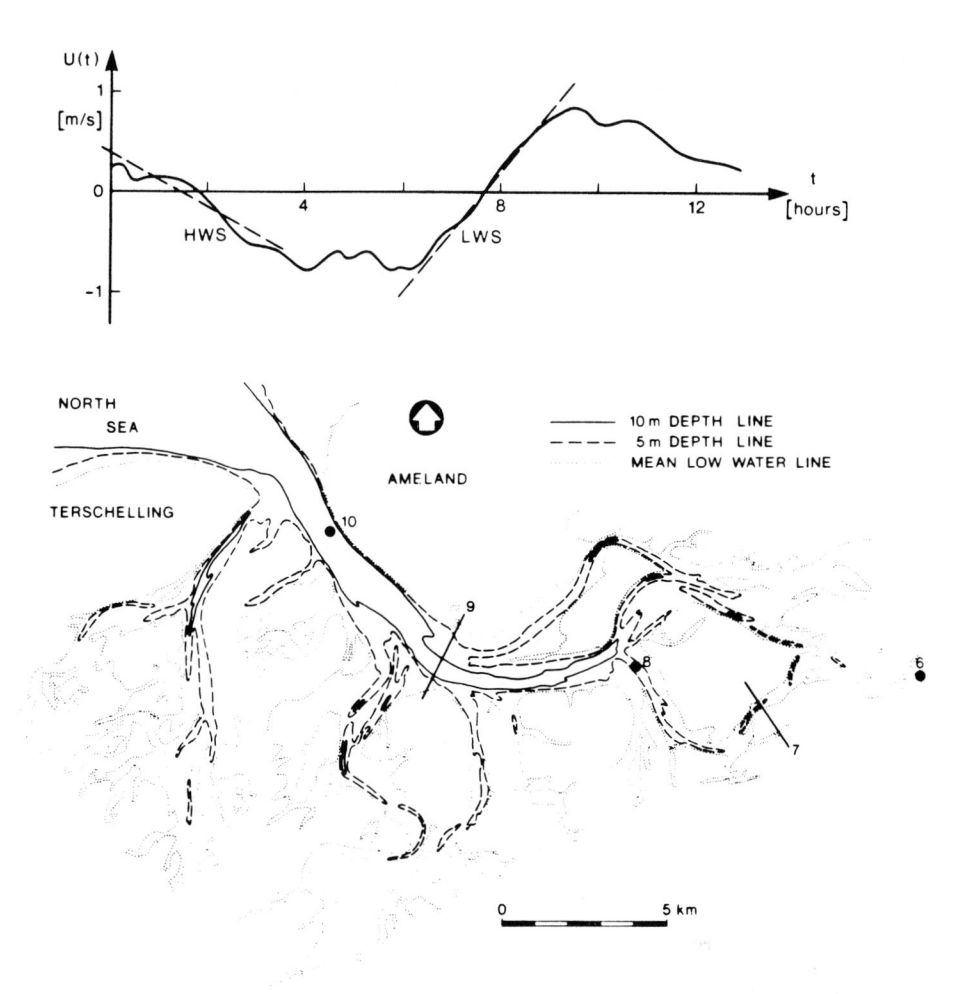

FIGURE 3. The Wadden Sea Ameland tidal system, with sampling stations 6, 8 and 10 and cross sections 7 and 9. Current velocity registration (from POSTMA 1961) in location 8 showing a slower variation of the current velocity at HWS than at LWS.

reached, then Eq. (3) is not valid, at least not for the bottom material which is brought in suspension when the current speed is close to its maximum. Thus in particular for coarse sediment Eq. (3) cannot be applied. The residual transport of coarse sediment depends more strongly on an ebb-flood asymmetry of the maximum current speed than on the asymmetry of the slack water periods. Eq. (3) only applies to fine sediment which is brought in suspension during a period that currents are on the order of half the maximum speed or less.

Fig. 1 finally shows that the suspended sediment concentration during flood already decreases shortly after maximum current speed while during ebb the decrease only starts just before slack water. This is partly explained by the higher fraction of coarse sediment eroded during flood than during ebb. Probably another process is also responsible for the above mentioned asymmetry. During flood the near bottom current velocity in a channel bend has a component directed towards the tidal flats at the inner bend. Water with a relatively high suspended sediment concentration is thus transported from the channel to the tidal flats. On the contrary during the ebb water with a relatively high sediment load is transported from the shallows to the channel. Such sediment fluxes between the channel and the channel boundaries may also contribute to a residual longitudinal transport. However, they are not incorporated in Eq. (3).

4. Comparison of model results and field data

In two tidal basins in The Netherlands the residual flux of sediment has been determined experimentally and compared with results given by the analytical expression, Eq. (3).

The geometry of these tidal basins, the Wadden Sea-Ameland area and the Eastern Scheldt is shown in Figs. 3 and 4 respectively. The geomorphological characteristics of these basins are quite different: the channels in the Ameland tidal system are shallow, especially in the landward part, the tidal flat area is relatively large and flooded soon after LW; the tidal flat bottom is mainly composed of fine grained material ($\leq 100\mu$). The channels in the Eastern Scheldt are deep, the tidal flat area is relatively less important. The major part of the tidal flats is situated above mean sea-level and the bottom contains only a small fraction ($\leq 10\%$) of fine sediment. In Figs. 3 and 4 also the tidal variation of the current velocity is shown. The behaviour around slack water is different in the two basins. Obviously

$$|dU/dt|_{t^+} < |dU/dt|_{t^-}$$

in the Wadden Sea, and the predicted tide-induced residual sediment flux is directed landward. The reverse holds for the major part of the Eastern Scheldt tidal basin.

In table 1 a more quantitative evaluation of the residual sediment flux in the Wadden Sea-Ameland area according to Eq. (3) is presented. The amount of sediment μ deposited in the period around slack water and the relative displacement λ have been estimated using suspended sediment and current velocity data collected during one tidal cycle by POSTMA (1961). The results are compared with the average bottom accretion determined from bottom soundings over a period of 15 years (DE BOER & VISSER 1981).

The bottom soundings show that, on the average, sedimentation slightly prevails over erosion. The sedimentation areas mainly coincide with the fine grained tidal flats. The average net bottom accretion due to the residual import of fine sediment is a few mm/year larger than the overall bottom accretion derived from the bottom soundings.

Bearing this in mind and also taking into account the possible systematic error of approximately 4 mm in the bottom soundings, the estimated tide-induced import through cross-section 9 is on the same order of magnitude as the import of fine sediment based on the bottom soundings. It also appears that the import of sediment is larger in the landward part of the tidal basin than the seaward part. This confirms that a net bottom erosion takes place in the central and seaward parts of the Ameland tidal system.

In the landward cross-section 7 the import evaluated from the formula for tide-induced transport is much larger than the import corresponding to the observed bottom accretion. Of course it may be argued that the tidal period considered is not representative for the average situation, and the value $U_e = 0.5$ ms^{-1} for the critical current velocity for erosion may also be questioned. More important however is the presence of wind waves which, particularly in the shallow landward part of the tidal system, contribute to the bottom shear stress and thus cause erosion even at current velocities much lower than U_e. In DRONKERS (1984) it is shown that the tide-induced residual sediment

TABLE 1. Estimate of tide-induced residual sediment transport in the Wadden Sea Ameland area (see Fig. 3) from experimental data (POSTMA 1961): current velocity and suspended sediment (particle size smaller than 64 micron) concentration at the locations 10, 8 and 6 during a tidal period. The distance between the locations approximately equals the tidal excursion. The cross sections 9 and 7 are situated halfway between the locations 10 and 8 and halfway between the locations 8 and 6 respectively.

location tidal phase		10 LWS	8 HWS	8 LWS	6 HWS
$\Delta C = C_{max} - C(x^\pm, t^\pm)$	$[kgm^{-3}]$	0.04	0.036	0.027	0.5
cross-section area	$[m^2]$	9 : 16500		7 : 1300	
deposited sediment per m					
$\mu = A \Delta C$	$[kgm^{-1}]$	660	594	35	650
$\lambda = \int_{t^-}^{t^+} \|U\| dt$	$[m]$	860	1330	940	3000
$\lambda \cdot \mu$	$[kg]$	5.7 10^5	7.9 10^5	0.3 10^5	19.5 10^5
cross section		9		7	
tide induced residual sediment transport in one tidal period	$[kg]$	2.2 10^5		19.3 10^5	
storage surface situated landward	$[m^2]$	143 10^6		18 10^6	
average landward bottom accretion (bottom sediment concentration = 1300 kgm^{-3})	$[mm/year]$	0.8		57	
average landward bottom accretion from bottom soundings	$[mm/year]$	0.1 ±4		7 ±4	
residual transport in one tidal period corresponding to average observed bottom accretion	$[kg]$	0.3 10^5		2.3 10^5	

import can be decreased by wave action. The decrease is much stronger in the landward part of the tidal basin than in the seaward region.

The large difference between the computed and observed long term sediment import through cross-section 7 cannot be explained entirely by the inclusion of the average wind influence in the computation of the bottom shear stress. From field observations it is known that an important erosion and seaward displacement of sediment takes place nearly every year as a result of strong wave and current action during a few storm surges (WINKELMOLEN & VEENSTRA 1980). During a storm surge large amounts of fine sediment are eroded at HW from tidal flats and marshes. This sediment is transported seaward by the ebb current when the surge is falling. The first flood tide after the storm surge penetrates the tidal basin less far: a net seaward displacement of fine sediment results.

The residual transport of fine sediment in the Eastern Scheldt tidal basin has been determined by direct transport measurements in two cross-sections A and B, see Fig. 5. In a number of locations in these cross-sections the current velocity and the suspended sediment concentration were measured during a tidal period.

The tidal transport is

$$M = P^F C_u^F - P^E C_u^E. \tag{7}$$

Here

$$C_u^{F,E} = \int_0^T \int\int_A uc\theta(\pm u)dA\, dt.$$

$$P^{F,E} = \int_0^T \int\int_A u\theta(\pm u)dA\, dt.$$

u is the velocity component perpendicular to the cross-section, A is the cross section area and θ is the Heavyside function.

The influence of flood or ebb surplusses is practically eliminated by adding to Eq. (7) a term

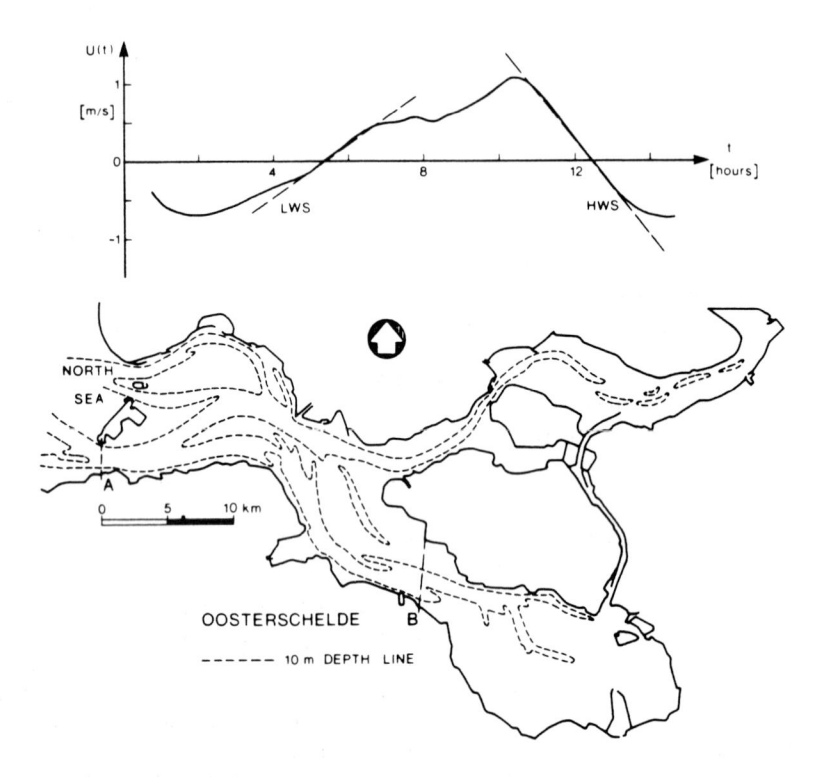

FIGURE 4. The Eastern Scheldt tidal basin and current velocity registration in transect B, showing a faster variation of the current velocity at HWS than at LWS

$$\frac{1}{2}(P^E - P^F)(C_u^E + C_u^F).$$

This yields

$$M = \frac{1}{2}(P^F + P^E)(C_u^F - C_u^E). \tag{8}$$

The transport measurements in both cross sections A and B were repeated every month for one year in order to gain insight into seasonal variations.

In order to reduce the sensitivity of the results to the sampling distribution in the cross-section, different types of measurements were performed:
- extensive transport measurement with a large number of stations in the cross-section,
- reduced transport measurements with two verticals, each representative for a part of the cross-section,
- velocity measurements in two verticals and sampling of suspended sediment in the whole transect at 5 m depth,
- extensive velocity measurements and sampling of suspended sediment in each vertical at 60% of the water depth from the surface. The sampling time is taken proportional to the discharge. Samples are cumulated over the ebb and flood periods separately.

The results of these different types of measurements do not yield systematic differences. In Fig. 5a the values for the residual transport of total suspended sediment are shown (ELGERSHUIZEN 1983).

TABLE 2. Estimate of tide-induced residual sediment transport through the transects A and B in the Eastern Scheldt tidal basin (see Fig. 7), using field data for the computation of μ and λ [*)].

location tidal phase		$A-L/2$ LWS	$A+L/2$ HWS	$B-L/2$ LWS	$B+L/2$ HWS
$\Delta C = C_{max} - C(x^{\pm}, t^{\pm})$	$[kgm^{-3}]$	0.017	0.014	0.016	0.012
cross-section area	$[m^2]$	A : 40,000		B : 35,000	
deposited sediment per m $\mu = A\,\Delta C$ $\lambda = \int_{t^{\pm}}^{t^{\pm}} \lvert U \rvert\,dt$ $\lambda \cdot \mu$	$[kgm^{-1}]$ $[m]$ $[kg]$	680 600 4.10^5	560 350 2.10^5	560 1100 $6.2\ 10^5$	420 600 $2.5\ 10^5$
cross section		A		B	
tide induced residual sediment transport in one tidal period computed yearly average (measured)	$[kg]$ $[kg]$	$-0.2\ 10^6$ $-1.3\ 10^6 \pm 2.10^6$		$-0.4\ 10^6$ $-0.9\ 10^6 \pm 10^6$	

[*)] Data of Fig.1 and other unpublished data of A. HOLLAND and H. ELGERSHUIZEN

Coulter Counter and microscope analyses reveal that the suspended sediment in the Eastern Scheldt contains a fraction (on the order of 30-70% of the weight) of small particles with an average diameter of approximately $10\,\mu$. The coarse fraction contains particles with a diameter between 100 and $200\,\mu$ The average fall velocity in still water is approximately $10^{-3}\ ms^{-1}$.

Most of the large particles are aggregates of fine particles. At maximum current speed sand may represent up to 40% of the weight of the suspended sediment load, under normal conditions. Residual import and residual export of suspended sediment both occur in the Eastern Scheldt. However, residual export dominates, especially in the landward cross-section B. In addition to the total suspended sediment concentration

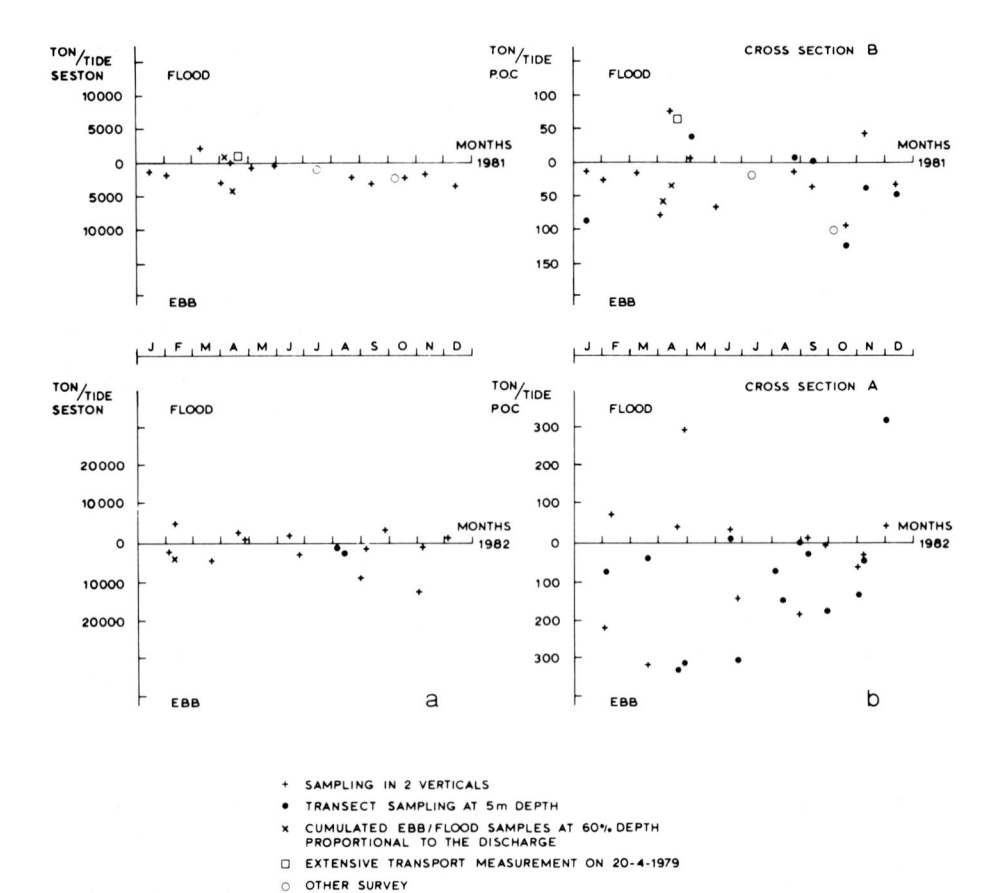

FIGURE 5a. Tidally averaged transport of suspended sediment through the transects A and B in the Eastern Scheldt (see Fig. 4), measured in different seasons during the years 1981 and 1982.

FIGURE 5b. Tidally averaged transport of particulate organic carbon (P.O.C.) in the Eastern Scheldt.

also the fraction of particulate organic carbon (P.O.C.) was determined. This constituent is more representative for the fine sediment fraction. The residual transport values for P.O.C. are shown in Fig. 5b. Here the dominance of residual export over residual import is stronger than for the total suspended sediment, especially in the seaward cross section *A*.

In Table 2 the yearly average residual seaward transport of suspended sediment is indicated for the cross sections A and B. This table also shows the residual transport in the Eastern Scheldt computed from the analytical solution Eq. (3) in a way similar to that used for the Wadden Sea. The conclusion is that the experimentally and theoretically determined transports have the same direction and agree within experimental accuracy.

5. Summary and conclusions

Estuaries and bays often behave as traps for fine marine sediment. Inside these semi-enclosed tidal systems fine sediment accumulates in regions which are sheltered from tidal currents and wave penetration (for example, harbours and tidal flats). Restricting attention to regions located further inward than the flood excursion, the accumulation rate depends on the ability of currents to induce a residual landward transport of fine sediment.

In estuaries a residual landward transport mechanism for fine sediment is provided by gravitational circulation. However, besides gravitational circulation the tidal motion alone may also produce an important net displacement of fine sediment. Indeed, during the propagation of the tidal wave in shallow irregularly shaped basins harmonic overtides are generated, causing an asymmetry in the velocity variation between ebb and flood. As a consequence the time interval during which sediment particles can settle at slack water and remain on the bottom until resuspension may be different for HWS and LWS. Therefore the temporarily deposited sediment experiences a net residual displacement during the tidal cycle. The magnitude and direction of this residual sediment flux is mainly determined by the difference in the variation of the current velocity around LWS and HWS.

This transport process only applies to fine sediment which is deposited and resuspended in a time interval in which the current speed is small compared to its maximum value. Most of the coarse sediment on the contrary is in suspension only in a small time interval around maximum current speed. The residual flux of coarse sediment therefore more strongly depends on an ebb-flood asymmetry of the maximum current speed than on the slack water velocity variation.

The residual flux of fine sediment predicted by the analytical expression which is derived in this paper has been compared with field observations in two different tidal systems, one of which behaves as a sediment trap and the other as a sediment source. The results are in agreement, at least as far as the order of magnitude and the direction of the sediment flux is concerned.

It must be noted that in the tidal system which behaves as a sediment trap, the influence of storm surges on the long term accumulation rate of sediment is very important. In those circumstances the major part of the fine sediment accumulated during calm weather circumstances can be eroded and returned to the coastal shelf.

References

ALLEN, G.P., SAUZAY, G., CASTAING, P. & JOUANNEAU, J.M. 1977. Transport and deposition of suspended sediment in the Gironde Estuary, France. In: Estuarine Processes II (M. WILEY ed.), Academic Press, 63-81.

ARIATHURAI, R., MACARTHUR, R.E. & KRONE, R.B. 1977. Mathematical model of estuarial sediment transport. Technical Report D-77-12; Environmental Effects Laboratory, U.S. Army Engineers Waterways Experiment Station.

BOER, M. DE & VISSER, G.C. 1981. Erosie en sedimentatie in de westelijke Waddenzee. Nota WWKZ-

80.H001, Rijkswaterstaat, The Netherlands (text in Dutch).

DRONKERS, J. 1984. Import of fine marine sediment in tidal basins. Proceedings of the International Wadden Sea Symposium, Texel 1983, Neth Inst. for Sea Res. Publ. Series 10 , 83-105.

EINSTEIN, H.A. & KRONE, R.B. 1962. Experiments to determine modes of cohesive transport in salt water. Journal of Geophysical Research, 67, 1451-1461.

ELGERSHUIZEN, J.H.B.W., Personal Communication.

FESTA, J.F. & HANSEN, D.V. 1978. Turbidity maxima in partially mixed estuaries: a two-dimensional numerical model. Estuarine and Coastal Marine Science 7 , 347-359.

GELFENBAUM, C. 1983. Suspended-sediment response to semi-diurnal and fortnightly tidal variations in a mesotidal estuary: Columbia river, U.S.A. Marine Geology, 52, 39-57.

KRONE, R.B. 1972. A field study of flocculation as a factor in estuarial shoaling process. Technical Bulletin 19, U.S. Army Corps of Engineers.

KRONE, R.B. 1974. Engineering interest in the benthic boundary layer. In: The Benthic boundary layer (MCCAVE ed.), 143-155.

McDOWELL, D.M. & O'CONNOR, B.A. 1977. Hydraulic Behaviour of Estuaries. Civil Engineering Hydraulics Series, MacMillan, London.

MIGNIOT, C. 1968. Etudes des propriétés physiques de différents sédiments très fins et de leur compartement sous des actions hydrodynamiques. La Houille Blanche, 7, 591-620.

ODD, N.V.M. & OWEN, N.W. 1972. A two layer model of mud transport in the Thames Estuary. Proceedings of the Institution of Civil Engineers. Suppl. 1972 (IX), paper 7517S, 175-205.

POSTMA, H. 1954. Hydrography of the Dutch Wadden Sea. Archives Néeerl. Zool. 10 , 405-511.

POSTMA, H. 1961. Transport and accumulation of suspended matter in the Dutch Wadden Sea. Netherlands Journal of Sea Research 1 , 148-190.

POSTMA, H. 1967. Sediment transport and sedimentation in the marine environment. In: Estuaries (G.H. LAUF ed.), 158-179.

WINKELMOLEN, A.H. & VEENSTRA, H.J. 1980. The effect of a storm surge on nearshore sediments in the Ameland-Schiermonnikoog area. Geologie Mijnbouw 59 , 97-111.

Appendix

The sum of the integrals $(a)+(b)+(c)$ equals

$$M_F = -\int_{t^-}^{t^+}\int_{X}^{x^+} - \int_{t_{1d}}^{t^-}\int_{x^-}^{x^+} + \int_{t_{1d}}^{t_{2d}^-}\int_{X_e^-}^{x^+}$$

The integrand has been omitted. As it contains the Heavyside function $\theta(U_d - |U(x,t)|)$ the first integral can be split into:

$$\int_{t_-}^{t^+} = \int_{t_-}^{t_{2d}^-} + \int_{t_{1d}}^{t^+}$$

Decomposing

$$\int_{x^-}^{x^+} = \int_{x^-}^{X} + \int_{X}^{x^+}$$

M_F contains five terms, three of which can be summed together:

$$-\int_{t^-}^{t_{2d}^-}\int_{X}^{x^+} - \int_{t_{1d}}^{t^-}\int_{X}^{x^+} + \int_{t_{1d}}^{t_{2d}^-}\int_{X_e^-}^{x^+} = -\int_{t_{1d}}^{t_{2d}^-}\int_{X}^{X_e^-}$$

Finally

$$M_F = -\int_{t_{1d}}^{t_{2d}^-}\int_{X}^{X_e^-} - \int_{t^+}^{t^+}\int_{X}^{x^+} - \int_{t_{1d}}^{t^-}\int_{x^-}^{X}$$

Similarly

$$M_E = \int_{t_{1d}^+ X_e^+}^{t_{2d}^+ X} + \int_{t_{1d} x^-}^{t^- X} + \int_{t_{1d}^- X}^{t^+ x^+}$$

Adding the sediment masses M_F and M_E passing respectively during flood and ebb through the moving plane $X(t)$, the result given by Eq. (2) is found.

Eq. (2) can be simplified to Eq. (3) in that region of the tidal basin (normally the major part) where

$$\Delta t_e^\pm, \Delta t_d^\pm \ll T$$

$$(X_e^- - x_0^-), (x_0^+ - X_e^+) \ll L.$$

Here $\Delta t_d^- = [t_{1d}^-, t_{2d}^-]$ is the time interval around LWS during which sediment is deposited; Δt_d^+ is the corresponding time interval around HWS.

For the derivation of the Eqs (4) and (5) it is further assumed that the sedimentation rate can be written in the form

$$\dot S = A_s \frac{w}{h} C,$$

where w is a constant settling velocity.

Following the simplified Eq. (1), $A_s \, dC/dt = -\dot S$, the concentration then will decrease during the settling time interval from its value C_{max} at $t = t_{1d}$ according to an exponential law:

$$C(t) = C_{max} e^{-\frac{w}{h}(t - t_{1d})}$$

In an equilibrium state, i.e. no net bottom erosion and no net change in the suspended sediment concentration, the concentration of suspended sediment at the position of the moving plane just after erosion ($t \geq t_e$) is approximately the same at ebb and flood. As the concentration in the moving frame is approximately constant from t_e^+ till t_{1d}^- and from t_e^- till t_{1d}^+, it follows that

$$C(X_e^-, t_e^-) \cong C(X_e^+, t_e^+) \cong C_{max}(x_0),$$

where C_{max} represents the suspended sediment concentration at the position x_0 at half tide, i.e. when the tidal velocity is highest.

Now consider the integral

$$I = \int_{t_{1d}}^{t_{2d}^-} dt \int_{X(t)}^{X_e^-} dx \, \theta(U_d - |U(x,t)|) \dot S(x,t)$$

and select $t = 0$ and $x_0^- = 0$ as time and space origins. Following the above assumptions and approximations

$$X(t) \cong t^2 X_e^- / (\Delta t_e^-)^2,$$

$$I \cong \int_{t_{1d}^-}^{t_{2d}^-} dt (X(t) - X_e^-) \dot S(x,t) \cong \frac{w}{h} A_s X_e^- C_{max} \int_{t_{1d}^-}^{t_{2d}^-} dt \left(1 - \frac{t^2}{(\Delta t_e^-)^2}\right) e^{\frac{w}{h}(t_{1d}^- - t)}.$$

Finally

$$I = \omega^- \lambda^- C_{max},$$

with

$$\omega^- = A_s(1-e^{-\frac{w}{h}\Delta t_d^-}),$$

$$\lambda^- = X_e^- \left[1 - \frac{\frac{w}{h}\int_{t_{1d}}^{t_{2d}} dt \frac{t^2}{(\Delta t_e^-)^2} e^{\frac{w}{h}(t_{1d}^- - t)}}{1-e^{-\frac{w}{h}\Delta t_d^-}} \right]$$

If $\Delta t_d^- < \Delta t_e^-$ the last term between brackets can generally be neglected:

$$\lambda^- \cong X_e^- \cong \frac{1}{2}\Delta t_e^- U_e.$$

Hydrodynamic Controls on Sediment Transport in Well-Mixed Bays and Estuaries

D.G. Aubrey

Woods Hole Oceanographic Institution, USA

ABSTRACT

Shallow bays and estuaries with negligible gravitational circulation respond to tidal forcing in a systematic manner determined primarily by channel geometry, bay geometry, and frictional effects. These characteristics impart a tidal distortion to the bay which can be represented by growth or decay of overtides and compound tides which experience down-channel evolution in amplitude, but which remain phase-locked to the parent tides throughout the bay. This systematic tidal response leads to either a flood-dominated bay with shorter, more intense flood tides, or an ebb-dominated bay with shorter, more intense ebb flows. Distortion of the tidal velocities leads to significant nonlinear growth of harmonics of the forcing tides, imparting an asymmetry to tidal velocity and near-bed shear stress time history. For coarse sediment transport, this shear stress asymmetry can lead to a net transport of sediment in the direction of tidal dominance. For flood-dominance, the bay will tend to fill; for ebb-dominance, the bay will be self-flushing. For fine sediment transported as suspended load, ignoring gravitational effects, the net transport is determined partly by the sediment fall velocity and time scale of tidal flow reversal, as well as the height to which sediment is suspended. A heuristic argument indicates that the net suspended transport can be either in the same, or opposite, sense as the tidal distortion. Improvements on estimates of net sediment transport require more complete descriptions of lateral variability in channel flow, biological and chemical effects, and bottom boundary layer modeling.

1. Introduction

Sediment transport in bays and estuaries influences their evolution and stability. In particular, the relative rates of sedimentation and rise of relative sea levels control the net shoaling, or infilling, rate of an estuarine system. These sedimentation and sea-level effects impact on the geological evolution of bays (for simplicity, the term bay includes those estuaries with little freshwater input), the navigability of such systems, as well as their chemical and biological health. While relative sea-level behaviour has been defined for many areas of the U.S. and foreign coasts (see, for instance, the various articles by Aubrey and Emery), and estuarine/bay sedimentation has been discussed by many investigators, the present paper addresses the distortion of tides in shallow, well-mixed bays, and its impact on sediment transport. The focus is not so much on the basic laws governing near-bed and suspended sediment transport, as the response of bay systems to the ocean tide and its potential influence on sedimentation. Numerical and field results pertaining to tidal distortion are presented in greater detail (without the emphasis on sediment transport) in Aubrey and Speer (in press) and Speer and Aubrey (in press). Related work can also be found in UNCLES (1981), describing estuaries with lower friction and less time-variable geometry.

This study addresses shallow bays with negligible freshwater inflow (so gravitational effects are minimized), well-mixed (so stratification is not dynamically important), and connected to the ocean by a narrow inlet, in which higher frequency tides are selectively attenuated. Although these restrictive assumptions limit the generality of the results, bays with these characteristics are common along the U.S. coasts, and other coasts of the world. Because of their relatively shallow depths, these systems tend to be highly non-linear (as indicated by the ratio of tidal range to water depth, a/h), and not studied as thoroughly as the much larger estuaries such as Chesapeake Bay.

Both near-bed and suspended load transport depend strongly on the flow field in an estuary, particularly when freshwater inflow is negligible, limiting riverine sediment sources and gravitational effects. When the riverine sediment influx is removed, sedimentation in estuaries depends largely on reworking of material within the estuary, influx of sediment through the tidal inlet from the ocean, and biological sources of sediment (marsh grasses, shell material, and peats, for instance). Local bottom shear stress will control much of this reworking, moving sediment either as near-bed or suspended load. Any mechanism imparting a sense of asymmetry to the tidal flows (assumed here to be the dominant transport process, although wind waves and storm surges can be important) will bias the sense of sediment transport such that the estuary will either fill (positive sediment flux) or be self-flushing (negative sediment flux).

This asymmetry, or distortion, takes the form of a difference in duration of ebb and flood flow, and a corresponding difference in magnitude of ebb and flood velocity. Since both water and sediment flux from rivers are purposely omitted from this discussion, these tidal distortions play the dominant role in sedimentation in these shallow, highly non-linear bays. Although channel bends and other flow complexities introduce considerable lateral variability in tidal flows, the primary interests in this study are how an entire bay/estuarine system responds to tidal forcing, what hydrodynamic effects control the characteristic tidal response of such systems, and how these affect near-bed and suspended load transport. The emphasis is therefore on the system response of these estuaries, rather than their more local flow and sediment behavior.

2. Hydrodynamics

The one-dimensional (depth-integrated) equations of motion for shallow bays with large ratio of tidal range/mean water depth (a/h), negligible freshwater inflow, and no Coriolis effects can be expressed as:

Momentum equation:

$$\frac{\partial U}{\partial t} + \frac{\partial}{\partial x}\frac{U^2}{A} = -gA\frac{\partial \zeta}{\partial x} - \frac{\tau_b}{\rho}P \tag{1}$$

Continuity equation:

$$\frac{\partial \zeta}{\partial t} + \frac{1}{b}\frac{\partial}{\partial x}U = 0 \tag{2}$$

where $\zeta(x,t)$ is the sea surface elevation, t is time, b is the channel width, $U(x,t)$ is the cross-sectional flux, τ_b is the average shear stress on solid boundaries, P is the wetted channel perimeter, A is the channel cross-sectional area, and ρ is the water density. The problem is closed mathematically by defining the average shear stress τ_b:

$$\tau_b = \rho f |U|U/(A \cdot A) \tag{3}$$

where f is a dimensionless friction factor. Principle non-linear effects in Eqs. (1), (2), and (3) enter through quadratic friction, advection of momentum, and tidal interactions with estuarine geometry in the continuity equation. These non-linearities are responsible for generation of overtides and compound tides, which cause distortion of the continental shelf tide as it enters these shallow bays and estuaries.

As an example of this non-linear tidal distortion, we can examine the M_2 tide and its compound tides. The M_2 sea surface variation can be represented as:

$$A_{M_2} = a_1\cos(w_1 t - \theta_1)$$

The principal overtides of M_2 are the quarter-diurnal tide (M_4) and the sixth-diurnal tide (M_6). Since tidal asymmetry is strongly dependent on the quarter-diurnal tide, we examine the effect of M_4:

$$A_{M_4} = a_2\cos(w_2 t - \theta_2)$$

The relative phase,

$$\phi_\zeta = 2\theta_1 - \theta_2 \tag{4}$$

determines the sense of tidal distortion, while the ratio A_{M_4}/A_{M_2} is an indicator of the degree of non-linearity and tidal distortion. As an example, if the magnitude of the relative sea-surface phase, ϕ_ζ, is between 0 and 180°, then falling tide exceeds rising tide in duration. Tidal distortion results not only in a duration asymmetry for sea surface, but also in a magnitude asymmetry for velocity or transport. When the phases of the M_2 and M_4 velocities are equal such that the velocity phase difference (ϕ_v-equation (4)) is zero, then one obtains shorter, enhanced flood currents relative to ebb currents. The cases of symmetrical and ebb-enhanced tidal currents can be illustrated by similar arguments.

This perspective of overtides and compound tides simplifies the representation of shallow water tides. In modelling, sea surface and velocity can be examined in terms of the magnitude and phases of the primary and forced harmonic tidal constituents. Use of the relative $M_2 - M_4$ phase (ϕ_ζ) provides an independent diagnostic indicator of the response of a body of water to tidal forcing. Similarly, relative phases of other overtides and compound tides (when there is more than a single primary tidal constituent present, as there will be in nature) provide insight into bay hydrodynamics. Used locally, this simple decomposition of the distorted tide can help separate the various terms in a local momentum balance (which could be applied to estimate bottom friction coefficients, for instance; see SWIFT & BROWN, 1983; AUBREY & SPEER, in prep.).

As an example of the above approach, both theoretical work and field observations have been examined (SPEER, 1984; AUBREY and SPEER, in press; Speer and Aubrey, in press.). SPEER (1984) and SPEER and AUBREY (in press) examined the one-dimensional equations of motion (Eqs. 1-3 above) for shallow ($a/h \sim 0.5$), short (length of bay a small fraction of tidal wavelength) bays with narrow connections to the adjacent continental shelf, and the response of the tide to various bay geometries. Briefly, they found that channels without tidal flats develop a time asymmetry characterized by a longer falling than rising tide. This behavior is enhanced by strong friction and large channel cross-sectional area variability over a tidal cycle. Resulting tidal currents have a shorter, intense flood and a longer, weak ebb (flood dominant). Tidal flats in channels can produce a longer rising tide and stronger ebb currents (ebb-dominated) if the area of tidal flats is large enough to overcome the effects of time-variable channel geometry. Weaker friction with flats can also produce this asymmetry.

This work demonstrates that shallow bays respond systematically to tidal forcing, leading to stable, uniform senses of tidal asymmetry (either flood- or ebb-dominated, due to phase-locking in forced tidal constituents), with down-channel evolution of magnitude of asymmetry. Overtides generated offshore and propagating into the bays are quickly swamped by the bay response to the incident M_2 tide. In areas where the offshore M_4 is particularly large, and where friction is so strong that the tide damps too quickly, this bay response and phase-locking may not be as evident. Observations by ROBINSON et al. (1983) in an English tidal lagoon may fall into this category, since they failed to document the phase-locking predicted by the above models.

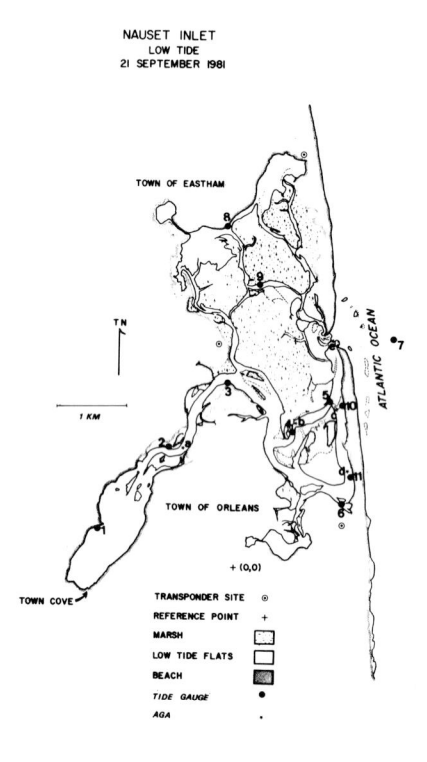

FIGURE 1. Location map for Nauset Inlet experiment of August through October 1982. Numbers refer to locations referenced in the text and other figures: 1-Goose Hummock (GH), 2-Mead's Pier (MP-location for figure 3), 3-Snow Point (SP), 4-Middle Channel West (MCW), 5-Middle Channel East (MCE), 6-Nauset Heights (NH), 7-Ocean site, 8-Nauset Bay (NB), and 9-North Channel (NC). Base map compiled from topography of 1981, referenced to high tide levels, showing primary tidal channels more clearly. AGA refers to locations of four-day long deployments of an electromagnetic current meter array.

The one-dimensional models provide a useful look at the system response, but they are clearly deficient in providing the details of the flow field. In particular, lateral gradients (which commonly occur in the field) are ignored, such as those generated by complex channel bends and resultant flow curvature. All stratification effects are lost because of the assumption of well-mixed bay. To this extent the one-dimensional models are of limited applicability, but they clearly illustrate in a diagnostic sense the response of the tide to varying bay geometries.

Field observations support the concept of a system response to incident tidal forcing (Aubrey and Speer, in press). In a detailed series of observations at Nauset Inlet, Massachusetts (Fig. 1), encompassing sea-surface meters recording for periods of one-to-three months and shorter duration (one-week) velocity measurements, the response of a shallow bay to incident tide forcing was clarified. Nauset Inlet is characterized by shallow water depths (average 2m) and a dominantly semi-diurnal (M_2) co-oscillating tide with an offshore range of 2m. Although the offshore tide decays rapidly within the bay (Fig. 2), the entire bay has a characteristic system response in which the rapidly growing forced tides are phase-locked to their parent constituents, consistently leading them by 60-70° in phase.

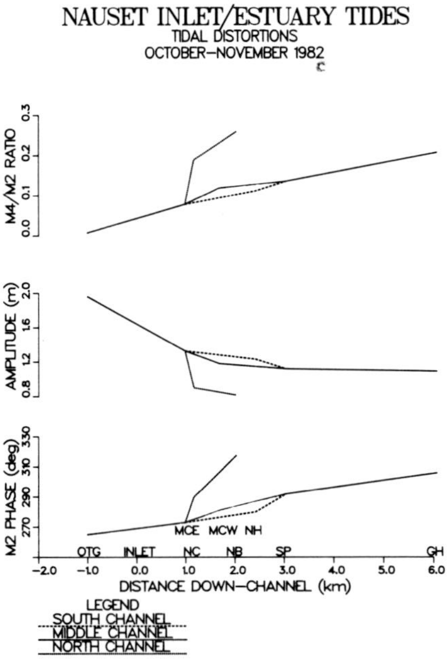

FIGURE 2. Tidal decay and evolution within Nauset Inlet, as a function of distance into the estuary (abscissa), for each of three primary drainage channels. Top panel shows M_4/M_2 ratio increasing with distance into the estuary, reflecting both a frictional dissipation of M_2 and real growth of M_4. The middle panel illustrates tidal range decreasing with distance down-channel. Tide-gauge stations are indicated on the abscissa, with the stations shown in plan form in Figure 1.

This phase-locking is characteristic of the entire estuary (the phase relationship for the small M_4 tide present offshore is different from that in the estuary), leading to a shorter, more intense flood and a longer, weaker ebb (flood dominant - Fig. 3). The ratio M_4/M_2 grows with distance into the estuary, as the first overtide amplifies through nonlinear interactions. The M_2 tide decreases with distance into the bay, losing energy to its overtides and to frictional decay. The M_6 overtide at Nauset is negligible, because dissipative effects overcome growth through non-linear interactions.

Observations at Wachapreague Inlet, Virginia (Fig. 4) document an ebb-dominated bay system (BOON and BYRNE, 1981). Although the measurement array was more sparse than that at Nauset Inlet, the sea surface data at Wachapreague show an M_4/M_2 ratio of 0.04, and a sea surface relative phase of 200°-220°, producing a longer rising tide and stronger ebb currents. Since the channels at Wachapreague are deeper (about 5m), and the tide is smaller, the non-linear parameter a/h is only 0.11, compared to Nauset Bay

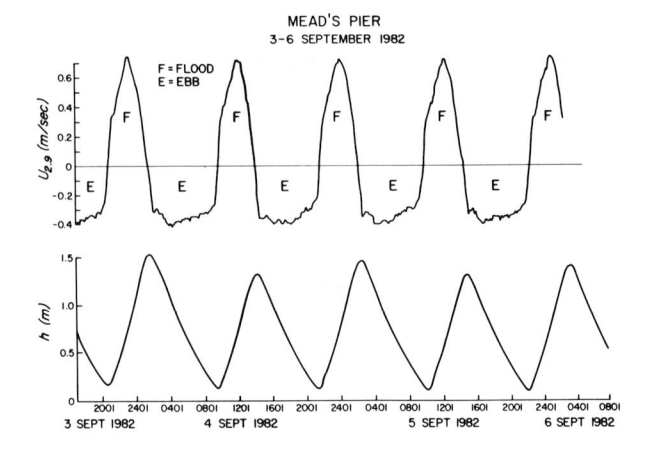

FIGURE 3. Three-day records of velocity (top panel - taken 2.9m above bottom in 3.5 meter water depth) and sea surface, showing the characteristic distortion of the tide at Mead's Pier (location 2, figure 1). Because flood tide is shorter, it is more intense, producing higher flood velocities. Ebb tides are longer, producing weaker tidal currents.

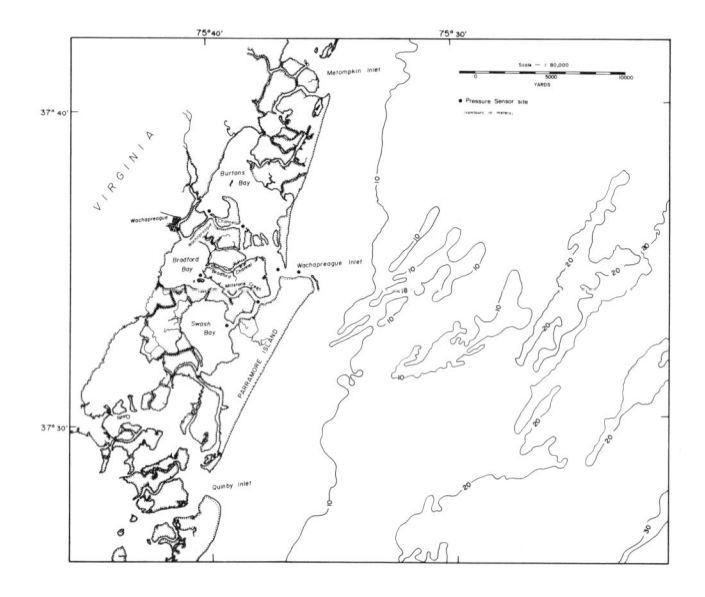

FIGURE 4. Wachapreague Inlet/estuary system, VA. The estuary is serviced by a deep (18m) stable inlet. Channel depths are 4-5 m and terminate in shallow bays consisting of extensive tidal flats.

where a/h reaches approximately 0.5-0.8. The deeper channels of Wachapreague empty into broad bays with extensive tidal flats, which control the sense of tidal asymmetry. In numerical models of Wachapreague Inlet (Speer and Aubrey, in press), even large values of friction are unable to reverse the sense of asymmetry in this system.

One-dimensional modeling of tidal distortion in channels and bays along with field observations strongly suggest that these bay systems respond in a uniform sense to tidal forcing. A harmonic representation of the hydrodynamic effects simplifies the description of bay response. However, simple kinematical quantities such as sea surface and velocity will not suffice to estimate sediment transport. Dynamical quantities of interest include the near-bottom stress (τ_{sf} - skin friction) or shear velocity (u_*):

$$u_* = \sqrt{\tau_{sf}/\rho} \tag{5}$$

These quantities depend on the friction coefficient, which is difficult to estimate in a tidal channel with large bedforms, since sediment transport is driven solely by the near-bed shear stress, $\tau_{s,f}$, and not the total bottom stress, τ_b. Total bottom stress can be estimated by careful measurements in a tidal channel through a local momentum balance (SWIFT and BROWN, 1983; AUBREY, and SPEER, in prep.), or from dissipation measurements. Most commonly, however, this stress is calculated from empirical equations and observations from open channel flows.

These calculations are complicated by the nature of the boundary layer in shallow channels. These boundary layers are generally segmented, with individual vertical segments responding to different roughness elements of the flow (see, for example, SMITH, 1977; SMITH and McLEAN, 1977a). Suspended sediment and near-bed transport also alter the boundary layer profile (e.g., SMITH, 1977a; SMITH and McLEAN, 1977b). Many channel boundary layers are depth-limited, where the water depth serves as the similarity length scale for the flow. NOWELL and CHURCH (1979) showed such boundary layers to have three layers in the vertical: an outer layer ($z/h > 0.35$), a wake layer ($0.35 > z/h > 0.10$), and an inner layer ($z/h < 0.10$). This latter study did not use bedform-scale roughness elements, but restricted itself to gravel-scale distributed roughness. The complexity of such channel boundary layers makes difficult interpretation of field observations of vertical profiles of velocity.

The vertical profile of horizontal velocity depends on all roughness scales (including a bedform density parameter) and discharge. Given these roughness and discharge estimates, a theoretical model must be invoked to define the velocity profile. For instance, SMITH and McLEAN (1977a) present models for stacked boundary layers responding to different scales of roughness in uniform, steady flow. These complexities make it difficult to extrapolate from a simple sea-surface or velocity asymmetry to estimate asymmetries in bottom shear stresses, or other dynamical terms. The channel roughness must be carefully determined, and a theory used to partition the total drag, τ_b, into a skin friction component, τ_{sf}. The skin friction is responsible for initiation of sediment motion (both near-bed and suspended load), but the vertical distribution of suspended sediment depends on the total shear stress at each level since this distribution depends on local mixing. With these caveats in mind, the ensuing discussion assumes that the friction coefficient does not vary strongly with flow rate, and is the same on flood and ebb. This approximation is for purposes of illustration only.

3. Sediment transport

Channel material can move in either suspended load or near-bed load (bed and bed-material load). In a shallow channel flow, with no freshwater inflow and complete vertical mixing, transport of sediment is dependent on the asymmetry and magnitude of the tidal flow, the time history of shear velocity (both near-bed and internal), and the fall velocity of the sediment (w_s). Fine sediments (with smaller fall velocities) will be transported in suspension longer than the more rapidly settling coarser sediment, so the asymmetry in tidal velocity will affect finer sediments more than coarser sediments.

Initiation of motion, including resuspension, will depend on the near-bed shear velocity.

The ratio of settling velocity to entrainment velocity, p_s, determines the range over which suspended transport is important (HUNT, 1969; SMITH, 1977):

$$p_s = w_s / \kappa u_*$$

where κ is von Karman's constant (0.4). For values of $p_s > 2$ (approximately), negligible suspended sediment transport occurs. For near-bed transport, the fall velocity is many times larger than the shear velocity (scaled on the local near-bed shear stress, τ_{sf}), while for suspended load transport the shear velocity is much larger than the fall velocity. This scaling allows one to differentiate regimes of suspended versus pure near-bed transport.

3.1 Near-bed transport

Near-bed transport occurs for coarse sediments when the fall velocity is large compared to the near-bed shear velocity. Modeling of near-bed transport assumes that the transport rate depends only on a single flow variable - shear velocity. Calculation of near-bed transport can be performed with many formulae, which separate into two distinct groups. On one hand, near-bed transport is calculated from energetics arguments (BAGNOLD, 1956, 1966, 1976), where grain-to-grain interactions are important in maintaining sediment transport. These equations may apply best under conditions of high volume transport rates. The second approach is a mechanistic one, with the interactions between individual sediment grains and the fluid emphasized, and particle-to-particle interactions neglected. This latter approach, which may be best for low rates of sediment transport, includes the work of MEYER-PETER/MÜLLER (1948), EINSTEIN (1950), YALIN (1963) and OWEN (1964), among others. Rather than compare each of these methods, we select one for illustration; although the method chosen (MEYER-PETER/MÜLLER, 1948) has been shown to be useful in nearshore applications (GOUD and AUBREY, 1985), no argument is made here for its indiscriminate, universal application.

For uniform, steady flow, the METER-PEYER/MÜLLER equations relate the volume transport rate of sediment (volume flux of sediment per unit of flow width) to the excess shear stress:

$$q = 8d \sqrt{(\frac{\rho_s}{\rho} - 1)gd} (\psi - \psi_c)^{\frac{1}{2}}$$

and

$$\psi = \tau_{sf} / (\frac{\rho_s}{\rho} - 1)\rho g d$$

where ψ is the Shield's parameter, ψ_c is the critical Shield's parameter at which sediment movement is initiated, ρ_s is the sediment density, and d is the grain size. Use of this equation requires knowledge of the critical shear stress for initiating sediment motion. This critical shear stress is normally determined from Shield's diagram, plotting a non-dimensional shear stress required for initiation of motion (ψ_c) versus a non-dimensional number representing fluid and sediment properties. These diagrams are empirical, and apply strictly only to abiotic, geometrically smooth beds with grains of a uniform size. Examples of such diagrams for uniform steady flow can be found in MADSEN and GRANT (1976) and SMITH (1977). For poorly sorted sediments with a mixture of grain sizes, the shape of the Shields relation is poorly understood.

The Meyer-Peter/Müller equation can be applied to distorted tidal flows in shallow bays. Using the simplest case of unimodal tidal forcing (M_2, for instance), one can compute the asymmetry in sediment transport from the parent tide and its first overtide (M_4). As discussed before, M_6 is relatively small in the field compared to the M_4 and does not impart an equal sense of distortion. Referring to the defining equations for M_2 and M_4 velocity:

$$V_{M_2} = v_1 \cos(w_1 t - \theta_1)$$

$$V_{M_4} = v_2 \cos(w_2 t - \theta_2)$$

The duration asymmetry for velocity is calculated from:

$$\frac{\cos w_1 t}{\cos(w_2 t - \theta_2)} = -\frac{v_2}{v_1} \qquad M$$

where θ_1 is arbitrarily set to zero. The magnitude asymmetry for velocity is V_{M_4}/V_{M_2}. The ratio of flood to ebb sediment transport is:

$$\frac{\int_f (\psi - \psi_c)^{\frac{3}{2}} dt}{\int_e (\psi - \psi_c)^{\frac{3}{2}} dt}$$

where the subscript e indicates integration over ebb, and the subscript f refers to flood. The flood/ebb duration asymmetry, plotted as a function of magnitude asymmetry and relative $M_2 - M_4$ phase (ϕ_v - Fig. 5), has the following features. For $|\phi_v|$ between 0° and 90°, flood duration is always shorter than ebb duration. For ϕ_v between 90° and 270°, flood duration exceeds ebb duration. For non-linear ratios (V_{M_4}/V_{M_2}) up to 0.25, which have been observed in the field, the maximum duration asymmetry is 7.2 hours versus 5.2 hours.

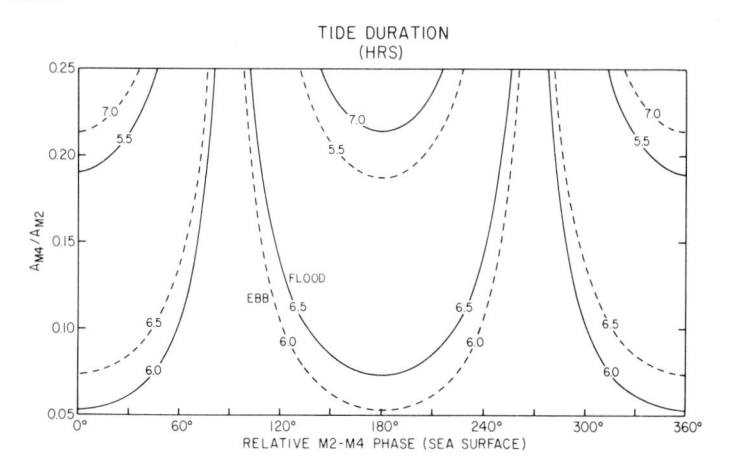

FIGURE 5. Duration asymmetries for non-linear tides in shallow channels, as a function of both relative phase between M_2 and M_4 velocities (abscissa), and nonlinear parameter relating the amplitude of the M_4 velocity (ordinate). For these limits of parameters, flood and ebb tide attain a maximum duration of 7 hours, and a minimum duration of approximately 5.5 hours.

Asymmetry in near-bed transport can be illustrated by the ratio $(\psi_f/\psi_e)^{\frac{3}{2}}$, integrated over a tidal cycle. This ratio is the relative flood-to-ebb transport over a tidal cycle assuming a threshold of motion equal to zero. This assumption leads to an underestimate of the transport asymmetry, but is a useful indicator of sediment transport system response (particularly when $\psi \gg \psi_c$ over most of the tidal cycle). The

RATIO OF FLOOD TO EBB NEAR BED TRANSPORT OVER A TIDAL CYCLE
($\psi_c = 0$)

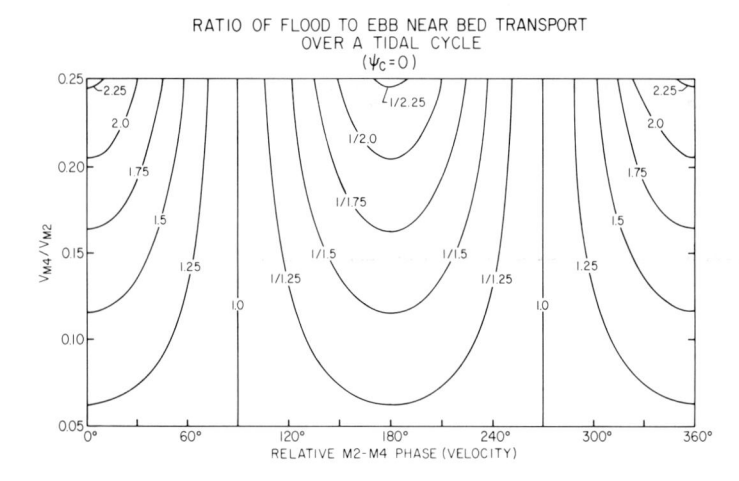

FIGURE 6. For the same parameters as figure 5, this depicts the flood-to-ebb tide near-bed sediment transport ratio over a tidal cycle, for the idealized case of zero threshold of motion. Although underestimating the maximum transport asymmetries in these systems, this diagram shows the dominant tide supports up to 2.25 times as much near-bed transport as does the non-dominant tide, depending on the nonlinear parameter and the sense of phase-locking.

formulation also assumes constant near-bed shear stress coefficients and uniform grain size. Results (Fig. 6) indicate a net flood transport for relative phases within 90° of zero, and net ebb transport elsewhere. Flood and ebb transport are equal only when relative phases in velocity are at 90° or 270°. For the magnitude of nonlinearity observed in the field (up to 0.25), maximum asymmetry ratios in velocity reach 2.25:1. This transport asymmetry plays a significant role in the evolution of an estuarine system. The sense of asymmetry reinforces itself, such that an ebb-dominated system will tend towards increasing ebb dominance as tidal channels deepen (until the ratio a/h becomes small), and a flood-dominated system will become increasingly so as tidal channels shoal (see Speer and Aubrey, in press.). This qualitative tendency does not address the formation of tidal flats, which exert a major control on flood/ebb asymmetry (see Speer and Aubrey, in press).

For any bottom roughness and known values of V_{M_4} and V_{M_2}, a graph similar to Fig. 6 can be generated. Fig. 6 cannot be made more general (without showing a family of curves) because of the threshold term, ψ_c. The example (Fig. 6) is used only for illustrative purpose.

3.2 Suspended sediment transport

Suspended sediment transport rates in well-mixed bays depend on the mixture of grain sizes present in the field, the shear velocity near the bed, the fall velocities of each sediment component, the velocity field above the bed, and temporal and spatial dependence of the momentum diffusion coefficient (e.g., SMITH, 1977), as well as biological and chemical factors (such as flocculation/aggregation). Suspended sediment can alter the velocity profile and the near-bed shear velocity by stratification effects; this self-stratification must be properly modeled by examining stable stratification profiles, rather than resorting to a non-physical change in von Karman's constant.

A coherent theory for noncohesive, abiotic suspended sediment transport in steady, uniform flow is presented by SMITH (1977), building on work by DOBBINS (1944), HUNT (1954, 1969) and SMITH and HOPKINS (1972), neglecting non-hydrodynamic effects. The equations for suspended sediment transport are complex, and require a known concentration at a specified reference level. Given this requirement it is difficult to calculate a general relationship for the ratio of suspended sediment transport on flood versus ebb. Instead, a more heuristic discussion of suspended transport under distorted tides can be presented, based on relations between the fall velocity and duration asymmetries of tidal flows in shallow bays and estuaries.

When estuarine waters have different suspended loads than coastal waters, then this concentration gradient will control whether the estuary acts as a source or sink of fine sediments to the coastal waters. When suspended sediment concentrations in estuaries are much higher than in adjacent coastal waters, the estuary will serve as a sink of suspended material to coastal waters, regardless of sense of tidal asymmetry within the bay or estuary.

Within the bay or estuary, sedimentation patterns will differ with location. For long, narrow estuaries, the following analysis holds. Neglecting river inflow and bed armoring, most sediment will be placed in suspension when the near-bed shear stress is greatest. At this time the scale height of the suspended material, δ_s, is the largest. If the discharge curve resembles the velocity curve in Fig. 3, then the bulk of sediment will be resuspended near peak velocity. The scale height for suspended sediment, δ_s, will also be greatest at this time. Since resuspension is related to a power of velocity greater than unity, this effect is accentuated by tidal velocity asymmetry.

The time required for the particles at any level δ_s, above the bed to fall can be estimated as:

$$t_s = \delta_s / w_s$$

While the sediment is falling, it is being advected in the free-stream. At some time (following the peak velocity), t_r, the tidal flow will reverse, and sediment will be transported in the opposite direction. If $t_s < t_r$, then the net suspended transport will be in the same direction as the maximum tidal shear stress (and velocity, in our simplified model). If the material remains in suspension past t_r, the direction of net suspended load transport will depend on the relative advection distances of the initial and reverse flows. If material exits the bay on ebb tide, it will not be free to re-enter the bay and the net flux may be seaward, regardless of the above scaling argument. This argument suggests that net suspended sediment transport may be either in the same sense as velocity domination, or in the opposite sense, for a suitably long channel (channel length much greater than tidal excursion distance).

This heuristic argument does not apply to areas with considerable two-dimensional flow structure, such as channel bends or locations where flow expands or contracts. It holds solely for a long uniform channel, with no appreciable gradients in near-bottom shear stress or water column shear velocities. It also requires that the rate of increase in tidal distortion not be too large. If the tide becomes non-linear at too rapid a rate downchannel, then the tendency for transport in the direction of maximum shear stress is emphasized.

POSTMA (1967) presented a qualitative model for suspended sediment transport, emphasizing up-channel retardation of the tidal velocity and resulting sedimentation. Two mechanisms were identified in the analysis: a settling lag due to the finite settling velocity, and a scour lag resulting from a difference between erosion velocity and transport velocity. These two physical effects lead to fine particle sedimentation within the landward end of the estuary, as observed in nature. Referring to a tidal flat region, POSTMA (1967) showed a similar landward shift in fine particles due to time-velocity asymmetry.

Both the POSTMA (1967) arguments and the newer argument presented in this paper are heuristic, addressing only limited aspects of the physics of suspended sediment transport. Along-channel variations in velocity result not only from reduced tidal range, but also from flow contractions and expansions. At Nauset Inlet, for example, the dominant southern channel empties into a wider, deeper basin of glacial origin where most suspended material settles out.

Resolution of the complete suspended sediment transport problem awaits detailed modeling of tidal flows, including their lateral variation, neglected in our heuristic argument. Sub-tidal motions must be correctly modelled, as well as accelerating/decelerating flows. Although these hydrodynamic considerations are essential, biological and chemical contributions to suspended sediment behaviour must also be resolved.

4. Discussion and conclusions

Spectral decomposition of tides in shallow embayments provides a useful framework for evaluation of sedimentation trends, and long-term behaviour of these systems. For bays forced by a semi-diurnal offshore tide, the primary contributors to duration and magnitude asymmetries in sea surface and velocity are the lowest harmonics of M_2, notably M_4. Higher harmonics are not as strongly coupled to the parent tide, and are dissipated more rapidly than the lower harmonics. Previous work (SPEER, 1984; AUBREY and SPEER, in press; and SPEER and AUBREY, in press.) showed that for highly non-linear shallow bays with negligible gravitational circulation, the sense of tidal distortion (flood-/or ebb-domination) is controlled largely by channel and bay geometry. In particular, time-variability in channel geometry and extent of tidal flats coupled with frictional effects control the duration and magnitude of flood and ebb tide. A system with a shorter flood tide will have a stronger flood tidal flow, leading to flood-dominance. A system with shorter ebb tide will have a stronger ebb tidal flow, leading to ebb-dominance. Field work shows overtides to be phase-locked to parent tides, leading to a uniform sense of tidal distortion throughout any particular bay.

For coarse near-bed sediment transport, tidal distortion and resulting velocity asymmetry directly affect the asymmetry in sediment transport. A flood-dominated system will have a near-bed transport asymmetry leading to net up-bay transport, while an ebb-dominated system will have an asymmetry leading to net ocean-ward coarse sediment transport. For typical nonlinear tidal responses characteristic of some of these shallow bays, the magnitude of this transport asymmetry reaches 2.25, indicating that the dominant tidal flow will transport 2.25 times as much sediment as the opposite tide. This asymmetry leads to filling of flood-dominated bays, such as Naulet Inlet, MA, where coarse sediment has migrated farther into the bay. For an ebb-dominant system, the bay/inlet will be self-flushing, leading to bay stability.

Since this simplified model of coarse sediment transport does not address some of the complexities of sediment transport, the results are necessarily qualitative. For instance, movable bed effects (where ripple height and length depend on the shear velocity) and bed armoring may alter sediment transport rates in some instances. Form drag over large bedforms may also affect transport rates over a tidal cycle. Nevertheless, this model illustrates the trends of coarse sediment transport in shallow bays and estuaries.

Fine sediment will behave more complexly than coarse sediment, since the sediment fall velocity increases in importance. A heuristic argument suggests that the ratio of sediment fall time to time scale of tidal reversal partly determines the direction of net suspended load transport (for bays with negligible gravitational circulation). Although the maximum near-bed shear stress will suspend sediments to the greatest height, δ_s, above the bed and transport it in the direction of tide dominance, if the flow reverses before the sediment reaches the bed the net suspended transport may be in the direction opposite that of the maximum near-bed shear stress. POSTMA's (1967) arguments relate to channels with uniformly decreasing velocities towards the landward end of the estuary. Complete resolution of the suspended-load problem requires more complete models of tidal velocity distribution, lateral gradients in entrainment rates, and sub-tidal circulation (including Lagrangian flow patterns). The heuristic model proposed in this study emphasizes the importance of the sediment fall velocity in determining the net suspended load transport, differing from the near-bed transport case. The distorted tide still dominates, imparting an asymmetry to near-bed shear stress and hence entrainment rates.

Evolution of bays and estuaries depends not only on net sedimentation, but also on rate-of-rise of relative sea levels. Since tidal distortion is a strong function of relative water depth (a/h), the balance between rising sea levels and sedimentation is critical. Along the U.S. east cost, a strong geographic trend dominates relative sea-level rise, with maximum relative sea-level rise near the Cape Hatteras-Virginia coast, reducing to either side towards Florida and Massachusetts (AUBREY and EMERY, 1983). Possible increases in rate of relative sea-level rise will accelerate the drowning of shallow bays and estuaries, perhaps increasing their stability. The effects of accelerated sea-level rise can be investigated in areas such as the south-east coast of the island of Honshu, Japan, where submergence rates of over 10 mm/yr exist along a deeply embayed coast (Aubrey and Emery, in press). Coupling improved models of bay/estuarine sediment transport with field studies of areas of more rapid submergence may help predict the future of many shallow estuaries and bays.

Acknowledgement

This work was supported by the NOAA National Office of Sea Grant under grant numbers NA79AA-D-0102 and NA80-AA-D-00077, and by the Woods Hole Oceanographic Institution's Coastal Research Center. W.D. Grant and J.D. Milliman suggested improvements to an early draft of this paper.

References

AUBREY, D.G. & SPEER, D.E., in press. A study of non-linear tidal propagation in shallow inlet/estuarine systems. Part I: Observations. Estuarine, Coastal and Shelf Science.

AUBREY, D.G. & K.O. EMERY, 1983. Eigenanalysis of recent United States sea levels. Continental Shelf Research, v. 2, pp. 21-33.

AUBREY, D.G. & EMERY, K.O., in press. Relative sea levels of Japan from tide-gauge records. GSA Bulletin.

AUBREY, D.G. & SPEER, P.E., in prep. Momentum balances and tidal bottom stress in a shallow estuary.

BAGNOLD, R.A., 1956. The flow of cohesionless grains in fluids. Phil. Trans. Royal Society of London, Series A, v. 249, pp. 235-297.

BAGNOLD, R.A., 1966. An approach to the sediment transport problem from general physics. U.S. Geol. Survey Prof. Paper 422I, 37 pp.

BAGNOLD, R.A., 1973. The nature of saltation and bed-load transport in water. Proc. Royal Society of London, Series A, v. 332, pp. 473-504.

BOON, J.D., II & BYRNE, R.J., 1981. On basin hypsometry and the morphodynamic response of coastal inlet systems. Marine Geology, v. 40, pp. 27-48.

DOBBINS, W.E., 1944. Effect of turbulence on sedimentation. Trans. Amer. Soc. Civil Engrs., v. 109, pp. 629-678.

EINSTEIN, H.A., 1950. The bed-load function for sediment transportation in open channel flows. U.S. Dept. of Agr. Soil Cons. Service, Tech. Bulletin 1026, 71 pp.

GOUD, M.R. & AUBREY, D.G., 1985. Theoretical and observed estimates of nearshore bedload transport rates. Marine Geology,v. ,pp.

HUNT, J.N., 1954. The turbulent transport of suspended sediment in open channels. Proc. Royal Society of London, Series A, v. 224, pp. 322-335.

HUNT, J.N., 1969. On the turbulent transport of a heterogeneous sediment. Quart. Journal of Mech. and Applied Math., v. 22, pp. 235-246.

MADSEN, O.S. & GRANT, W.D., 1976. Sediment transport in the coastal environment. Ralph M. Parsons Laboratory Report No. 209, M.I.T., 105 pp.

MEYER-PETER, E. & MÜLLER R., 1948. Formulas for bedload transport. Proc. Second Meeting, IAHR, Stockholm, pp. 39-64.

NOWELL, A.R.M. & CURCH, M., 1979. Turbulent flow in a depth-limited boundary layer. Jour. Geophysical Research, v. 84, pp. 4816-4824.

OWEN, P.R., 1964. Saltation of uniform grains in air. J. Fluid Mechanics, v. 20, pp. 225-242.

POSTMA, H., 1967. Sediment transport and sedimentation in the estuarine environment. In: G.G. Lauff (editor), Estuarines, Amer. Ass. Advancement of Science, Washington, D.C., pp. 158-179.

ROBINSON, I.S., WARREN, L. & LONGBOTTOM, J.F., 1983. Sea-level fluctuations in the Fleet, an English tidal lagoon. Estuarine, Coastal and Shelf Science, v. 16, pp. 651-668.

SMITH, J.D. & MCLEAN, S.R., 1979a. Spatially averaged flow over a wavy surface. Jour. of Geophysical Research, v. 82, pp. 1735-1746.

SMITH, J.D. & MCLEAN, S.R., 1979b. Boundary layer adjustments to bottom topography and suspended sediment. In J.C.J. Nihoul (ed.), Bottom Turbulence, Elsevier, pp. 123-151.

SMITH, J.D. & HOPKINS, T.S., 1972. Sediment transport on the continental shelf off of Washington and Oregon in light of recent current measurements. In D.J.P. Swift, D.B. Duane and O.H. Pilkey (eds.), Shelf Sediment Transport, Dowden, Hutchinson and Ross, Stroudsburg, PA, pp. 143-180.

SMITH, J.D., 1977. Modeling of sediment transport on continental shelves. In Goldberg, E.D. (ed.), The Sea: Ideas and Observations on Progress in the Study of the Seas, v. 6, John Wiley and Sons, Inc., pp. 539-577.

SPEER, P.E. & AUBREY, D.G., in press. A study of non-linear tidal propagation in shallow inlet/estuarine systems. Part II: Theory. Estuarine, Coastal and Shelf Science.

SPEER, P.E., 1984. Tidal distortion in shallow estuarines. Ph.D. thesis, WHOI/MIT Joint Program in Oceanography, Woods Hole, MA, 210 pp.

SWIFT, M.R. & BROWN, W.S., 1983. Distribution of tidal bottom stress in a New Hampshire estuary. Report no. UNH-MP-T/DR-SG-83-2, Univ. of New Hampshire, 40 pp.

UNCLES, R.J., 1981. A note on tidal asymmetry in the Severn Estuary. Estuarine, Coastal and Shelf Science, v. 13, pp. 419-432.

YALIN, M.S., 1973. An expression for bed load transportation. J. Hydraul. Div., Proc. Amer. Soc. Civil Engrs., v. 89 (HY3), pp. 221-250.

Sediment-Driven Density Fronts in Closed End Canals

Chung-Po Lin

Ashish J. Mehta

University of Florida, USA

ABSTRACT

Laboratory tests were conducted to investigate the motion of density fronts, or turbidity currents, caused by suspended, fine sediment in a 9 m long slip representing a closed-end canal. The open end of the slip was connected to a flume carrying sediment-laden flow. Motion of the front was recorded following gate opening at the entrance to the slip which initially contained fresh water. Out of the thirteen tests conducted, the front was arrested in three tests before reaching the closed-end. In all cases, a shear-induced gyre occurred near the entrance, and a stratified front occurred beyond. The front velocity is shown to be dependent on the densimetric velocity, densimetric Reynolds number, Re_Δ, and particle Reynolds number, Re_p. The arrested front length, L_0, normalized by the flow depth, appears to depend on Re_Δ and Re_p. Suspension concentration in the front decreased from its value at the entrance to nearly zero at the toe of the front. Likewise, the sediment became finer with distance from the entrance. This sorting trend was represented in the bottom deposit.

1. Introduction

Sedimentation in closed-end residential canals, tidal docks and pier slips is a common phenomenon. Such a canal is connected to the main channel by an ungated or sometimes gated entrance or lock (Fig. 1a). Given sediment-laden flow in outside waters in the channel, there are five principal causative factors which result in circulation and sediment intrusion in the canal (MEHTA et al., 1984). These are: water level variations, such as due to astronomical tides; flow shearing and water exchange at the entrance between the flow outside and waters in the canal; salinity- and temperature-induced density gradients; wind-induced shear at the water surface; and finally, sediment-induced density gradient between the channel and canal waters. The last is perhaps the least investigated mechanism, and is the subject matter of present interest. This mechanism becomes particularly important relative to the others, for example, in upstream estuarial locations where the tidal range is small, the flow is predominantly uni-directional and wind and salinity effects are not significant.

Although sedimentation in the prototype is usually complicated by the simultaneous occurrence of the five aforementioned causative factors, measurements obtained in closed-end residential canals and in docks with locked entrances tend to suggest the importance of the role played by sediment-driven density flows in causing a significant degree of sedimentation (HALLIWELL and O'DELL, 1970; WANLESS, 1974).

From the point of view of understanding sediment dynamics, the gated entrance case poses several interesting problems. If the gate (initially closed with clear, quiescent water in the canal) is opened instantaneously, a sediment-laden density front, or turbidity current, will propagate into the canal. With increasing penetration, the front speed will slow down and the sediment will settle out. If the canal is relatively short, the front will reach the closed-end and reflection will generate a front wave in the opposite direction. On the other hand, if the canal is long, front penetration will terminate at a distance

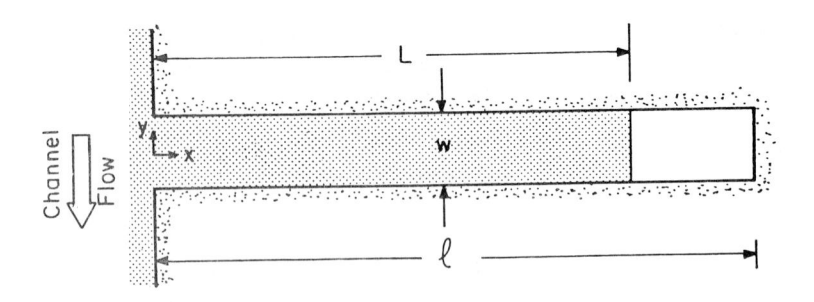

FIGURE 1a. Schematic plan view of a closed-end canal or slip.

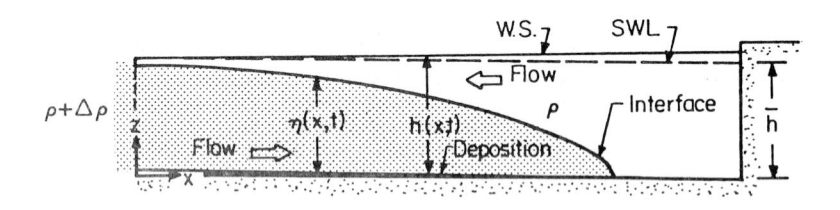

FIGURE 1b. Schematic side view of a closed-end canal or slip.

from the entrance where the smallest depositable particles in suspension settle out.

The phenomenon has some similarities with the formation of salinity-induced density fronts; the driving force is gravity, and there is an interface between fluids having different effective densities. The densimetric Froude number is important in characterizing interfacial stability and mixing. Differences between the two types of phenomena are, however, significant. In the case of salinity-induced flow, an arrested wedge is eventually formed when the net horizontal pressure force due to the hydrostatic head resulting from the density difference is balanced by the pressure head and the interfacial shear force caused by fresh water from upstream sources flowing out (KEULEGAN, 1966; PARTHENIADES, DERMISSIS and MEHTA, 1975). On the other hand, for the sediment front to be arrested, no upstream fresh water outflow is necessary, and a state of equilibrium is eventually attained in which an inward sediment-laden flow occurs through the lower portion of the water column, the sediment settles out, and sediment-free water flows out through the upper portion of the water column, with volumetric flow continuity maintained at every vertical section along the length of the canal. A somewhat similar phenomenon occurs in closed-end sidearms of cooling ponds. In this case, hot water from the pond penetrates the sidearm and forms an arrested front in the upper layer, with cooler water below (STURM, 1976; BROCARD and HARLEMAN, 1980).

In order to investigate the sediment-induced density front behavior, a limited number of laboratory experiments were recently conducted. Results and a preliminary analysis of the data are given in this paper.

2. Background

Figures 1a,b are schematic, definition sketches showing the instantaneous position of the density front in a canal of length, ℓ, and width, w, as well as counter flows in the two layers, bottom deposit, still water level (SWL) and the instantaneous water surface (W.S.) resulting from front penetration. McDowell (1971) gave consideration to the characteristics of the arrested front based upon certain simplifying assumptions concerning momentum and mass conservation in the two-layered flow. One of his conclusions was that the mean velocity, u_e, of the inflow at the entrance in the lower layer (front) would be equal to the densimetric velocity, $[(\Delta \rho / \rho_m) g h_1]^{1/2}$, where ρ_m = mean density between sediment-laden waters and clear water, $\Delta \rho$ = density difference between sediment-laden waters and clear water, h_1 = depth of lower layer at the entrance, and g = acceleration due to gravity. No experimental evidence was presented to support this conclusion. Later, Gole et al. (1973) conducted a limited number of laboratory tests in a 3.7 m long slip. These investigators measured the propagating front profiles and vertical velocities within the front. Sediment concentrations in the slip were, however, not reported. In some of their tests arrested fronts are reported to have occurred. Furthermore, it was found that the inflow velocity, u_e, was approximately 50% of that postulated by McDowell (1971).

Following the work of McDowell (1971), Mehta et al. (1984) considered flow continuity, momentum and mass conservation for the arrested front using width-averaged equations for the two layers. In order to obtain an analytic solution to these equations the following assumptions were made: 1) steady, viscous flow, 2) flow velocity variation with distance along the canal small in both layers, 3) vertical fluid velocity negligible compared with the particle settling velocity, 4) sediment confined to the lower layer, and 5) constant sediment concentration in the lower layer. Using appropriate fluid mechanical conditions for the surficial, interfacial and bottom boundaries, solutions were obtained for the arrested front shape, distance of penetration, velocity distributions in the two layers, and the steady rate of sediment transport through the entrance. As would be expected in phenomena of this nature, the particle Reynolds number, $\text{Re}_p = w_s h / \nu$ and the densimetric Reynolds number, $\text{Re}_\Delta = u_\Delta h / \nu$, where the densimetric velocity, $u_\Delta = (gh\Delta \rho / \rho_m)^{1/2}$, h = depth of flow, w_s = particle settling velocity and ν = kinematic viscosity of the fluid, were found to be significant dimensionless parameters. In addition, a characteristic velocity, $u_0 = u_\Delta (\text{Re}_p)^{1/2}$, was determined to be the normalizing parameter in the solution for the depth-varying horizontal fluid velocity, u, in the two layers. The predicted fluid velocity agreed well with limited measurements of Gole et al. (1973). It was further found that the length, L_o, of the arrested front is equal to $\alpha_1 \text{Re}_\Delta \text{Re}_p^{-\frac{1}{2}}$, and the sediment flux, F (mass transported per unit flow area per unit time), into the slip is equal to $\alpha_2 C_o u_o$, where C_o = sediment concentration outside, and α_1, α_2 are coefficients with theoretical values of 0.05 and 0.06, respectively. Further details of this approach are not given here because the experimental results to be described were found to differ considerably from the analytic solutions in the entrance region where a channel flow-induced turbulent gyre occurred. The influence of such a gyre, which is as well present under real conditions, was not accounted for in the approach considered by Mehta et al. (1984). Nevertheless, the aforementioned parameters are useful in organizing results on the front behavior, as corroborated by dimensional analysis. Thus, for example, the front speed, u_f, may be represented in the functional form as

$$u_f = f(h, \Delta \rho, \rho_m, g, w_s, \nu) \tag{1}$$

from which, the well-known pi-theorem will yield four dimensionless groups: Re_p, $\Delta \rho / \rho_m$, gh^3 / ν^2 and $u_f h / \nu$. These, upon rearrangement, yield: u_f / u_Δ, $\text{Re}_p, \text{Re}_\Delta$ and

$\Delta \rho / \rho_m$, from which the importance of Re_p and Re_Δ is immediately recognized.

3. Experiments

3.1. Apparatus and procedure

The main experiments were conducted in a 9.1 m long and 0.23 m wide plexiglass closed-end slip connected at the open end to a 100 m long and 0.23 m wide plexiglass flume at the U.S. Army Waterways Experiment Station, Vicksburg, Mississippi. An easily removable gate was placed at the entrance. Characteristics of the 100 m flume have been described by DIXIT et al. (1982). The slip was placed at right angles to the flume at a location which was approximately half-way along the length of the flume. Desired flow in the flume was generated by discharging water through a large headbay, and the depth of flow was controlled by an underflow-type tail gate. The flume and the slip were subject to changes in water temperature resulting from temperature variations in the source water as well as the ambient air. Sediment was introduced into the flow at the desired rate by injecting a slurry of sediment in water at a point upstream of the entrance to the slip. Sediment-laden flow was discharged out of the flume without recirculation. Each experiment consisted in initially establishing a steady, sediment-laden flow in the flume with the gate at the entrance closed, and with fresh, sediment-free water inside the slip. The gate was then opened and the behavior of the density front was investigated. A mini-flow current meter was used to measure velocity profiles in the flume, and a series of vacuum pressure taps at different elevations and locations were used to obtain suspended sediment concentrations. By dyeing the sediment slurry with Rhodamine-B, it was possible to record the progression of the front at various instants of time. A dye injection technique and floating tracer particles were used to record vertical and lateral velocity profiles in the slip. Point gages attached to the flume and slip were used to measure water surface elevations before and after gate opening. Fluid temperature was measured in the slip and in the flume. Deposited sediment was sampled at the end of a given test at various distances from the entrance.

3.2. Material

Sediments used were: a commercial kaolinite, silica flour, two types of flyashes (I and II) and a loess from Vicksburg, Mississippi. Sediment densities were: kaolinite 2.50 g/cm^3, silica flour 2.65 g/cm^3 flyash (I) 2.45 g/cm^3 and flyash (II) 2.37 g/cm^3. The fluid was tap water whose chemical composition has been given by DIXIT et al. (1982). The settling velocities of the sediments were determined at various initial suspension concentrations from tests in a 1.1 m tall column using a procedure described by MCLAUGHLIN (1959).

Two series of tests were conducted, one without and one with sediment. The series without sediment consisted of tests with three objectives: 1) to determine the influence of gate opening in causing transient flows, 2) to investigate flow circulation set up by naturally occuring thermal gradients in the slip, and 3) to examine the development of the shear-induced gyre near the entrance.

A preliminary analysis of data from the first series and comparison of the results with those from the second series indicated the following: 1) transient flows developed as a result of gate opening lasted for relatively short durations, on the order of a minute, and did not otherwise influence the front propagation phenomenon. 2) For flume sediment concentrations less than approximately 200 mg/ℓ, the fluid density difference between the slip and the flume arising from corresponding differences in the water temperature (which occurred during the course of the experiment) were found to be on the order of 50 to 80% of the density difference due to sediment concentration gradient. For flume sediment concentrations exceeding approximately 1500 mg/ℓ, thermally-induced density effects were comparatively negligible. 3) The shear-induced gyre at that entrance without sediment was not significantly different from that with sediment.

A summary of conditions for the thirteen tests in the second series is given in Table 1. Magnitudes of the mean velocity, U, in the flume indicate that the flume flow was fairly turbulent, with Reynolds numbers in excess of 1.6×10^4. The depth-averaged concentration, C_o, of suspended sediment (which was comparatively well-mixed over depth in the flume) was varied from 50 to 2000 mg/ℓ, while the still water depth, h, ranged from 5.0 to 12.7 cm. In test 1, an initial hydrostatic head difference of 2.6 cm was present at the time of gate opening. This was not the case in the remaining twelve tests in which the water levels were equal on both sides of the gate. The settling velocities, w_s, ranged from 2.1×10^{-2} to 7.8×10^{-1} mm/sec. In all but three tests, the front reached the closed-end and thus was reflected. In test 2 with kaolinite at a relatively low concentration of 80 mg/ℓ, the front was arrested before reaching the closed-end. Likewise in tests 10 and 11 employing flyash (I), the fronts were arrested.

At the time of gate opening, the depth in the slip was 10.1 cm while in the flume it was 12.7 cm.

TABLE 1. Summary of test conditions

Test No.	Sediment	Flume Vel., U (cm/sec)	Flume Conc., C_o (mg/ℓ)	Still Water Depth, $\overline{h}$ (cm)	Settling Vel., w_s (mm/sec)	Front Behavior
1	kaolinite	22	180	12.7[a]	6.7×10^{-2}	reflected
2	kaolinite	18	80	7.9	4.0×10^{-2}	arrested
3	kaolinite	18	260	8.0	8.8×10^{-2}	reflected
4	kaolinite	18	292	8.3	9.5×10^{-2}	reflected
5	kaolinite	19	850	8.5	2.1×10^{-2}	reflected
6	kaolinite	19	143	8.4	4.2×10^{-2}	reflected
7	kaolinite	19	346	8.2	1.1×10^{-1}	reflected
8	kaolinite	16	2000	5.0	3.1×10^{1}	reflected
9	silica flour	19	213	8.8	1.9×10^{-1}	reflected
10	flyash (I)	19	50	8.9	7.8×10^{-1}	arrested
11	flyash (I)	19	365	8.8	7.8×10^{-1}	arrested
12	flyash (II)	16	1056	5.7	3.3×10^{-1}	reflected
13	loess	16	350	5.8	3.4×10^{-1}	reflected

4. Results

4.1. Entrance characteristics

Figure 2a shows the interfacial elevation, η, as a function of distance, x, from the entrance for the propagating front at various instants of time after gate opening in test 12, in which, as observed, the front reached the closed-end and was then reflected. Figure 2b shows similar results from test 2 in which the front was arrested at an approximate distance of 2.7 m from the entrance. In all tests, stable stratification occurred with a negligible degree of upward sediment transport across the interface. The concentration of suspended sediment in the upper layer was essentially nil.

A characteristic feature of all the tests was the presence, near the entrance, of a gyre

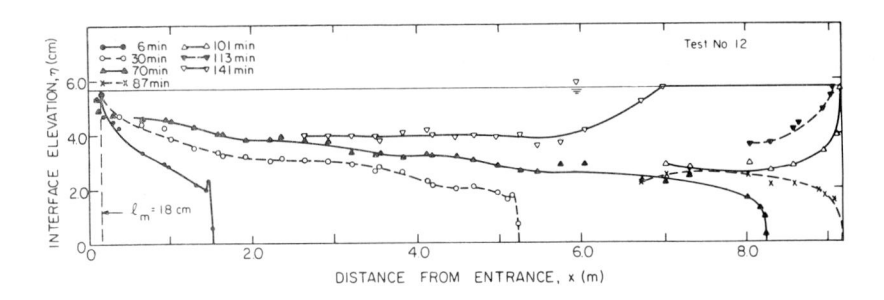

FIGURE 2a. Instantaneous front profiles in test no. 12. Front reached closed end at 87 minutes. Profiles at 101, 113 and 141 minutes indicate front propagating towards the open end.

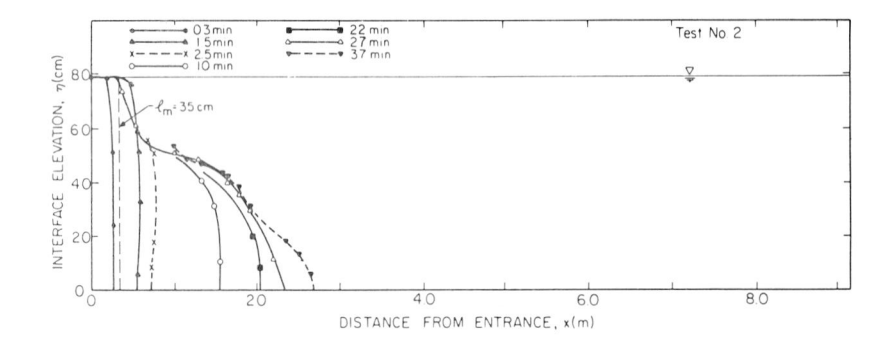

FIGURE 2b. Instantaneous and arrested (at 37 min) front profiles in test no. 2.

induced by flow shear between the flume and the slip waters. Flow in this gyre appeared to be turbulent, and there was a considerable amount of vertical, lateral and longitudinal mixing of the suspended sediment, so that a stratified front did not occur here. Figure 3 shows isovels from a near-steady gyre from test 7 corresponding to a mean flow velocity $U = 19$ cm/sec in the flume. The maximum tangential velocity, u_θ, in the gyre was on the order of 3.5 cm/sec, and the mean angular velocity was 0.26 sec^{-1}. The gyre center, (x_m, y_m), was at $x_m = 13$ cm, $y_m = 13$ cm. The slip centerline was at distance of 11.5 cm from the wall, so that the gyre center was slightly downstream (with reference to the direction of U) of the slip centerline. Such an asymmetric location of the center is characteristic of shear-induced gyres (ASKREN, 1979). The length, ℓ_m, considered to be a characteristic dimension of the gyre, was 31 cm. This was determined from front profiles such as those shown in Figs. 2a,b, where ℓ_m is shown as the zone of mixing corresponding to a distance from the entrance beyond which the front began to develop. Beyond ℓ_m a less well-defined and weak, secondary gyre with direction of rotation opposite to that of the primary gyre was formed, as observed in Fig. 3. Maximum velocities, u_θ, in the secondary gyre were on the order of 1 cm/sec. In the primary gyre, the maximum value of the Reynolds number, $u_\theta \bar{h} / \nu$, where $\bar{h}$ is still water depth, was on the order of 1800, while in the secondary gyre the Reynolds number dropped to 500. In other words, flow turbulence decreased significantly in the secondary gyre in relation to the primary gyre, and, in fact, viscous effects became dominant, thus allowing for the formation of a stable, stratified front.

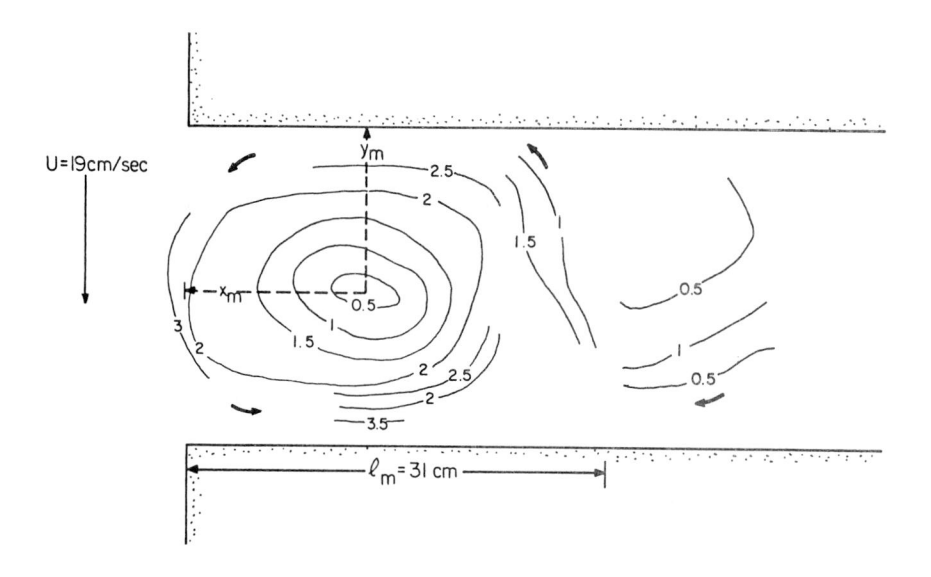

FIGURE 3. Isovels from primary and secondary gyres near the entrance from test no. 7.

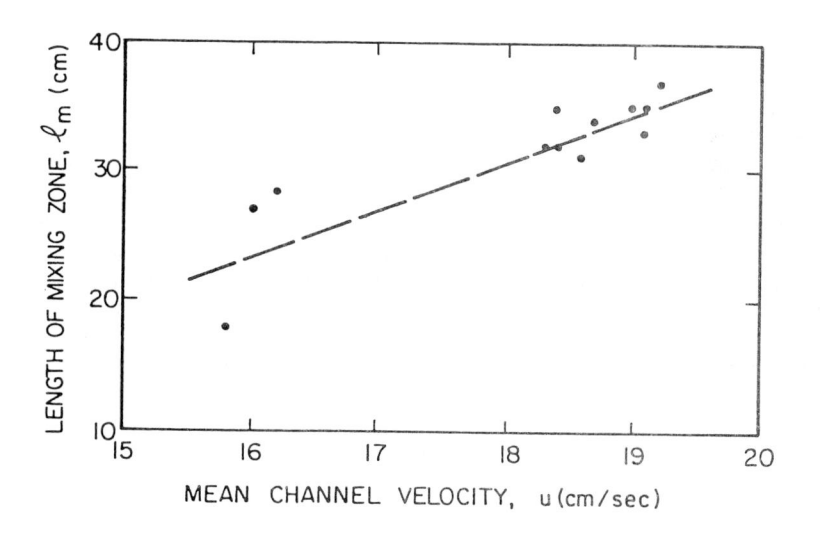

FIGURE 4. Characteristic length, ℓ_m, of the entrance gyre against mean flume velocity, U.

In Fig. 4 the gyre length, ℓ_m, is plotted against the mean channel velocity, U, from twelve tests excluding test 1, in which, as noted earlier, the initial flow conditions were somewhat different (see Tables 1 and 2). The dashed line indicates a qualitatively increasing trend of ℓ_m with U. As a rule of thumb, ℓ_m is typically found to be of the same order of magnitude as the slip width, w, in this case 23 cm (ASKREN, 1979). The magnitude of ℓ_m in fact cannot be expected to increase indefinitely with U, although the strength of the secondary gyre tends to increase with increasing U, particularly up to a distance on the order of $2w$ from the entrance (ASKREN, 1979).

TABLE 2. Characteristic test parameters

Test No.	Gyre Length, ℓ_m (cm)	Densimetric Vel., u_Δ (cm/sec)	Re'_Δ $=u_\Delta R / \nu$[b]	Re_p $=\overline{w_s h} / \nu$
1	[a]	2.05	1.23×10^3	9
2	35	0.79	3.69×10^2	3
3	32	1.43	6.73×10^2	7
4	32	2.42	1.16×10^3	8
5	35	3.21	1.56×10^3	18
6	34	1.44	6.97×10^2	4
7	31	1.90	9.06×10^2	9
8	29	3.32	1.16×10^3	16
9	33	2.06	1.03×10^3	17
10	35	1.67	8.33×10^2	69
11	37	2.14	1.07×10^3	69
12	18	2.71	1.03×10^3	19
13	27	1.82	7.0×10^2	20

[a] ℓ_m was determined.
[b] Re_Δ is densimetric Reynolds number based on the hydraulic radius, R.

4.2. Front characteristics

Figure 5 shows an example of the instantaneous depth-variation of the horizontal velocity, u, from test 12, at a distance $x = 1.1m$ from the entrance at times $t = 27$ and 130 min. At equilibrium, i.e. when the front is arrested, the flows become steady and the discharge in the lower layer towards the closed-end would be equal to the discharge in the upper layer towards the entrance. However, since the front was not arrested in test 12, equilibrium conditions were not attained. Furthermore, during the measurement of these velocities using a dye injection procedure, some errors appeared to have been introduced as a result of the presence of secondary flows. These errors are likely to be inherent in the data shown in Fig. 5.

The horizontal velocity, u, varied laterally to a measurable degree, as typified by the variation of the surface velocity, u_s, from test 9, presented in Fig. 6. The slight asymmetry of the data about the slip centerline is believed to have resulted from secondary flows. Notwithstanding this effect, it may be noted that u_s approximately varied with $\delta = 2.8$ power of the lateral coordinate, y, as seen from the empirical fit equation given in Fig. 6. In this equation, u_{sm} is the value of u_s at the slip centerline ($y = 0$), and ϵ is an empirical coefficient. The relatively strong dependence of u_s on y is indicative of the viscous nature of the flow.

In Fig. 7, the front speed, u_f, at the toe is plotted against distance, x, from the entrance from tests 2 and 12. For test 12, the speed of the front after reflexion is not shown. The reflected front wave in fact appeared to have propagated more slowly. The

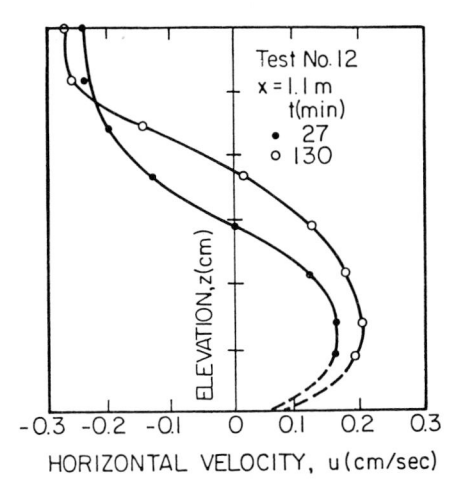

FIGURE 5. Instantaneous variation of the horizontal velocity, u, with elevation, z, from test no. 12, at $x = 1.1$ m and $t = 27$ and 130 min.

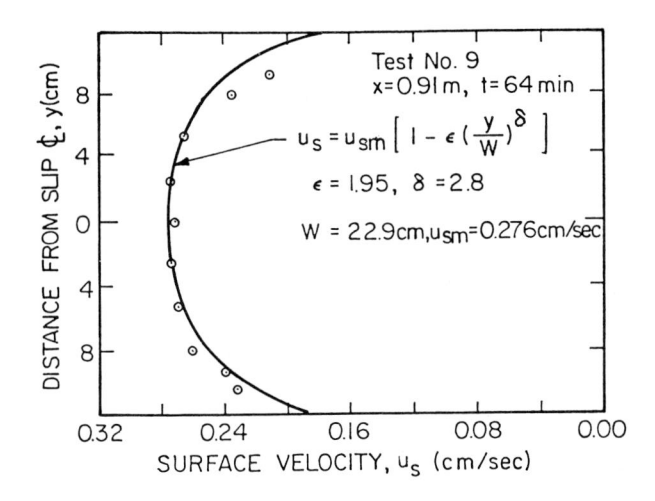

FIGURE 6. Lateral Variation of the Horizontal Surface Velocity, u_s, from test no. 9.

speed of the reflected front at time $t = 95$, 107 and 127 min was measured to be 0.0035, 0.013 and 0.13 cm/sec, respectively. In test 2 the front was arrested at a distance $x \approx 2.7m$ and time $t \approx 40$ min. The mean lines through the data points for both tests in Fig. 7 suggest a monotonically decreasing value of u_f with distance and time. While such a decreasing trend was found in all tests, the data were believed to have been influenced, at least in some cases, by density gradients induced by temperature changes during the course of the tests. Such an influence is observed, for instance, in tests 12, in which the data point corresponding to $u_f = 0.29$ cm/sec at $x = 1.2$ m shows a measurable deviation from the mean trend given by the curve.

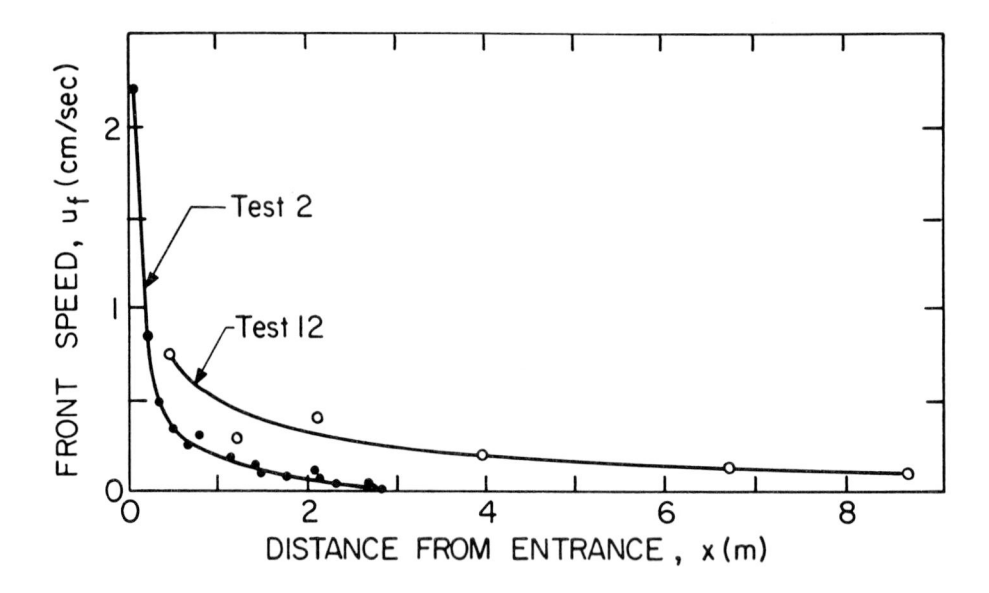

FIGURE 7. Variation of front speed u_f, with distance, x, from entrance in test nos. 2 and 12.

Knowing ℓ_m (Table 2), plots such as Fig. 7 were used to determine the initial front speed, u_{fo}, at $x = \ell_m$, which corresponds to the distance from the entrance where the front first developed. In Fig. 8, u_{fo} is plotted against the densimetric velocity, $u_\Delta = (g\bar{h}\Delta\rho/\rho_m)^{\frac{1}{2}}$, where it should be noted that in computing u_Δ (given in Table 2), the density difference, $\Delta\rho$, used is the combined difference resulting from gradients (between the flume and the slip) in sediment concentration as well as fluid temperature. The density difference due to sediment concentration was calculated using the flume concentration, C_o (Table 1). The corresponding mean difference due to temperature (ranging from 0 to 4.2×10^{-4} g/cm^3) was derived from temperature measurements during the tests. The two differences were then added algebraically for each test. It is observed from Fig. 8 that the mean line drawn through the data points yields the relationship:

$$\frac{u_{fo}}{u_\Delta} = 0.43 \qquad (2)$$

The proportionality constant 0.43 is close to 0.46 obtained by KEULEGAN (1957) and 0.44 of MIDDLETON (1966). Keulegan's laboratory tests dealt with the motion of saline density fronts from locks having the same width as the canal to which the lock was connected. Middleton conducted two series of tests, one with saline water and another using fine plastic beads in suspension. The close similarity between the saline and turbidity current phenomena at the initial stages is quite evident from the closeness of results.

The front speed before front reflection was further investigated by considering the relationship for saline fronts proposed originally by KEULEGAN (1958) and modified by WU (1969):

$$\frac{u_\Delta}{u_f} = \frac{u_\Delta}{u_{fo}} + \beta\frac{L}{R} \qquad (3)$$

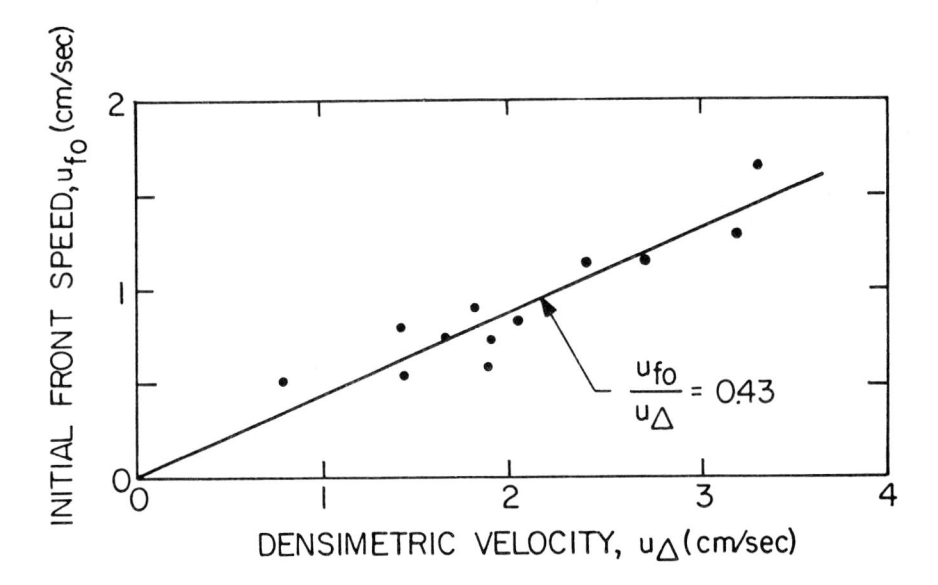

FIGURE 8. Initial front speed, u_{fo}, against densimetric velocity, u_Δ.

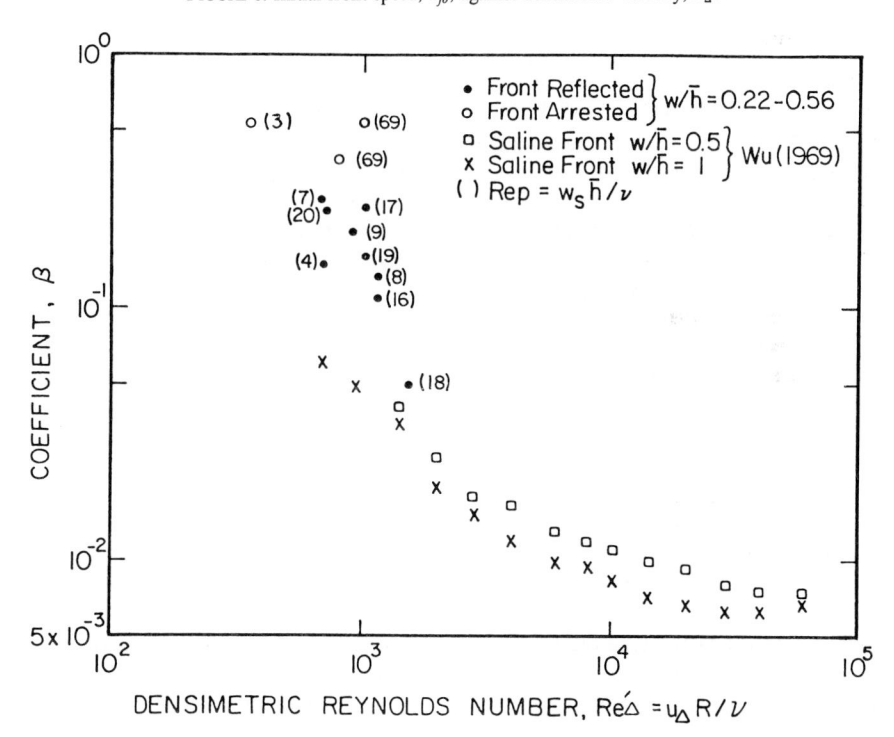

FIGURE 9. Variation of coefficient, β, from Eq. 3 with densimetric Reynolds number, Re'_Δ.

where L is the instantaneous front length (Fig. 1a), β is an empirical coefficient, and R is the hydraulic radius. Equation 3 was fitted to the data such as those given in Fig. 7, and

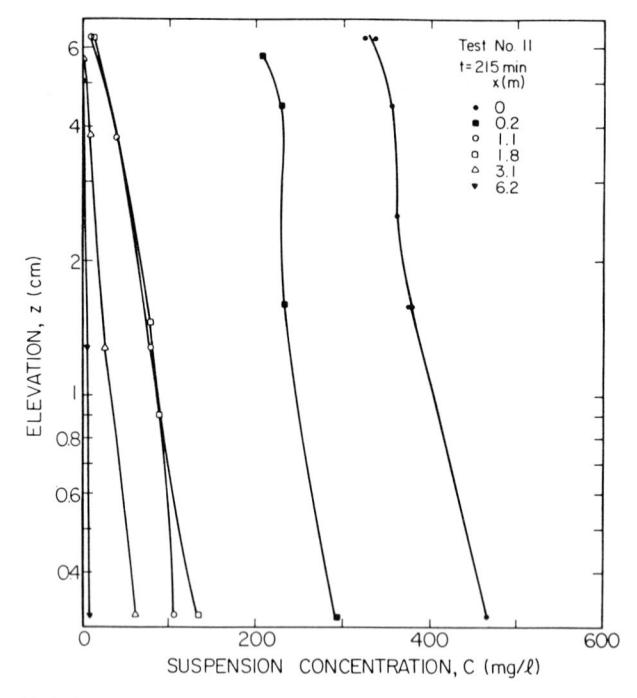

FIGURE 10. Variations of suspension concentration, C, with elevation, z, above the bottom from test no. 11.

in Fig. 9, β is plotted against the hydraulic radius-based densimetric Reynolds number $Re'_\Delta = u_\Delta R / \nu$ (using $\nu = 1 \times 10^{-6}$ m²/sec for clear water). Data points for three tests in which the front was eventually arrested are shown separately from tests in which the front was eventually reflected at the closed-end. Included also are values of the particle Reynolds number, $Re_p = w_s \bar{h} / \nu$, in parantheses. Comparison is made with the data of KEULEGAN (1958) for saline fronts, reanalyzed by WU (1969) for aspect ratios, $w/\bar{h} = 0.5$ and 1. The same ratio in the present tests varied from 0.22 to 0.56. In general, β is observed to decrease with increasing Re'_Δ. Furthermore, there appears to be a general agreement between the trends for β from Wu's plot and the present tests, in spite of the fact that the sediment intrusion phenomenon is additionally complicated by the ability of the particles to settle out of suspension. In general, in the range of Re_Δ covered in the tests, β appears to be insensitive to Re'_Δ in comparison with the values plotted by Wu, which are for higher values of Re_Δ. Although the number of tests in the present study were not sufficient to clearly identify the influence of Re_p on the results, there is a possible trend of increasing β with Re_p. It is further noted that the β values from tests in which the front was eventually arrested were generally higher than those from tests in which the front was not arrested.

As an example of the type of suspended sediment concentration profiles in the lower layer obtained in the tests, Fig. 10 shows depth-variations of concentration, C, at various distances, x, from the entrance at 210 min after gate opening from test 11. In Fig. 11, the depth-averaged concentrations, $\bar{C}$, from test 11 (using data from Fig. 10) and test 12 are plotted against distance x from the entrance. In test 11, $\bar{C}$ decreased from $C_o = 365$ mg/ℓ at the entrance relatively rapidly at first and then more slowly, to nearly zero at $x = 6.0$ m, where the toe of the arrested front occurred. A similar trend is observed for test 12. In their theoretical considerations, MCDOWELL (1971) and MEHTA ET AL. (1984) assumed the concentration to be constant throughout the arrested front, equal to C_o. However, in view of the observed trend of decreasing value of $\overline{C}$ with x, the assumption of constant concentration throughout the front appears to be difficult to justify.

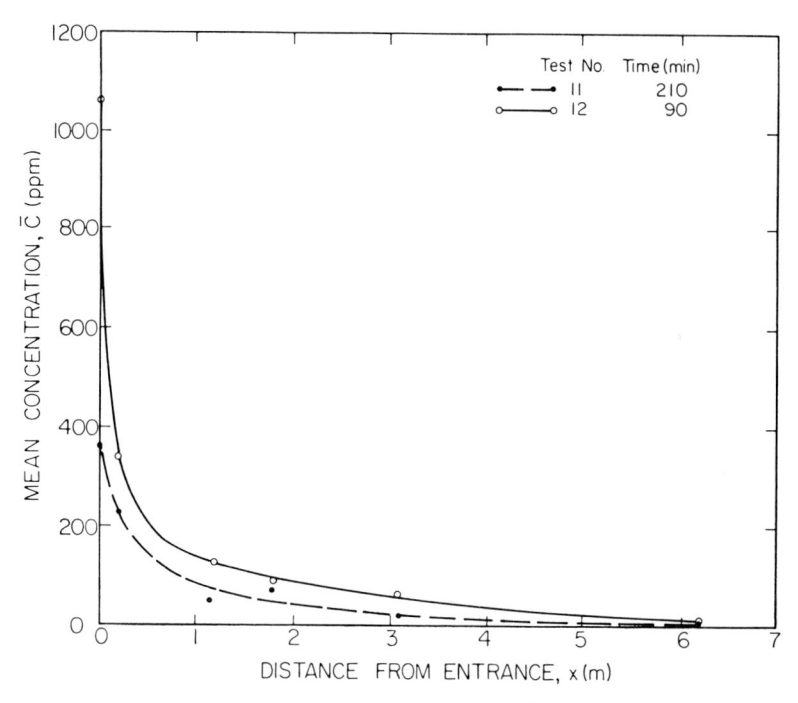

FIGURE 11. Variation of depth-averaged suspension concentration, $\overline{C}$, with distance, x, from entrance in tests 11 and 12.

4.3. Arrested front

For tests 2, 10 and 11 in which an arrested front eventually occurred, Table 3 gives the front length, L_o. In an effort to relate L_o to the governing parameters Re_Δ and Re_p, the following relationship is proposed, following previous considerations by MEHTA et al. (1984) noted earlier:

$$\frac{L_o}{h} = \alpha_3 \left(\frac{u_\Delta h}{\nu}\right)^{\frac{1}{2}} \left(\frac{w_s}{u_\Delta}\right)^{-\frac{1}{2}} = \alpha_3 Re_\Delta Re_p^{-\frac{1}{2}} \quad (4)$$

where α_3 is considered here to be an empirical coefficient. This relationship is similar to the dimensionless expression for the arrested saline wedge derived by KEULEGAN (1966). The latter relationship is obtained if w_s in Eq. 4 is replaced by the fresh water velocity multiplied by two, and the exponent $-\frac{1}{2}$ is changed to $-5/2$. Since an arrested front occurred in only three tests, it is not feasible to determine if any relationship other than Eq. 4 would be more suitable; however, computed α_3 values for the three tests appear to be of the same order of magnitude, ranging from 0.10 to 0.32. It is conceivable that this spread of values arises out of likely experimental error, and that this error represents a deviation from a constant value of α_3 contained within the measured range. It is interesting to note that for the arrested saline wedge, KEULEGAN (1966) obtained 0.23 as the corresponding coefficient, given the densimetric Reynolds number on the order of 10^5.

4.4. Sediment deposition

In tests 8, 11, 12 and 13, the thickness of the deposited sediment, ΔH, was estimated at various distances from the entrance using data on deposited mass and dry density, as shown in Fig. 12. These deposits occurred at the end of each test. The total run duration for each test and the dry density, ρ_D, of each deposit (measured in a separate series of static cylinder tests) are given in the figure. The variation of ΔH with distance was, in general dependent upon initial test conditions, type of sediment and duration of the experiment. In tests 8, 12, 13, the front was not arrested. In test 11, the front propagated for 204 min before being arrested. Shortly after this time, at 210 min, the test was terminated. In all four tests, therefore, the deposit thickness observed in Fig. 12 resulted from time-varying rate of sediment intrusion through the entrance. The trend of rapidly decreasing deposit thickness with distance from entrance in tests 11, 12 and 13 is similar to that observed by KUENEN and MIGLIORINI (1950) in experiments on turbidity deposit from surges of poorly sorted sand-mud suspensions.

TABLE 3. Arrested Front Length Parameters

Test No.	Densimetric Reynolds Number, Re_Δ	Arrested Front length, L_o (m)	Coefficient α_3
2	6.24×10^2	2.7	0.10
10	1.48×10^3	6.9	0.32
11	1.89×10^3	6.0	0.30

Primarily for the purpose of comparing qualitative trends, the data of test 8 (using kaolinite) from Fig. 12 are replotted in Fig. 13 together with those of MIDDLETON (1967) who, as noted earlier, used fine plastic beads in place of sediment. The bottom deposit thickness, ΔH, is normalized by the corresponding maximum value, ΔH_m, and the distance from the entrance, x, is normalized by channel length, ℓ ($=5$ m in Middleton's tests). The normalized profiles indicate a remarkable similarity, at least in this case. While it would be speculative to offer any physical explanation of the observed trends given the limited data, the results provide encouragement for further examining similarity conditions for sediment deposition in the channels.

Measurable sorting (with distance) of sediment in the deposit was found to have occurred during the tests. Sorting would be expected to become significant in the slip beyond the gyre (characterized by ℓ_m) near the entrance; the degree of sorting depending on, among other factors, the spread of the particle settling velocity, w_s, about the median value. As the suspended matter entered the slip, particles with larger w_s settled out first, leaving finer material in suspension which settled out subsequently. A consequence of this behavior was reflected in the variation of the (dispersed) particle size, d_{75}, of the deposited sediment with distance, x, from the entrance, shown in Fig. 14. As observed, d_{75} decreased rapidly with distance in tests 8, 11 and 12. In test 13, using loess, only one data point corresponding to a location just beyond the gyre could be obtained. In general, the significance of the observed sorting trend, when compared with the previously noted trend in suspension concentration variation with distance (such as in Fig. 11), is that not only the suspension concentration decreased with distance, but that the material in suspension became finer as well. In as much as the settling velocity varies with sediment concentration, particularly for cohesive materials (DIXIT et al. 1982), the coupled processes of decreasing suspension concentration and particle size with distance make it difficult to assign a unique, characteristic value of the settling velocity for describing

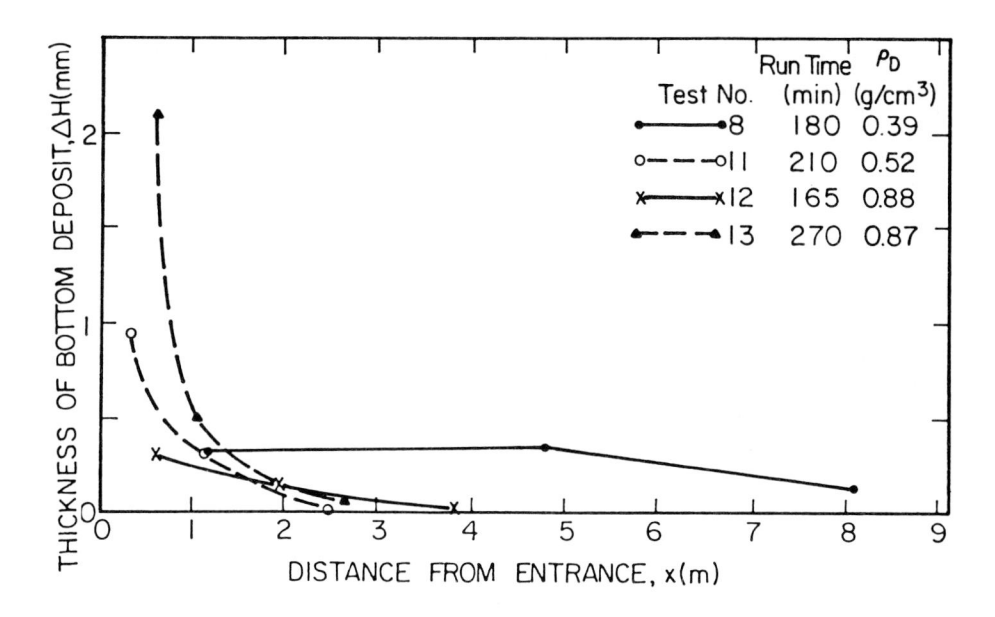

FIGURE 12. Bottom deposit thickness, ΔH, against distance, x, from entrance: tests 8, 11, 12 and 13.

details of sediment transport within the front.

5. Conclusions

The following main conclusions can be derived:

1. Near the slip entrance a zone of mixing occurred as a result of a gyre resulting from flow-induced shear at the entrance. Flow in the gyre was turbulent, whereas beyond the end of the gyre, at a characteristic distance ℓ_m from the entrance, viscous effects become progressively dominant. Within the range of mean flow velocities, U, in the flume covered in the tests, ℓ_m increased with U.

2. There was a considerable amount of vertical, lateral and longitudinal mixing of the suspended sediment within the gyre. A stratified front occurred in the slip at distances greater than the length, ℓ_m, of the gyre.

3. The initial front speed, u_{fo}, at a distance, ℓ_m, from the entrance was found to be proportional to the densimetric velocity, u_Δ. The proportionality constant, 0.43, was close to the value, 0.46, obtained by KEULEGAN (1957), in studies of the motion of saline density fronts, and 0.44 obtained by MIDDLETON (1966) in studies on salt water motion as well as turbidity currents.

4. The propagating front speed, u_f, before front reflection, was described by Eq. 3 in which the empirical coefficient, β, varied with the densimetric Reynolds number, Re_Δ. In general, the trend of decreasing β with increasing Re_Δ is in agreement with β values plotted by WU (1969) using the data of Keulegan (1958) for saline fronts. In addition, however, the β values from the present tests suggested a likely dependence on the particle Reynolds number, Re_p. Also, arrested fronts appeared to correspond to comparatively higher values of β.

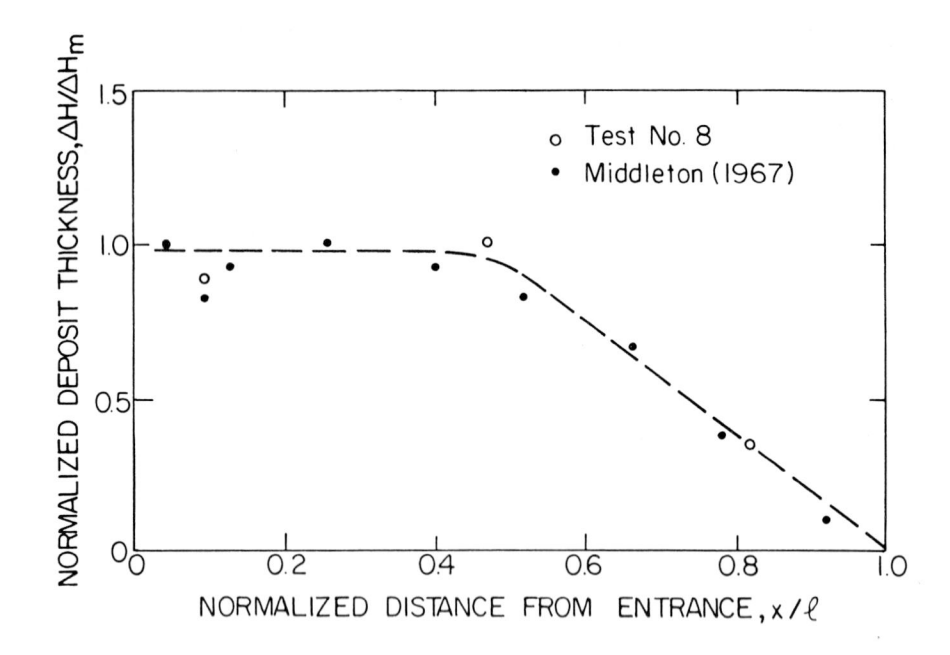

FIGURE 13. Plot of normalized deposit thickness $\Delta H / \Delta H_m$ against normalized distance from entrance, x / ℓ. Comparison between test 8 and data of MIDDLETON (1967).

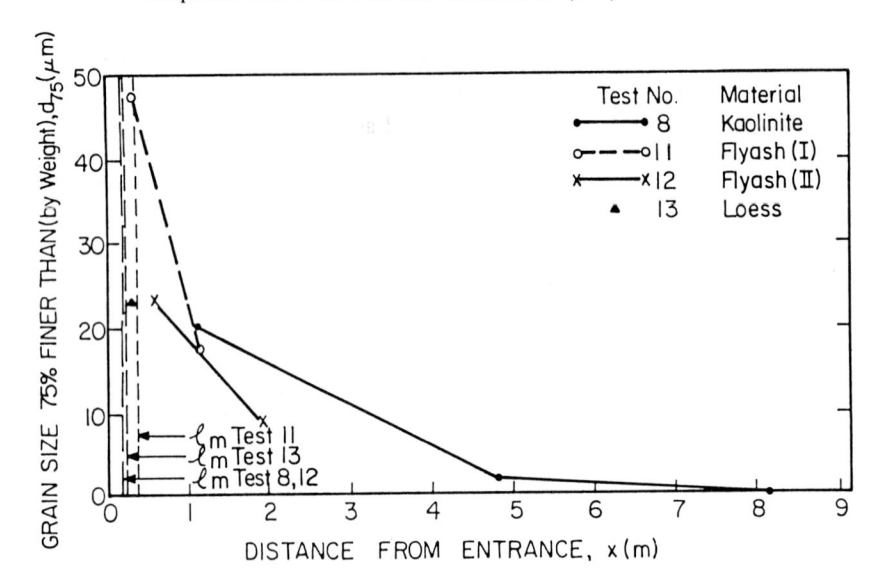

FIGURE 14. Grain size, d_{75}, 75% finer than (by weight), against distance, x, from entrance: tests 8, 11, 12 and 13.

5. For both propagating as well as arrested fronts, vertical sediment concentration profiles indicated increasing concentration with decreasing elevation above the bed in the lower layer. The depth-averaged concentration decreased with distance from the entrance, relatively rapidly at first and then more slowly to nearly zero at the toe of the front. In view of this variation of concentration, it is unrealistic to assume the concentration to be constant, as considered by McDOWELL (1971) and MEHTA et al. (1984).

6. The arrested front length, L_o, may be approximately estimated through Eq. 4. Additional data are, however, required for further verification of this equation.

7. Sediment sorting due to deposition occurred along the length of the slip. Essentially, the particle size in the deposit (and, therefore, in suspension) decreased with distance from the entrance.

8. In future experiments, it would be essential to conduct the tests under controlled water and ambient air temperatures, in order to minimize the influences of uncontrolled air and water temperature variations on test results.

Acknowledgement

Assistance provided by Messrs. E.C. McNair, Jr., A. Teeter and S. Heltzel at the U.S. Army Waterways Experiment Station, Vicksburg, Mississippi, is sincerely acknowledged.

References

ASKREN, D.R., 1979. Numerical simulation of sedimentation and circulation in rectangular marina basins. NOAA Technical Report NOS 77, U.S. Department of Commerce, Rockville, Maryland.

BROCARD, D.N. & D.R.F. HARLEMAN, 1980. Two-layer model for shallow horizontal convective circulation. Journal of Fluid Mechanics, Vol. 100, part 1, pp. 129-146.

DIXIT, J.G., A.J. MEHTA & E. PARTHENIADES, 1982. Redepositional properties of cohesive sediments deposited in a long flume. UFL/COEL-82/002, Coastal and Oceanographic Engineering Department, University of Florida, Gainesville, Florida.

GOLE, C.V., Z.S. TARAPORE & M.R. GADRE, 1973. Siltation in tidal docks due to density currents. Proceedings of the 15th Congress of IAHR, Vol. 1, Istambul, Turkey, pp. 335-340.

HALLIWELL, A.R. & M. O'DELL, 1970. Density currents and turbulent diffusion in locks. Proceeding of the 12th Coastal Engineering Conference, ASCE, Vol. III, Washington, D.C., pp. 1959-1977.

KEULEGAN, G.H., 1957. An experimental study of the motion of saline water from locks into fresh water channels. Thirteenth Progress Report on Model Laws from Density Currents (NBS Report 5168), National Bureau of Standards, Washington, D.C.

KEULEGAN, G.H., 1958. The motion of saline fronts in still water. Twelfth Progress Report on Model Laws for Density Currents (NBS Report 5831), National Bureau of Standards, Washington, D.C.

KEULEGAN, G.H., 1966. The mechanism of arrested saline wedge. in Estuary and Coastline Hydrodynamics, A.T. Ippen Editor, McGraw-Hill, New York, pp. 546-574.

KUENEN, Ph. H. & C.J. MIGLIORINI, 1950. Turbidity currents as a cause of graded bedding. Journal of Geology, Vol. 58, pp. 91-127.

McDOWELL, D.M., 1971. Currents induced in water by settling solids. Proceedings of the 14th Congress of IAHR, Vol. 1, Paris, France, pp. 191-198.

McLAUGHLIN, R.T., 1959. The settling properties of suspensions. Journal of the Hydraulics Division, ASCE, Vol. 85, No. HY12, pp. 9-41.

MEHTA, A.J., R. ARIATHURAI, P. MAA & E.J. HAYTER, 1984. Fine sedimentation in small harbors. UFL/COEL-84/007, Coastal and Oceanographic Engineering Department, University of Florida, Gainesville, Florida.

MIDDLETON, G.V., 1966. Experiments on density and turbidity currents, I. Motion of the head. Canadian Journal of Earth Sciences, Vol. 3, pp. 523-546.

MIDDLETON, G.V., 1967. Experiments on density and turbidity currents, III. Deposition of sediment. Canadian Journal of Earth Sciences, Vol. 4, pp. 475-505.

PARTHENIADES, E., V. DERMISSIS & A.J. MEHTA, 1975. On the shape and interfacial resistance of arrested saline wedges. Proceedings of the 16th Congress of IAHR, Vol. 1, Sao Paulo, Brazil, pp. 157-164.

STURM, T.W., 1976. An analytical and experimental investigation of density currents in sidearms of cooling ponds. Ph.D. Dissertation, University of Iowa, Iowa City, Iowa.

WANLESS, H.R., 1974. Sediments in natural and artificial waterways, Marco Island area, Florida, UM-RSMAS No. 74032, Rosenstiel School of Marine and Atomspheric Science, University of Miami, Miami, Florida.

WU, J., 1969. Laboratory results of salt water intrusion. Journal of the Hydraulics Division, ASCE, Vol. 95, No. HY5, pp. 2209-2213.

INDEX

Aggregation 239,254

Bottom shear stress 7,251,257
    critical erosion 220,223,230,252
    quadratic 49,73,75

Circulation 22,23,24,25,71,80,114,120,121,122,
        147,154,161,199,200,201,202,203,207
    gravitational 7,11,12,20,22,23,24,28,58,65,
        228,234,245,257
    lateral 24,25,64,65,69
    transverse 22,25,27
    thermohaline 114
Coastal boundary layer 28,199,200,201,202,
        203,204,205,206,207
Compound tides 246,247
Coral reefs 165,197
Coulter counter 239

Densimetric velocity 261,268,273
Diffusion 75,158,223
    turbulent, of salt 222
    turbulent, of suspended sediment 222
    coefficient 215,216,217,218,222,254
Dispersion 25,26,125,234
    numerical 215
    coefficient 12,210,212

Eddy viscosity 32,34,35,49,52,68,71,73,75,77,
        78,79,121,122,155,173,207,223
Ekman number 199
    transport 143,147
    boundary layer 161,173
    depth 207
    suction 158,161
    pumping 173
Empirical orthogonal function 133,135,
        136,137
Estuary number 10
Erosion
    rate of 230
    function 230
    threshold 231,253
Exchange 168,177,228,229
    tide-induced 6,143,145,146,151,172,173
    density-induced 6,7,15,16
    sub-tidal 132,143,144,145,147,151

Fall velocity,
    see settling velocity
Flocculation 210,254
Frequency domain regression 137,142

Front velocity 259,261,266,268,273
Fronts 32,37
    density 36,259,260,268,270,273
    suspended sediment 259,261,262,264,265,
        266,267,268,270,272,273,275
Froude number
    densimetric 260

Karman vortex street 158

Landsat 168,173
Large scale mixing 6,13

Mean Lagrangian velocity,
    see Lagrangian residual current
Momentum diffusion,
    see also eddy viscosity

Overtides 246,247

Particulate organic carbon 240
Phase locking 245,247,248,249,256

Residence time 23,179,196,197
Residual circulation 42,160
    due to freshwater flow 42,51,56
    due to surface wind 42,51,56
    due to density gradients 42,51,56

Residual current 49,50,51,53,61,62,65,69,71,73,
    75,77,78,79,81,104,114,121,126,127,128
    tide-induced 75,114,119,121,177,199
    vertical structure of 49,54,56,58
    Eulerian 61,65,69,71,72,73,75,78,79,102,
        103,105,107,110,111,112,113
    Langrangian 61,71,72,73,75,78,79,102,103,
        105,106,107,108,109,110,111,112,113
    second order 102
Residual ellipse 102,110,111,112,113
Residual flow,
    see Residual flux
Residual flux 39,50,58,59,60,64,65,75,77,81,82,
    84,87,91,92,95,98,100,113,120,121
    of salt 58,59,63
    of suspended sediment 58,59,63,64,228,
        229,230,231,235,236,237,239,240,241
    stream function of 82
Residual transport,
    see Residual flux
Residual velocity,

see Residual current
Residual vorticity 127
Reynolds number 264
  densimetric 259,261,270,272,273
  particle 259,261,270,273
Richardson number 212,219
  local 34,75,219

Salt balance 23
Salt distribution
  longitudinal 6,211
Salt flux 22,25,26,28
  advective 22,25,26,27
  dispersive 22,26
Salt intrusion 6,7,10,12,13,15,19,42,49,53,210
  effect of harbour basins on 6,15,17,18,
    19,20
  effect of tidal phase on 6,7,13,14,18
  effect of tidal amplitude on 13
  length of 12,53
  wedge type 51,53,56,163,222,260,271
Salt transport,
  see Salt flux
Sedimentation
  rate of 243
  function 230
Settling time interval 243
Settling velocity 210,213,215,217,222,230,234,
    239,243,251,252,257,261,262,272
Shear velocity 252,254
Shields parameter 252
Skin friction 251
Stokes drift 58,61,65,66,67,69,71,72,73,78,102,
    103,105,106,107,110,111,112,114
  vertical structure 58,67,69
Stratification 10,12,13,20,23,55,58,254
  vertical 184
  diagram 53,54,56
  number 55
Strouhal number 49,56

Tidal currents 30,46
  Eulerian 104
  Lagrangian 104,105
  vertical structure of 48,56,58,63
Tidal exchange ratio 173
Tidal distortion 245,246,247,255,256,257
Tidal ellipse 49
Tidal jets 154,163,166,169
Tidal pumping 58,59,63,64,69
Tidal rectification 120,121,122,126,128
Tidal sloshing 26,27

Turbulent entrainment 172
Turbulent mixing 6.12,32
Turbidity
    currents 259,273
    maximum 58,59,64,69,210,211,213,222,
        225,228

Vortex pair 163,166,168,169
Vorticity 122,123,128,158,161,163,173,200

Water level elevation
    sub-tidal 28,132,146
        residual 105